SCHAUM'S A–Z

CHEMISTRY

SCHAUM'S A-Z

CHEMISTRY

ANDREW HUNT

SCHAUM'S A-Z SERIES

McGraw·Hill

New York   Chicago   San Francisco   Lisbon   London   Madrid   Mexico City
Milan   New Delhi   San Juan   Seoul   Singapore   Sydney   Toronto

Originally published by Hodder & Stoughton Educational, a division of Hodder Headline Plc, 338 Euston Road, London NW1 3BH.

2 3 4 5 6 7 8 9 0   2 1 0 9 8 7 6 5 4 3

ISBN 0-07-141935-7

HOW TO USE THIS BOOK

The *Schaum's A–Z Chemistry* is an alphabetical textbook designed to help you at every stage of an advanced level chemistry course. Entries begin with one sentence definitions which explain why the terms are important. These are followed by fuller explanations and examples, together with diagrams, tables and equations. Words in italics are cross-references to other entries in the book.

You can use the book as a course companion and examination aid.

Course companion

The book covers all the main ideas in intermediate and advanced level chemistry courses. This concise, alphabetical book makes it easy for you to find the explanations, information and examples that you need when starting to study a new topic. Worked examples show you how to tackle the calculations.

Entries include many aspects of practical and applied chemistry, including both analytical chemistry and organic preparations. Furthermore, the book explains some of the terms in more specialist options so it will help you to get started on topics such as spectroscopy, environmental chemistry and biochemistry.

Andrew Hunt

A
B
C
D
E
F
G
H
I
J
K
L
M
N
O
P
Q
R
S
T
U
V
W
X
Y
Z

ACKNOWLEDGEMENTS

Firstly I thank Geoffrey Barraclough who both read and checked the whole book and contributed to the appendices. His experience of chemistry teaching and detailed understanding of examining at the advanced level have made sure that the key term entries in this book will help students to improve the quality of their examination answers and avoid common errors.

Secondly I thank Ian Marcousé, the series editor, and three experienced chemistry teachers: Rod Clough, Jeffrey Hancock and Ted Lister, who read early drafts. Their wise comments helped me to develop the content and style of this book.

Thirdly I thank Alexia Chan of Hodder & Stoughton, David Mackin of GreenGate Publishing Services and all the people who work with them.

Finally I thank all the people I have worked with while contributing to the Nuffield Advanced Chemistry, SATIS 16–19, Salters Advanced Chemistry and other Nuffield Science projects. Writing and editing for these projects has clarified my understanding of chemical ideas and taught me much about the importance of chemistry in our lives.

Andrew Hunt

A$_r$ is the symbol for *relative atomic mass*.

absolute zero (0 K): the temperature at which atoms and molecules in crystals are effectively motionless. It is the lowest temperature on the absolute or *Kelvin* temperature scale. A plot of the *volume* of a sample of gas against temperature at constant pressure is a straight line which on extrapolation cuts the temperature axis at −273.15°C.

Gases appear to behave as if they would have zero volume at absolute zero. This would be true of a non-existent *ideal gas* but in practice real gases turn to liquids and solids and occupy a definite volume before absolute zero is reached.

Absolute zero is now defined precisely by setting the *triple point* of water at 273.16 K.

absorption of liquids and gases happens when a *fluid* soaks into the pores of a material, like water soaking into a sponge. Absorption should be carefully distinguished from *adsorption*.

absorption of radiation: when *electromagnetic radiation* passes through a material, some or all of the wavelengths of the radiation may be absorbed. When white light passes through a red filter, for example, all the wavelengths are absorbed except for red which passes through – so it looks red.

The gases in the *atmosphere* absorb most of the radiation reaching the Earth from the Sun.

The bands of wavelengths in the electromagnetic spectrum of sunlight which can pass through the Earth's atmosphere to reach the surface of the Earth. It is as if there are two "windows" letting through bands of radiation while all other wavelengths are shut out (by absorption).

Ozone in the upper atmosphere helps to protect living things by absorbing harmful *ultraviolet radiation*. Chemists use *spectroscopy* to study the absorption of radiation. From an *absorption spectrum* they can make deductions about the composition and structure of the sample.

absorption spectrum: a plot showing how strongly a sample absorbs radiation over a range of frequencies. Absorption spectra from *infrared spectroscopy* and *ultraviolet spectroscopy* give chemists valuable information about the composition and structure of chemicals.

accuracy of data is determined by the agreement between a measured quantity and the correct value. In chemical analysis the correct value is often not known, so chemists need to estimate the *uncertainty of the measurements* (see also *errors of measurement*).

acetominophen: a mild painkiller (*analgesic*) which can also help to lower the body temperature during fever. It is readily available as an over-the-counter drug. Unlike aspirin it does not reduce inflammation.

Acetominophen is made by *acylating* 4-aminophenol with acetic anhydride.

The advantage of acetominophen over aspirin is that it does not irritate the lining of the stomach and it has few side effects.

The great danger is that an overdose with *medicines* containing acetominophen causes severe liver damage which can be fatal. The symptoms do not appear until after about two days. Unfortunately with acetominophen, the dose that causes poisoning is not many times the dose required to relieve pain or lower fever.

acetone (propanone): the simplest *ketone*, CH_3COCH_3, which is widely used as a solvent. Three to four million metric tons of acetone are manufactured worldwide each year. Acetone is also used to make monomers for the manufacture of *polymers*, including *acrylics*, polycarbonates and *epoxyresins*. Most acetone is made in the cumene process which also produces *phenol*.

acids: compounds with the following characteristic properties:

- in water they form solutions with a *pH* below 7
- they change the colors of *acid–base indicators*
- they react with *metals* such as magnesium to produce hydrogen gas:

  $$Mg(s) + 2HCl(aq) \longrightarrow MgCl_2(aq) + H_2(g)$$

- they react with carbonates such as calcium carbonate to form carbon dioxide gas:

  $$CaCO_3(s) + 2HCl(aq) \longrightarrow CaCl_2(aq) + CO_2(g) + H_2O(l)$$

- they react with *basic oxides* to form salts and water:

  $$CuO(s) + H_2SO_4(aq) \longrightarrow CuSO_4(aq) + H_2O(l)$$

Pure acids may be solids (such as citric and tartaric acids), liquids (such as *sulfuric*, *nitric* and *acetic* acids) or gases (such as hydrogen chloride which becomes *hydrochloric acid* when it dissolves in water).

Definitions of acids are based on the theories used to explain why they have similar properties. Chemists use different theories in different contexts. When considering acids in solution in water they still sometimes refer to the theory which the Swedish chemist Svante Arrhenius (1859–1927) suggested. He explained the behavior of acids in terms of hydrogen ions. What acids have in common, according to this theory, is that they produce hydrogen ions when they dissolve in water.

$$HCl(g) \longrightarrow H^+(aq) + Cl^-(aq)$$

The definition of an acid generally used today is based on the *Brønsted–Lowry theory*. Another definition is based on the *Lewis acid/base theory*.

acid anhydrides are related to *carboxylic acids*. The *functional group* in an anhydride is formed by eliminating a molecule of water from two carboxylic acid groups. Sometimes this happens simply on heating the acid but generally anhydrides are made in other ways.

With water, alcohols and amines, acid anhydrides react in a similar way to *acyl chlorides*. Chemists call this type of reaction *acylation*.

Anhydrides are less reactive than acyl chlorides and so are often preferred for laboratory and industrial syntheses. Their reactions may require heating in a flask fitted with a *reflux condenser*.

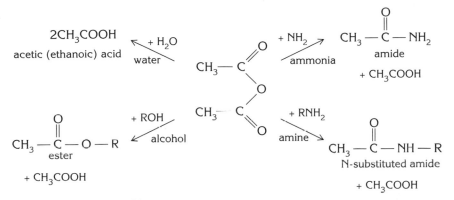

Reactions of acetic anhydride

acid–base equilibria: equilibrium systems involving acid–base reactions. Acid–base reactions are *reversible*. This is illustrated by a solution of an *ammonium salt* in water.

$$NH_4^+(aq) + H_2O(l) \rightleftharpoons NH_3(aq) + H_3O^+(aq)$$
$$\text{acid 1} \qquad \text{base 2} \qquad\quad \text{base 1} \qquad \text{acid 2}$$

There is competition for *protons* (hydrogen ions) between ammonia molecules and water molecules. On the left-hand side of the equation the protons are held by *lone pairs of electrons* on the ammonia molecules. On the right-hand side they are held by lone pairs on water molecules.

The *equilibrium law* applies. The position of equilibrium is determined by the value of the equilibrium of constant for the reaction. For the example in the equation, the relevant equilibrium constant is the *acid dissociation constant* for the ammonium ion.

The equilibrium involves two conjugate acid–base pairs:

- NH_4^+ and NH_3
- H_3O^+ and H_2O.

An acid turns into its conjugate base when it loses a proton. A base turns into its conjugate acid when it gains a proton.

acid–base indicators are used to show changes in *pH* of solutions. Indicators are *weak acids* or *bases* which change color when they lose or gain hydrogen ions. When added to a solution an indicator gains or loses protons depending on the pH of the solution. It is conventional to represent a weak acid indicator as HIn where In is shorthand for the rest of the molecule. In water:

$$HIn(aq) + H_2O(l) \rightleftharpoons H_3O^+(aq) + In^-(aq)$$
un-ionized indicator color 1   indicator after losing a proton color 2

3

The structures of methyl orange in acid and alkaline solutions. In acid solution the added hydrogen ion (proton) localizes two electrons to form a covalent bond. In alkaline solution the removal of the hydrogen ion allows the two electrons to join the other delocalized electrons. The change in the number of delocalized electrons causes a shift in the peak of the wavelengths of light absorbed, so the color changes and the molecule acts as an indicator.

Indicators such as phenolphthalein, methyl orange and bromothymol blue change color over a range of pH values and are used to detect the *end-point* in *acid–base titrations*.

The pH range over which an indicator changes color is determined by its strength as an acid (or base). Typically the range is given roughly by $pK_a \pm 1$ (see the *Henderson–Hasselbalch equation* for an explanation).

Indicator	pK_a	Color change HIn/In⁻	pH range over which color change occurs
methyl orange	3.6	red/yellow	3.2–4.2
methyl red	5.0	yellow/red	4.2–6.3
bromothymol blue	7.1	yellow/blue	6.0–7.6
phenolphthalein	9.4	colorless/red	8.2–10.0

acid–base reactions, according to the *Brønsted–Lowry theory*, are reactions involving the transfer of protons from an acid to a base.

acid–base titration: a practical technique used to determine the *concentration* of an *acid* or an *alkali*. A *titration* measures the volume of a *standard solution* of alkali or acid needed to react exactly with a measured volume of the unknown solution. The procedure for an acid–base titration is the same as for any other titration but the method for finding the end-point is distinctive: the analyst either adds an *acid–base indicator* or uses a *pH meter*.

Worked example

Limewater is a saturated solution of calcium hydroxide in water. 25.0 cm³ of 0.04 mol/L hydrochloric acid from a buret neutralized 20.0 cm³ of limewater. What was the concentration of the limewater?

Notes on the method

Always start by writing the equation for the reaction. See *titration* for a general method for the calculations.

Remember to convert volumes in cm³ to volumes in liters by dividing by 1000.

In any titration there is one unknown – in this case the concentration of the limewater, c_A.

Answer

The equation for the reaction is:

$$Ca(OH)_2(aq) + 2HCl(aq) \longrightarrow CaCl_2(aq) + 2H_2O(l)$$

The volume of calcium hydroxide in the flask, $V_A = \dfrac{20.0}{1000}$ L

Let the concentration of calcium hydroxide be c_A.

The volume of hydrochloric acid added from the buret, $V_B = \dfrac{25.0}{1000}$ L

The concentration of hydrochloric acid, $c_B = 0.04$ mol/L

$$\frac{V_A \times c_A}{V_B \times c_B} = \frac{n_A}{n_B}$$

$$\frac{20.0/1000 \times c_A}{25.0/1000 \times c_B} = \frac{1}{2}$$

Therefore $\quad c_A = \dfrac{25.0 \times 0.04}{2 \times 20.0} = 0.025$ mol/L

The concentration of the limewater was 0.025 mol/L.

acid catalysis: any reaction speeded up by an acid *catalyst*. One example is the acid-catalysed *hydrolysis* of an *ester* to give a *carboxylic acid* and an *alcohol*. All the chemicals are in the same solution so this is an example of *homogeneous catalysis*.

acid chlorides: see *acyl chlorides*.

acid dissociation constants measure the strength of acids. They show the extent to which acids dissociate into ions in solution. Acid dissociation constants are used to compare the strengths of relatively *weak acids* which are only partly ionized in solution.

For a weak acid represented by the formula HA:

$$HA(aq) + H_2O(l) \rightleftharpoons H_3O^+(aq) + A^-(aq)$$

According to the *equilibrium law*, the equilibrium constant,

$$K_c = \frac{[H_3O^+(aq)][A^-(aq)]}{[HA(aq)][H_2O(l)]}$$

In dilute solution the concentration of water is effectively constant, so the expression can be written in this form:

$$K_a = \frac{[H_3O^+(aq)][A^-(aq)]}{[HA(aq)]}, \text{ where } K_a \text{ is the acid dissociation constant}$$

Acid	K_a in mol/L	
nitrous acid, HNO_2	4.7×10^{-4}	stronger
formic (methanoic) acid, $HCOOH$	1.6×10^{-4}	
acetic (ethanoic) acid, CH_3COOH	1.7×10^{-5}	
hypochlorous acid, $HOCl$	3.7×10^{-8}	
phenol, C_6H_5OH	1.3×10^{-10}	weaker

A B C D E F G H I J K L M N O P Q R S T U V W X Y Z

Worked example

Calculate the hydrogen ion concentration and the pH of a 0.01 mol/L solution of acetic acid. K_a for the acid is 1.7×10^{-5} mol/L.

Notes on the method

Two approximations simplify the calculation.

1   The first assumption is that $[H_3O^+(aq)] = [A^-(aq)]$. In this example A^- is the acetate ion CH_3COO^-. This assumption seems obvious from the equation for the ionization of a weak acid but it ignores the hydrogen ions from the ionization of water. Water produces far fewer hydrogen ions than most weak acids so its ionization can be ignored.

2   The second assumption is that so little of the acetic acid ionizes in water that $[HA(aq)] \approx 0.01$ mol/L. Here HA represents acetic acid. This is a riskier assumption which has to be checked because in very dilute solutions the degree of ionization may become quite large relative to the amount of acid in the solution.

Answer

$$K_a = \frac{[H_3O^+(aq)][A^-(aq)]}{[HA(aq)]} = \frac{[H_3O^+(aq)]^2}{0.01 \text{ mol/L}} = 1.7 \times 10^{-5} \text{ mol/L}$$

Therefore $[H_3O^+(aq)]^2 = 1.7 \times 10^{-7}$ (mol/L)^2

So $[H_3O^+(aq)] = 4.12 \times 10^{-4}$ mol/L

$pH = -lg\,[H_3O^+(aq)] = -lg\,[4.12 \times 10^{-4}] = 3.39$

Check the second assumption: in this case about 0.0004 mol/L of the 0.0100 mol/L of acid (4%) has ionized. In this instance the degree of ionization is just about small enough to justify the assumption that $[HA(aq)] \approx$ the concentration of un-ionized acid.

acidic oxide: an oxide of a *non-metal* which reacts with water to form an *acid*. Some acidic oxides are insoluble but they can be recognized because they react directly with *basic oxides* to form *salts*. Note that acidic oxides are not themselves *acids* as defined by the Brønsted–Lowry theory because they do not contain ionizable hydrogen atoms, so they cannot act as proton donors.

Oxide	Acid formed with water
carbon dioxide, CO_2	carbonic acid, H_2CO_3
sulfur dioxide, SO_2	sulfurous acid, H_2SO_3
sulfur trioxide, SO_3	sulfuric acid, H_2SO_4
phosphorus pentoxide, P_2O_5	phosphoric acid, H_3PO_4

Silica or silicon dioxide (SiO_2) is an acidic oxide which is insoluble in water. Silica reacts with basic metal oxides at high temperatures in furnaces during glass making and in steelmaking. Calcium oxide (quicklime) is added to a *blast furnace* to remove silica and other impurities. The calcium silicate is a liquid at the temperature of the furnace; it runs towards the bottom where it floats on top of the molten iron as a slag which can be tapped off separately.

$$CaO(s) + SiO_2(s) \longrightarrow CaSiO_3(l)$$

acid rain: a type of pollution produced by burning fuels or wastes or when industrial processes release *acidic oxides* into the air. Sulfur dioxide (SO_2) and *nitrogen oxides* (NO_x) form during *combustion* in engines, furnaces, incinerators and power stations. These primary pollutants are converted to secondary pollutants by chemical reactions in the air. Among the secondary pollutants are sulfuric acid, nitric acid and ammonium sulfate. The pollutants cause acidification by being deposited in the environment as gases or particles (dry deposition) or in the form of rain or mist (wet deposition).

acid salt: a compound formed by partly neutralizing *acids* such as *sulfuric acid*, H_2SO_4, or *phosphoric acid*, H_3PO_4. These are acids with two or three ionizable hydrogen atoms. Sodium hydrogensulfate, $NaHSO_4$, and sodium dihydrogenphosphate, NaH_2PO_4, are examples of acid salts.

The negative ion in an acid salt can act either as an acid by giving away a further proton, or as a *base* by accepting a proton. The pH of an *aqueous solution* of an acid salt depends on the *acid strength* of the parent acid.

A solution of sodium hydrogensulfate, a salt of a *strong acid*, is acidic with a pH below 7 because the hydrogensulfate ion is also an acid, giving protons to water molecules:

$$HSO_4^- + H_2O \rightleftharpoons SO_4^{2-} + H_3O^+$$

A solution of sodium hydrogencarbonate, a salt of the *weak acid* called carbonic acid, is alkaline with a pH above 7 because the hydrogencarbonate ion is a base, taking protons (H^+) from water molecules:

$$HCO_3^- + H_2O \rightleftharpoons H_2CO_3 + OH^-$$

acid strength: see *strong acid*, *weak acid* and *acid dissociation constant*.

acid strength of organic hydroxy compounds: the order of acid strength for organic compounds with —OH groups is:

carboxylic acids > phenols > water > alcohols.

Carboxylic acids are *weak acids* which ionize to a significant extent by giving protons to water forming a solution with pH 3–4. *Phenol* is acid enough to lower the pH of a solution in water to 5–6 but it ionizes significantly only when mixed with a stronger base such as the hydroxide ions in a solution of sodium hydroxide. Ethanol and other *alcohols* are such weak acids that they do not ionize to a significant extent in water or in a solution of a strong alkali.

acrylics: a group of *addition polymers* based on derivatives of acrylic acid, $H_2C = CH — COOH$. The most widely used acrylic is the polymer of acrylonitrile, $H_2C = CH — CN$. Often a small amount of chloroethene is added to the monomer. *Copolymerization* produces a polymer which, when spun into *fibers*, feels similar to wool.

actinides (or actinoids) are the *f-block elements* in period 7 of the *periodic table*. They are the 14 elements from thorium to lawrencium which lie between actinium (element 89) and element 104, rutherfordium.

activated complex: a combination of reacting atoms, molecules or ions at that point during a chemical reaction when they are at the top of the *activation energy* barrier between reactants and products. An activated complex is a combination of reacting particles at a higher energy because chemical bonds are stretched. The activated complex is the *transition state* for the reaction step. (See the diagram on the next page.)

Reaction profile to show the activated complex formed in the course of a nucleophilic substitution reaction

activation energy: the minimum energy needed in a collision between molecules if they are to react. The activation energy is the height of the energy barrier separating reactants and products during a chemical reaction (see *activated complex*).

Activation energy is an important idea in the *collision theory* of reaction rates. It accounts for the fact that reactions go much more slowly than would be expected if every collision between atoms and molecules led to a reaction. Only a very small proportion of collisions bring about chemical change. Molecules react only if they collide with enough energy between them to overcome the energy barrier. At around room temperature only a small proportion of molecules have enough energy to react.

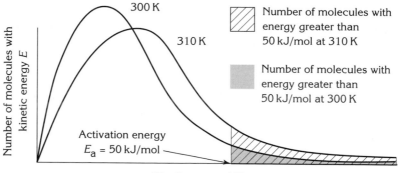

Maxwell–Boltzmann distribution curve for the kinetic energies of molecules at about room temperature, 300 K and 310 K. The shaded area shows the proportion of molecules having at least the activation energy for a reaction. This area is bigger at a higher temperature.

Activation energies account for the way in which reaction rates vary with temperature, T. At a higher temperature there are more molecules with enough energy to react when they collide.

The activation energies for many reactions in biochemical systems are around 50 kJ/mol and for these reactions the rate of reaction doubles for each 10 K rise in temperature.

It is possible to determine the activation energy for a reaction by measuring its rate over a range of temperatures and then using the *Arrhenius equation*.

active site: the part of a *catalyst* which is responsible for its catalytic activity. *Zeolites* are examples of inorganic catalysts, for example, which owe their catalytic activity to active sites in the crystal structure which can adsorb specific molecules so that they can react to produce the required products.

Enzymes are biological catalysts. An enzyme molecule consists of a coiled *protein* chain. The coiling gives rise to an active site with a precise three dimensional shape. The enzyme can act only on molecules with the right shape to fit into the active site. This accounts for the specificity of enzymes.

A model illustrating an enzyme with an active site splitting a molecule into two smaller molecules

acyl chlorides are made from *carboxylic acids* by replacing the hydroxyl (— OH) group with a chlorine atom.

Formation of ethanoyl (acetyl) chloride from ethanoic (acetic) acid

Acyl chlorides are very reactive; they are powerful acylating agents. Most are quickly *hydrolyzed* back to the acid by cold water. Acyl chlorides react rapidly with *alcohols* and *phenols* to form *esters*. These reactions are all examples of *acylation*.

$$CH_3 - C\overset{O}{\diagdown}_{Cl} \quad + \quad C_2H_5OH \quad \longrightarrow \quad CH_3 - \overset{O}{\overset{\|}{C}} - O - C_2H_5 \quad + \quad HCl$$

ethyl acetate

Formation of an ester

They react with *ammonia* and *amines* to form *amides*.

Formation of an amide and an N-substituted amide

Acyl chlorides can also acylate *benzene* by the *Friedel–Crafts* reaction.

acylation: a reaction which substitutes an acyl group for a hydrogen atom. The H atom may be part of an — OH group, an — NH_2 group or a *benzene* ring. Acylating agents are either *acyl chlorides* or *acid anhydrides*.

The acetyl group – an acyl group

Acylation is the reaction which converts salicylic acid to *aspirin*.

Formation of aspirin by acylation of salicylic acid with acetic anhydride

Acylation of benzene is an example of the *Friedel–Crafts reaction*.

addition–elimination reactions take place when two molecules first add together and then immediately split off a small molecule such as water or hydrogen chloride. They are sometimes called *condensation reactions*.

The reactions take place in three stages:

• nucleophilic addition

• gain and loss of hydrogen ions (protons)

• elimination of water.

The reactions of *acyl chlorides* with the *nucleophiles* water, alcohols, ammonia and amines are all addition–elimination reactions.

Mechanism of the reaction of an alcohol (ethanol) with an acyl halide (acetyl chloride) to form an ester (ethyl acetate)

addition polymerization: a process for making *polymers* from compounds containing *double bonds*. Many molecules of the *monomer* add together to form a long chain polymer. Ethene, for example, polymerizes to form poly(ethene).

$$n\text{CH}_2 = \text{CH}_2 \longrightarrow [-\text{CH}_2-\text{CH}_2-]_n$$

Monomer	Polymer	Notes
ethene (ethylene)	poly(ethene) or polythene	A high-pressure, high-temperature process in the presence of a peroxide initiator produces low-density poly(ethene) with branched chains. The mechanism involves a *radical chain reaction*. A low-pressure, low-temperature process with a *Ziegler–Natta catalyst* produces high-density poly(ethene) in which the polymer chains pack closer because they have no side branches.
propene (propylene)	poly(propene) or polypropylene	*Isotactic* poly(propene) is made using a Ziegler–Natta catalyst. This strong polymer is used to make pipes, shrink-wrap, carpet fibers and ropes.
phenylethene (styrene)	poly(phenylethene) or polystyrene	Expanded polystyrene has low density and is an excellent thermal insulator. It is used for packaging because it absorbs shocks.
chloroethene (vinyl chloride)	poly(chloroethene) or polyvinylchloride (PVC)	Unplasticized uPVC is a rigid polymer suitable for guttering and window frames. PVC with a *plasticizer* is flexible and used for packaging, flooring and cable insulation.

A B C D E F G H I J K L M N O P Q R S T U V W X Y Z

11

tetrafluorethene

poly(tetrafluorethene) or PTFE

Engineers use PTFE to provide low-friction surfaces to allow bridges and other engineering structures to move slightly as the metals expand and contract with temperature changes. It is the polymer used to coat nonstick pans.

addition reaction: a reaction in which two molecules combine to form a single product. Bromine, for example, adds to ethene to form the addition product 1,2-dibromoethane.

1,2-dibromoethane

Structure diagram to show addition of bromine to ethene

Addition reactions are characteristic of *unsaturated compounds* such as the *alkenes* and *carbonyl compounds*.

adsorption: a process in which atoms, molecules or ions are held on the surface of a solid. Adsorption processes are important in *heterogeneous catalysis* and in some types of *chromatography*. Adsorption should be carefully distinguished from *absorption*.

aerobic respiration: see *respiration*.

aerosols: *colloids* in which particles of a solid, or droplets of a liquid are finely dispersed in a gas. Smoke is an aerosol in which the dispersed particles are solid. Examples of aerosols with liquid droplets are mist, clouds and insecticide or paint sprays.

agrochemicals include *pesticides* which destroy the organisms that damage crops such as insects (insecticides), fungi (fungicides) and weeds (*herbicides*). Other agrochemicals are the growth regulators which can stimulate or inhibit plant growth. The chemical industry distinguishes agrochemicals from *fertilizers*, which provide the chemicals needed for healthy plant growth.

air: a mixture of gases as shown in the table below. The proportion of water *vapor* in the air varies widely according to the weather conditions. Air pollution adds other gases such as *nitrogen oxides*, sulfur dioxide, *ozone* and *hydrocarbons* (see also *acid rain* and *photochemical smog*). The gases are separated on a large scale by *fractional distillation* of liquid air.

Gas	Percentage by volume in dry air	Boiling point/K
nitrogen	78.0	77
oxygen	21.0	90
argon	0.9	87
carbon dioxide (about 0.04%) and small traces of neon, helium and krypton	0.1	

air condenser: a glass tube without a water jacket used for condensing *vapors* during *distillation* or refluxing (see *reflux condenser*). An air condenser is fitted when the temperature of the vapor is so high that the surrounding air is cool enough to condense the liquid. Normally this is when the boiling point of the liquid being distilled is above about 150°C.

alcohols: compounds with the formula R—OH where R represents an *alkyl group*. The hydroxy group —OH is the *functional group* which gives the compounds their characteristic reactions.

The chemical industry makes alcohols by the *hydration* of *alkenes* in the presence of an acid catalyst.

Equation for the industrial production of ethanol. Ethanol is also produced by fermentation.

Alcohols are named by changing the ending of the corresponding *alkane* to –ol. So ethane becomes ethanol.

Names and structures of two alcohols

$$CH_3 — CH_2 — CH_2 — CH_2 — OH$$ 1-butanol, a primary alcohol

$$CH_3 — CH_2 — CH — CH_3$$
$$| $$
$$OH$$ 2-butanol, a secondary alcohol

$$CH_3$$
$$|$$
$$CH_3 — C — CH_3$$
$$|$$
$$OH$$ 2-methyl-2-propanol, a tertiary alcohol

Primary, secondary and tertiary alcohols

Even the simplest alcohols such as methanol and ethanol are liquids at room temperature because of *hydrogen bonding* between the hydroxy groups. For the same reason alcohols with relatively short hydrocarbon chains mix freely with water.

The reactions of ethanol are typical of the alcohols in general.

Reactions of ethanol

Oxidation reactions distinguish primary alcohols, secondary alcohols and tertiary alcohols (see *primary, seconday and tertiary organic compounds*).

primary alcohol $\xrightarrow[\text{warm}]{Cr_2O_7{}^{2-},\ H^+(aq)}$ aldehyde $\xrightarrow[\text{reflux}]{\text{excess } Cr_2O_7{}^{2-},\ H^+(aq)}$ carboxylic acid

secondary alcohol $\xrightarrow[\text{reflux}]{Cr_2O_7{}^{2-},\ H^+(aq)}$ ketone

tertiary alcohol   no reaction

Use of oxidation with an acidified solution of dichromate ions to distinguish primary, secondary and tertiary alcohols

aldehydes: *carbonyl compounds* in which the carbonyl group is attached to two hydrogen atoms or an *alkyl group* and a hydrogen atom. So the carbonyl group is at the end of a carbon chain. The — CHO group is the *functional group* which gives aldehydes their characteristic reactions. They are named systematically after the alkane with the same carbon skeleton by changing the ending "e" to "al" (so, for example, ethane becomes ethanal).

formaldehyde   acetaldehyde   propanal
(methanal)   (ethanal)

Structures and names of aldehydes

Methanal is a gas at room temperature. Ethanal boils at 21°C so it may be a liquid or gas at room temperature depending on the conditions.

Aldehydes are formed by *oxidation* of primary *alcohols* on heating with a mixture of dilute sulfuric acid and potassium dichromate under conditions which allow the aldehyde to distil

14

off as it forms. Unlike *ketones*, aldehydes can easily be oxidized further to *carboxylic acids* by longer heating with an excess of the reagent.

$$CH_3CH_2CH_2OH \xrightarrow[\text{heat}]{Cr_2O_7^{2-}, H^+} CH_3CH_2C{\overset{\displaystyle O}{\diagdown}}_H \xrightarrow[\text{reflux}]{Cr_2O_7^{2-}, H^+} CH_3CH_2C{\overset{\displaystyle O}{\diagdown}}_{OH}$$

1-propanol   propanal   propanoic acid

Two stage oxidation of 1-propanol

Fehling's solution and *Tollen's reagent* (ammoniacal silver nitrate) are mild oxidizing agents, used to distinguish aldehydes from ketones. Preparing a crystalline derivative with Brady's reagent (2,4-dinitrophenylhydrazine) makes it possible to identify aldehydes.

Sodium borohydride ($NaBH_4$) reduces aldehydes to primary alcohols. An alternative reducing agent is *lithium aluminum hydride* ($LiAlH_4$) but it is not so easy to use.

$$CH_3CH_2 - C{\overset{\displaystyle O}{\underset{H}{\diagdown}}} + 2[H] \xrightarrow{Na^+BH_4^-} CH_3\,CH_2\,CH_2OH$$

Reduction of propanal to 1-propanol. The 2[H] comes from the reducing agent. This is a shorthand way of balancing a complex equation involving reduction.

Thanks to the double bond in the carbonyl group, aldehydes (like ketones) undergo *addition reactions*. These are *nucleophilic addition reactions*.

Addition reactions of acetaldehyde

2-hydroxypropanenitrile
(which hydrolyzes to
2-hydroxypropanoic acid)

alicyclic hydrocarbons: *hydrocarbons* with rings of carbon atoms (but no *benzene* rings). Examples are cycloalkanes and cycloalkenes.

cyclohexane C_6H_{12} cyclohexene C_6H_{10}

Structures of cyclohexane and cyclohexene

aliphatic hydrocarbons: *hydrocarbons* with no rings of carbon atoms. The chains of carbon atoms may be branched or unbranched. *Alkanes*, *alkenes* and *alkynes* are all aliphatic compounds.

aliquot: a measured volume taken from a liquid sample for analysis. Typically aliquots are taken by a pipet for *titration*.

alkalis: *bases* which dissolve in water. The common laboratory alkalis are the hydroxides of sodium and potassium, calcium hydroxide (in limewater) and *ammonia*. Alkalis form solutions with a *pH* above 7 so they change the colors of *acid–base indicators*. What alkalis have in common is that they dissolve in water to produce hydroxide (OH^-) ions. Sodium hydroxide (Na^+OH^-) and potassium hydroxide (K^+OH^-) contain hydroxide ions in the solid as well as in solution. Ammonia produces hydroxide ions by reacting with water. Ammonia acting as a base takes *protons* from water molecules.

$$NH_3(g) \quad + \quad H_2O(g) \rightleftharpoons NH_4^+(aq) \; + \; OH^-(aq)$$

alkali metals: the elements in *group I* of the periodic table. All the elements react with water to form alkaline solutions of the metal hydroxide. Sodium, for example, reacts to form *sodium hydroxide*. It is important to remember that the hydroxides are alkaline because of the presence of OH^- ions. Alkali metal ions themselves, such as sodium ions, Na^+, do *not* make solutions alkaline.

alkaline cell: an *electrochemical cell* in which the electrolyte is the alkali, potassium hydroxide. Alkaline cells are suitable for toys and cassette recorders with a motor drawing a heavy or continuous current.

alkaline earth metals: the elements in *group II* of the periodic table. The alkaline earths are the *oxides* of these metals which are much less soluble in water than the corresponding group I oxides.

alkaloids: organic nitrogen compounds extracted from plants. Alkaloids can have a powerful effect on the human nervous system and can be very poisonous. Caffeine, the stimulant in tea and coffee, is an alkaloid. Other alkaloids are the *drugs* quinine, morphine and codeine.

Structure of caffeine

alkanes: the *hydrocarbons* which make up most of crude oil and natural gas. Alkanes are *saturated compounds* with the general formula C_nH_{2n+2}. The carbon atoms in alkane molecules may be in straight chains or branched chains but all the bonds are single bonds.

The names of branched alkanes are based on the longest straight chain in the molecule with the positions of the side chain alkyl groups identified by numbering the carbon atoms.

Name	Molecular formula	Structure
methane	CH_4	
ethane	C_2H_6	
propane	C_3H_8	

Alkane molecules are non-*polar* so they do not mix with, or dissolve in, *polar solvents* such as water. The molecules are only held together by weak *Intermolecular forces* (*van der Waals forces*). The longer the molecules the greater the attraction between them. The boiling points rise as the number of carbon atoms per molecule increases. Alkanes in the range C_1 to C_4 are gases at room temperature and pressure. Under the same conditions, alkanes in the range C_5 to C_{17} are liquids while those with more than 17 carbon atoms per molecule are solids. Liquid alkanes with longer chain lengths are *viscous liquids* used as lubricants.

2,2-dimethylbutane   2,3-dimethylbutane

Names and structures of some branched alkanes

The *bond energies* for C — C and C — H bonds are high so the bonds are relatively hard to break. Also the bonds are not polar. This means that alkanes are very unreactive towards reagents in water such as acids and alkalis, as well as oxidizing and reducing reagents. Three important reactions of alkanes are: *combustion*, *halogenation* and *cracking* (see *thermal cracking* and *catalytic cracking*). Cracking and the halogenation of alkanes are examples of *radical chain reactions*.

Reactions of alkanes with ethane as the example

alkenes, such as ethene and propene, are products of *cracking* oil fractions. They are important starting points for making other chemicals because of the reactivity of the double bonds in their molecules.

Alkenes are *unsaturated compounds*. They are hydrocarbons with the general formula C_nH_{2n}. The characteristic *functional group* of the alkenes is a carbon–carbon double bond. The presence of the double bond makes alkenes more reactive than *alkanes*.

The name of an alkene is based on the name of the corresponding alkane with the ending change to "ene." Where necessary, a number in the name shows the position of the double bond in the structure, as in the two structural *isomers* 1-butene and 2-butene. Counting starts from the end of the chain that will give the lowest possible number in the name. The number in the name shows the first of the two atoms connected by the double bond. In 1-butene, for example, the double bond is between the first and the second atoms in the carbon chain.

The *pi (π) bonds* in alkenes prevent free rotation about the double bonds. This means that some alkenes can show *geometrical isomerism*.

1-butene, C_4H_8

2-butene, C_4H_8

propene, C_3H_6
(propylene)

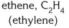

ethene, C_2H_4
(ethylene)

Names and structures of alkenes

The boiling points of alkenes increase as the number of carbon atoms in the molecules increase. Ethene, propene and the butenes are gases at room temperature. Alkenes with more than four carbon atoms are liquids or even solids. Alkenes, like other hydrocarbons, do not mix with or dissolve in water.

The characteristic reactions of alkenes are *addition reactions* (see below). The reactions with reagents such as bromine and hydrogen bromide are *electrophilic addition reactions*. The high pressure process for making poly(ethene) involves a *radical chain reaction*.

Addition reactions of ethene

Markovnikov's rule helps to predict the main product of an addition reaction when an unsymmetrical molecule adds to an unsymmetrical alkene.

Potassium permanganate oxidizes alkenes. The products depend on the conditions. A dilute, acidified solution of potassium permanganate converts an alkene to a diol at room temperature.

1,2-ethanediol

The reaction of ethene with dilute, acidified permanganate ions producing
1,2-ethanediol

A solution of permanganate ions is purple. The color disappears as it reacts with an alkene. So the reaction with cold $MnO_4^-(aq)$ ions can be used to distinguish unsaturated and saturated hydrocarbons. Another test for unsaturated hydrocarbons uses a solution of bromine (see *organic analysis*).

A hot, concentrated solution of acidic potassium permanganate breaks apart the double bond in an alkene.

cyclohexene   1,6-hexanedioic acid

Using hot, acidic potassium permanganate to break a carbon chain at a double bond. In this example the reaction converts cyclohexene to 1,6-hexanedioic acid.

alkoxides: metal *salts* of alcohols. Ethanol, for example, reacts with *sodium* to produce sodium ethoxide.

$$C_2H_5OH(l) + Na(s) \longrightarrow C_2H_5O^-Na^+(s) + H_2(g)$$

The ethoxide ion is a very *strong* base. It rapidly gains a proton if water is added to turn back to ethanol.

alkyl group: a group of carbon and hydrogen atoms which forms part of the structure of a molecule. The simplest example is the methyl group CH_3 — which is methane with one hydrogen atom removed. In general, alkyl groups are *alkane* molecules minus one hydrogen atom.

Alkyl group	Formula
methyl	CH_3-
ethyl	CH_3CH_2-
propyl	$CH_3CH_2CH_2-$
butyl	$CH_3CH_2CH_2CH_2-$

A useful shorthand for any alkyl group is the capital letter R. Further alkyl groups are then represented by R' or R". So, for example a tertiary *amine* with three different alkyl groups attached to the nitrogen atom can be written as:

General formula for a tertiary amine

$$\begin{array}{c} R \\ | \\ R'—N: \\ | \\ R" \end{array}$$

alkylation: a reaction which introduces an *alkyl group* into a molecule by *addition* or *substitution*. Industrially the alkylation of alkenes produces branched alkanes needed to raise the octane number of *gasoline*.

The *Friedel–Crafts reaction* adds alkyl groups to benzene rings in *arenes*.

methylpropene   methylpropane   2,2,4-trimethylpentane

Production of a branched alkane by alkylation of an alkene

20

alkynes: hydrocarbons with a triple bond between two carbon atoms. They are *unsaturated hydrocarbons* with the general formula C_nH_{2n-2}. The best known example is ethyne (historically known as acetylene) which is the first member of the series.

The structure of ethyne $H — C \equiv C — H$

allotropes: different forms of the same element in the same physical state. *Oxygen* has gaseous allotropes with different molecular structures; normal diatomic oxygen (O_2) and *ozone* (O_3). *Sulfur* has solid allotropes with the same molecular structure (S_8) but different crystal structures; rhombic and monoclinic sulfur.

Carbon has allotropes consisting of different extended networks of atoms; diamond and graphite. Carbon also exists in molecular form as the *fullerenes*.

Fragment of a
diamond network structure

Fragment of a
graphite network structure

C_{60} molecule

Allotropes of carbon

Other elements with allotropes are *phosphorus* and *tin*.

alloys: usually mixtures of metals, but *steel* is an alloy of iron with the nonmetal carbon. Alloys are often more useful than pure metals. Controlling the composition of an alloy makes it possible to vary its properties. Examples of alloys are:

- *mild steel* – an alloy of iron with 0.12–0.25% carbon; it is relatively cheap, easily rolled into sheets and can be pressed into shape

- *brass* – an alloy of copper with less than 20% zinc; it is easily worked, has a gold color and does not corrode

- *bronze* – an alloy of copper with up to 12% tin; it is strong, hardwearing and resistant to corrosion

- *solder* – an alloy of lead and tin which melts at a low temperature and can be used to join metals

- *duralumin* – an alloy of 95% aluminum with 4% copper and smaller amounts of magnesium, iron and silicon; it is stronger, harder and more resistant to corrosion than pure aluminum.

alpha decay: the emission of an alpha particle from the nucleus of an *atom* of a *radioactive* element. Alpha decay produces an atom of an element two places to the left in the periodic table because the *atomic number* of the atom decreases by two. At the same time the *mass* *number* decreases by four. When an atom of uranium-238 decays by loss of an alpha particle it turns into an atom of thorium-234.

$$^{238}_{92}U \longrightarrow \,^{234}_{90}Th + \,^{4}_{2}He$$

alpha (α) particles consist of two protons and two neutrons and are identical with the nuclei of helium atoms. They are emitted from the nuclei of *radioactive* elements during alpha decay.

Alpha particle

An alpha particle emitted from a radioactive nucleus is usually represented as $^{4}_{2}He$ but chemically it is $^{4}_{2}He^{2+}$

aluminum (Al): a silvery metal. It is very useful because it has a relatively low density while being resistant to *corrosion* and is a good conductor of thermal energy and electricity.

Aluminum is the second element in group III, coming below boron in the periodic table. The *electron configuration* of aluminum is $[Ne]3s^23p^1$. Aluminum has a cubic *close packed giant structure* with the atoms held together by *metallic bonding*.

Aluminum is a reactive metal with a *standard electrode potential* $E° = -1.66$ V for the $Al^{3+}(aq) \mid Al(s)$ half-cell. This puts aluminum above zinc in the *electrochemical series*. A thin, tough and transparent layer of aluminum oxide on the surface of the metal stops most reactions. As a result aluminum does not corrode in air or dissolve in acids as might be expected. The oxide layer is so important that it is often thickened by *anodizing*.

The oxide layer protects the metal from dilute acids but aluminum is attacked by fairly concentrated hydrochloric acid.

$$2Al(s) + 6HCl(aq) \longrightarrow 2AlCl_3(aq) + 3H_2(g)$$

Concentrated nitric acid is an oxidizing agent which thickens the oxide layer making the metal less likely to react ("passive").

The oxide layer and the metal itself dissolve in concentrated, strong alkalis.

$$2Al(s) + 2OH^-(aq) + 6H_2O(l) \longrightarrow 2[Al(OH)_4]^- + 3H_2(g)$$

On heating, aluminum reacts directly with nonmetals, including oxygen, sulfur, nitrogen and the halogens. Aluminum combines very strongly with oxygen and is used to reduce the oxides of other metals in the *thermit reaction*.

Important aluminum compounds include *aluminum oxide* and hydroxide, *aluminum chloride* and salts containing the hydrated *aluminum(III) ion*. In all its compounds aluminum is in the *oxidation state* +3. The relatively high charge and small size of the Al^{3+} ion state helps to account for the differences between aluminum compounds and the compounds of metals in group I and group II. In some ways aluminum compounds have properties more characteristic of nonmetals than metals. Compared with the larger Na^+ ions, Al^{3+} ions have a strong tendency to polarize neighboring anions, giving rise to *polar covalent* rather than *ionic bonding*. Only the fluoride and oxide are ionic. (See also *polarization of ions*.)

aluminum extraction: *aluminum* is obtained by *electrolysis* of a solution of aluminum oxide in molten cryolite, Na_3AlF_6. Pure aluminum oxide for the process is obtained by purifying *bauxite*.

The discovery that aluminum oxide dissolves in molten cryolite was essential to the development of the process because the pure oxide melts at 2015°C – much too high for economic industrial processing.

Electrolysis takes place in carbon-lined steel tanks called "pots." The carbon lining is the *cathode* of the cell. The *anodes* are blocks of carbon. The currents used are high, of the order of 200 000 A, so the process is generally carried out where electricity is relatively cheap, often close to a source of hydroelectric power.

Reduction at the cathode: $Al^{3+} + 3e^- \longrightarrow Al$

The aluminum is liquid at the temperature of the molten electrolyte (970°C) and it collects at the bottom of the pot. The molten metal is tapped off from time to time.

Oxidation at the anode: $2O^2 \longrightarrow O_2 + 4e^-$

Molten aluminum oxide in cryolite

Cross-sectional diagram of an electrolysis cell for extracting aluminum

Much of the oxygen reacts with the carbon of the anodes, forming carbon dioxide. The anodes burn away and have to be replaced regularly. The energy from the burning anodes helps to heat the electrolyte, keeping it at the working temperature at around 960°C.

Waste gas from the cell contains fluorides and has to be thoroughly cleaned to avoid pollution of the region surrounding the plant.

aluminum oxide (Al_2O_3): a white solid with a very high melting point. Aluminum oxide exists naturally as a group of minerals called corundum. Corundum is an important abrasive; its hardness is 9 on the *Mohs scale*. Emery is a greyish-black variety of corundum made of aluminum oxide mixed with iron oxide minerals. Some gemstones are varieties of corundum including rubies which are red because of the presence of some chromium(III) ions in place of aluminum(III) in the crystal structure. Sapphires are blue because some of the aluminum ions are replaced by a mixture of titanium(IV) and cobalt(II) or iron(II) ions.

Aluminum oxide is widely used as an engineering *ceramic*. It is used to make the insulator in spark plugs. A translucent, polycrystalline form of the compound encloses high-pressure, high-temperature sodium street lamps which give a much whiter light than the yellow low-pressure lamps.

Aluminum oxide is *amphoteric* but the pure solid is insoluble, and this property is more easily demonstrated by adding alkali to a solution of aluminum(III) ions. The amphoteric behavior is used in the production of the pure oxide from *bauxite*.

aluminum chloride ($AlCl_3$): an off-white solid which *sublimes* on heating and fumes in moist air because of *hydrolysis* to hydrogen chloride and the hydroxide. Aluminum chloride is manufactured on a large scale by injecting a stream of chlorine gas into molten aluminum at 800°C. At this temperature the product vaporizes and is collected as a solid in condensers.

Aluminum atoms in $AlCl_3$ have only six electrons and have a strong tendency to accept two more. As a result, aluminum chloride vapor contains Al_2Cl_6 *dimers.*

Solid $AlCl_3$ has a layer lattice with bonding intermediate between ionic and covalent.

The Lewis electron dot formula *for $AlCl_3$ and the bonding in an Al_2Cl_6 dimer. Lone pairs on chlorine atoms form* coordinate bonds *with aluminum atoms.*

$AlCl_3$ is a strong acceptor of electron pairs, making it a powerful *Lewis acid*.

Aluminum chloride is the catalyst for *Friedel–Crafts* and related reactions. In industry this means that it is an important catalyst for the manufacture of a wide range of useful products including *polymers*, pigments, pharmaceuticals, *dyes* and *detergents*.

aluminum(III) ions are hydrated in aqueous solution and are present as a *complex ion* $[Al(H_2O)_6]^{3+}$. Six water molecules form *coordinate bonds* with each metal ion (see the diagram on page 25).

The electrons in the water molecules are pulled towards the highly polarizing aluminum ion, making it easier for the water molecules linked to the aluminum ion to give away protons. The hydrated aluminum ion is as strong an *acid* as acetic acid. The hydrated ion gives protons to water molecules, forming *hydronium ions*. The solution is acidic enough to release carbon dioxide when added to sodium carbonate.

$$[Al(H_2O)_6]^{3+} + H_2O \rightleftharpoons [Al(H_2O)_5OH]^{2+} + H_3O^+$$

$$[Al(H_2O)_5OH]^{2+} + H_2O \rightleftharpoons [Al(H_2O)_4(OH)_2]^+ + H_3O^+$$

Adding a base such as hydroxide ions to a solution of aluminum ions removes a third proton, producing an uncharged complex. The uncharged complex is much less soluble in water and precipitates as a white jelly-like precipitate – hydrated aluminum hydroxide.

$$[Al(H_2O)_4(OH)_2]^+ + OH^- \rightleftharpoons [Al(H_2O)_3(OH)_3](s) + H_2O(l)$$

Adding excess alkali removes yet another proton, now producing a negatively charged ion which is soluble in water so that the precipitate redissolves. This demonstrates the *amphoteric* properties of aluminum hydroxide.

$$[Al(H_2O)_3(OH)_3](s) + OH^- \rightleftharpoons [Al(H_2O)_2(OH)_4]^-(aq) + H_2O(l)$$

All these changes are reversed by adding a solution of a strong acid such as hydrochloric acid which turns the hydroxide ions in the complex back to water molecules.

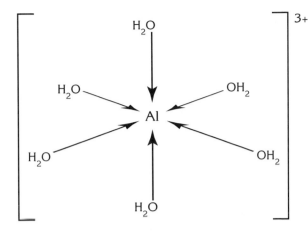

Structure of a hydrated aluminum ion showing the octahedral arrangement of water molecules. This is the hexaaquoaluminate(III) ion.

alums: *double salts* which crystallize easily from solutions in water forming attractive octahedral crystals. Alums have the general formula $M(I)M(III)(SO_4)_2.12H_2O$. $M(I)$ is an ion with a 1+ charge such as Na^+, K^+ or NH_4^+ and $M(III)$ is a metal ion with a 3+ charge such as Al^{3+}, Cr^{3+} or Fe^{3+}.

amalgams: solutions of metals in *mercury*. Dentists make amalgams for filling teeth. They grind up the compound Ag_3Sn with mercury. This produces an amalgam which is workable so that it can be shaped easily to fill a cavity. The amalgam soon hardens as the mercury reacts with the silver and the tin. As it hardens the mixture expands to fit tightly in the tooth cavity.

amides: organic nitrogen compounds derived from *carboxylic acids* by replacing the —OH group with an —NH_2 group.

Amides are white crystalline solids at room temperature, apart from methanamide which is a liquid. The simple amides are soluble in water.

acetamide
(ethanamide)

benzamide

Names and structures of amides

Amides form rapidly at room temperature when *acyl chlorides* or *acid anhydrides* react with ammonia. *Esters* react more slowly to form amides.

Reactions which produce amides

Another way of making amides is to convert a carboxylic acid to an ammonium salt and then to dehydrate the salt by heating it under reflux for several hours.

$$CH_3COOH + NH_3 \longrightarrow CH_3COO^- NH_4^+ \longrightarrow CH_3CONH_2 + H_2O$$

carboxylic ammonia   ammonium salt   amide
acid

Unlike *amines*, amides are very weak bases. They give neutral solutions in water and do not form salts with acids such as hydrochloric acid. The lone pair of electrons on the nitrogen atom is delocalized with the carbonyl group making it unavailable to form a dative bond to a proton.

Delocalization of electrons in the amide group

Reduction converts an amide to an amine. A suitable procedure is to treat the amide with *lithium aluminum hydride* in dry ether and then to add water.

$$CH_3CONH_2 \xrightarrow[\text{2. } H_2O]{\text{1. LiAlH}_4 \text{ in ether}} CH_3CH_2NH_2$$

Heating with phosphorus(v) oxide dehydrates an amide to a *nitrile*.

Heating with dilute acid or dilute alkali hydrolyzes amides to the corresponding acid

Hydrolysis of acetamide

(See also the *Hofmann degradation*.)

amines: nitrogen compounds in which one or more of the hydrogen atoms in ammonia (NH_3) is replaced by an *alkyl group* or *aryl group*. The number of alkyl groups determines whether the compound is a primary amine, a secondary amine or a tertiary amine.

The names of amines are based on the nature and number of alkyl groups attached to the nitrogen atom.

Amines can be very smelly. Ethylamine has a fishy smell. Animal flesh smells putrescent because it gives off diamines.

methylamine (primary)   dimethylamine (secondary)   trimethylamine (tertiary)

Names and structures of primary, secondary and tertiary amines. Note the meaning of primary, secondary and tertiary here – it is not the same as for alcohols and haloalkanes.

Two methods of making primary amines are:

* the reduction of a nitrile
* the substitution reaction of ammonia with a haloalkane.

Primary amines, like ammonia, can acts as *bases*, form *complex ions* with metal ions and react as *nucleophiles*.

Acid anhydrides and acyl chlorides both acylate primary amines.

$CH_3CH_2 NH_2$

$+ H_2O(l)$ → $CH_3CH_2NH_3^+(aq) + OH^-(aq)$
alkaline solution

$HCl(aq)$ → $CH_3CH_2NH_3^+ Cl^-(aq)$
salt

$Cu(H_2O)_6^{2+}(aq)$ → $[Cu(CH_3CH_2NH_2)_4(H_2O)_2]^{2+}(aq)$
deep blue solution of complex ion

$CH_3CH_2Br(l)$ heat → $(CH_3CH_2)_2NH(l)$
secondary amine

Reactions of the primary amine ethylamine

Primary alkyl amines are stronger bases than ammonia (see *base dissociation constants*). This can be explained in terms of the *inductive effect*.

For the properties of aryl (aromatic) amines see *aniline*.

amino acids: the compounds which join together in long chains to make *proteins*. They are carbon compounds with two *functional groups:* a basic amino group and a carboxylic acid group. Proteins consist of long chains of amino acids. There are about 20 different amino acids which link together to make proteins. Chemists sometimes call them α-amino acids because the amino group is attached to the alpha carbon atom – the one next to the carboxylic acid group.

Amino acids are crystalline solids. They are very soluble in water and crystallize as *zwitterions*.

glycine   alanine   serine

Examples of amino acids in proteins. Note that the amino group is attached to the carbon atom next to the carboxylic acid group.

Each of the amino acids in a protein has a central carbon atom attached to four other groups so, except for glycine which has two hydrogen atoms, these are *chiral* molecules which can exist in mirror image forms. The natural amino acids all take the L-form.

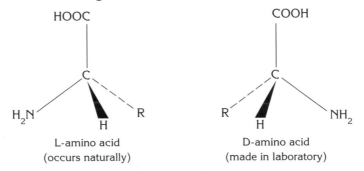

L-amino acid   D-amino acid
(occurs naturally)   (made in laboratory)

Mirror image forms of an amino acid

amino group: the $-NH_2$ functional group found in primary *amines*, and *amino acids*. As in *ammonia*, there is a *lone pair of electrons* on the nitrogen atom. The amino group, like ammonia, can act as a *base*, form *complex ions* with metal ions and react as a *nucleophile*.

ammonia (NH_3): the simplest compound of nitrogen and hydrogen. It is a colorless gas with a pungent smell. It is the only common alkaline gas. The molecule has a pyramidal shape:

Dot and cross diagram for ammonia and the shape of an ammonia molecule. There are three bonding pairs and one lone pair in the outer shell of the nitrogen atom. The four electron pairs point to the vertices of a distorted tetrahedron. The bond angle is less than the regular tetrahedral angle (109.5°) because repulsion between the lone pair and the bonding pairs is greater than the repulsion between bonding pairs (see shapes of molecules).

The properties of ammonia are affected by *hydrogen bonding*. Ammonia molecules can form hydrogen bonds with each other and with water molecules. As expected for a compound with small molecules, ammonia is a gas at room temperature but because of hydrogen bonding it is quite easily liquefied by cooling or increasing the pressure. Ammonia is also very soluble in water because of hydrogen bonding.

Many of the important properties of ammonia involve the lone pair of electrons on the nitrogen atom. The lone pair means that ammonia is:

- a weak base: $NH_3(g) + HCl(g) \longrightarrow NH_4^+Cl^-(s)$

- a ligand in complex ions:

  $[Cu(H_2O)_6]^{2+}(aq) + 4NH_3(aq) \longrightarrow [Cu(NH_3)_4(H_2O)_2]^{2+}(aq) + 4H_2O(aq)$

- a nucleophile: $CH_3CH_2CH_2Br + NH_3(aq) \longrightarrow CH_3CH_2CH_2NH_3^+Br^-(aq)$.

Other reactions of ammonia involve the N—H bonds. In particular, the reactions in which ammonia acts as a *reducing agent*. Ammonia burns in oxygen to form nitrogen and steam. It reacts with chlorine to form nitrogen and ammonium chloride. It reduces metal oxides, such as copper(II) oxide.

$$3CuO(s) + 2NH_3(g) \longrightarrow 3Cu(s) + N_2(g) + H_2O(l)$$

The catalytic oxidation of ammonia is important in *nitric acid manufacture*.

ammonia manufacture: the Haber process for the synthesis of ammonia combines nitrogen with hydrogen in the presence of an iron catalyst.

$$N_2(g) + 3H_2(g) \rightleftharpoons 2NH_3(g) \qquad \Delta H = -92.4 \text{ kJ/mol}$$

Hydrogen for the process comes from natural gas and steam (see *steam reforming*). The nitrogen comes from the air.

The reaction is very slow at room temperature. Raising the temperature increases the rate of reaction but the *reversible* reaction is *exothermic* so, according to *Le Chatelier's principle*, the higher the temperature the lower the yield of ammonia at equilibrium. A catalyst makes it possible for the reaction to go fast enough without the temperature being so high that the yield is too low. Also, according to Le Chatelier's principle, increasing the pressure raises the percentage of ammonia at equilibrium.

Flow diagram for the synthesis of ammonia

The process typically operates at a pressure of 150 atmospheres with temperatures in the range 650–720 K.

Graph showing how the equilibrium yield of ammonia varies with pressure and temperature

The main uses of ammonia are in making:

- *fertilizers* (80%)

- nylon (7%)

- *nitric acid* (5%).

ammonium hydroxide: the traditional name for "ammonia solution" or "aqueous ammonia." The older name was used because a solution of *ammonia* in water contains some ammonium ions and hydroxide ions.

$$NH_3(aq) + H_2O(l) \rightleftharpoons NH_4^+(aq) + OH^-(aq)$$

However ammonia is a *weak base*, so most of the ammonia in the solution is not ionized. Aqueous ammonia is a better name, because almost all the ammonia is present as dissolved molecules linked to water molecules by *hydrogen bonds*.

ammonium salts contain the ammonium ion (NH_4^+), formed when *ammonia* reacts with acids. Examples are ammonium chloride, ammonium sulfate and ammonium nitrate.

$$2NH_3(aq) + H_2SO_4(aq) \longrightarrow (NH_4)_2SO_4(aq)$$
$$NH_3(aq) + HNO_3(aq) \longrightarrow NH_4NO_3(aq)$$

An ammonium ion forms when the *lone pair* on the nitrogen atom in ammonia forms a dative bond with a hydrogen ion (proton) from an acid. The ammonium ion has four bonding electron pairs around the nitrogen atom so it has a tetrahedral shape.

Ammonium salts such as ammonium sulfate and ammonium nitrate are soluble nitrogen compounds manufactured and used on a large scale as *fertilizers*. Ammonium nitrate is also used as an *explosive*.

Formation of an ammonium ion and the shape of the ammonium ion. Once formed, all four bonds are the same.

amorphous solids have no regular crystal structure; their particles are in a random jumble. *Glasses are amorphous solids.*

amount of substance: the measurement which allows chemists to find formulas, write equations and make fair comparisons between equal numbers of atoms, molecules or ions. Amount of substance is a *physical quantity* (symbol *n*) which is measured in the unit *mole* (symbol mol). In chemistry the word "amount" has this precise meaning. There is no measuring instrument for determining amounts directly, unlike balances for determining masses in kilograms or graduated glassware for measuring volumes in liters. Instead, chemists first measure masses or volumes and then calculate the amount in moles given the *molar mass, concentration* or *molar volume* of the specified substance or *entity.*

For any pure substance:

$$\text{amount of substance (mol)} = \frac{\text{mass of substance (g)}}{\text{molar mass (g/mol)}}$$

For a solution:

$$\text{amount of substance (mol)} = \text{volume of solution (L)} \times \text{concentration of solution (mol/L)}$$

For a gas:

$$\text{amount of substance (mol)} =$$
$$\frac{\text{volume of a gas at a specified temperature and pressure (m}^3)}{\text{molar volume of any gas under the same conditions (m}^3\text{/mol)}}$$

(See also the *Avogadro constant.*)

ampere: the SI unit of current. A current of one ampere is a flow of one *coulomb* of charge per second. 1 A = 1 C/s.

amphetamines: *drugs* which stimulate the central nervous system and make people feel more alert. For a while amphetamines, such as benzedrine, were used medicinally until it was found that they cause strong psychological dependence.

Molecular structure of benzedrine

amphoteric compounds: substances which can behave both as *acids* and as *bases*. Water is an example. Water can both accept *protons* from acids forming *hydronium ions* and donate protons to stronger *bases* forming hydroxide ions.

Water acting as a base (taking protons from an acid):

$$H_2O(l) + H^+ \rightleftharpoons H_3O^+(aq)$$
$$\text{from acid}$$

Water acting as an acid (giving protons to a base):

$$H_2O(l) \rightleftharpoons OH^-(aq) + H^+$$
$$\text{to base}$$

Alternatively, these substances are described as being amphiprotic. Other examples are the hydrogencarbonate ion (HCO_3^-) and *amino acids*.

amphoteric oxides: oxides which react both like *acidic oxides* and *basic oxides*. Examples are: ZnO, Al_2O_3, Cr_2O_3, SnO, SnO_2, PbO, PbO_2. The anhydrous oxides are often relatively *inert* to aqueous reagents so that it is easier to demonstrate their amphoteric behavior on a test-tube scale with freshly formed samples of the metal hydroxides. Zinc hydroxide, for example, dissolves in acids to form *salts* (so acting as a *base*); it also dissolves in a solution of a strong alkali, forming zincate ions (thus acting as an *acid*).

$$Zn(OH)_2(s) + 2H^+(aq) \rightleftharpoons Zn^{2+}(aq) + 2H_2O(l)$$
$$Zn(OH)_2(s) + 2OH^-(aq) \rightleftharpoons ZnO_2^{2-}(aq) + 2H_2O(l)$$

The reaction of an amphoteric hydroxide with acids and bases can also be described in terms of the addition or removal of protons from a complex of the metal ion, such as $Al(H_2O)_3(OH)_3$. (See *aluminum(III) ions*.)

analgesics: drugs which relieve pain. There are mild analgesics such as *aspirin, acetominophen* or *ibuprofen*. Other analgesics are powerful, like the *narcotic* morphine from the opium poppy.

analytical chemistry is concerned with determining the *qualitative* and *quantitative* composition of substances.

anerobic respiration: see *respiration*.

anesthetics: total anesthetics are used in medicine to induce pain-free sleep during surgery. Early anesthetics were trichloromethane (chloroform), diethyl ether and dinitrogen oxide (laughing gas). After 1956 the most commonly used anesthetic was halothane ($CF_3CHBrCl$). Unfortunately the liver metabolizes halothane producing toxic products which can cause hepatitis. To avoid these side effects new anesthetics were developed based on *ethers*. Ethers are highly flammable but can be converted to safe and effective anesthetics by replacing hydrogen atoms with fluorine atoms. The latest anesthetics were introduced in the mid-1990s. They include Sevoflurane, $FCH_2 - O - CH(CF_3)_2$.

Local anesthetics, such as Novocaine, for minor surgery and dentistry, prevent pain just in the region of the body where they are applied or injected.

anhydrides: see *acid anhydrides*.

anhydrous salts: the compounds left after removing the water from a *hydrated* salt. Hydrated copper(II) sulfate, for example, is blue. Heating drives off the *water of crystallization* as steam leaving a white solid, anhydrous copper(II) sulfate.

$$CuSO_4.5H_2O(s) \longrightarrow CuSO_4(s) + 5H_2O(l)$$

aniline: a primary *amine* with an —NH$_2$ group attached to a benzene ring. There is a two-step laboratory route from benzene to aniline. First *nitration* then reduction.

Formation of aniline from benzene

The chemical industry makes aniline from phenol and ammonia with an alumina catalyst. Aniline and other arylamines are important intermediates in the manufacture of *azo dyes*. This is because they react with nitrous acid below 10°C to produce *diazonium salts*.

Aniline is a weaker base than ammonia, unlike alkyl amines which are stronger bases than ammonia. The lone pair on the nitrogen atom interacts with the delocalized electrons of the benzene ring. This extended delocalization stabilizes the free amine and lessens the tendency of the molecule to form a coordinate bond with a proton.

anions: negative ions attracted to the *anode* during *electrolysis*.

Charge	Anion	Symbol
1–	bromide	Br$^-$
	chloride	Cl$^-$
	acetate	CH$_3$COO$^-$
	hydrogencarbonate	HCO$_3^-$
	hydroxide	OH$^-$
	iodide	I$^-$
	nitrate	NO$_3^-$
	nitrite	NO$_2^-$
	permanganate	MnO$_4^-$
2–	carbonate	CO$_3^{2-}$
	oxide	O^{2-}
	sulfate	SO$_4^{2-}$
	sulfide	S^{2-}
	sulfite	SO$_3^{2-}$
	thiosulfate	S$_2$O$_3^{2-}$
	dichromate	Cr$_2$O$_7^{2-}$
3–	nitride	N^{3-}
	phosphate	PO$_4^{3-}$

anion tests are used in *qualitative analysis* to identify the negative ions in salts. The tests use reagents which can produce color changes, gases and precipitates. Knowledge of the chemistry makes it possible to interpret the changes and identify the ions. See the table on the next page.

Anion tests (see page 33)

Test	Observations	Inference
Test for carbonate, sulfite and nitrite Add dilute hydrochloric acid to the solid salt. Warm gently if there is no reaction at first.	Gas which turns limewater milky white.	carbon dioxide from a carbonate
	Gas which is acidic, has a pungent smell and turns acid-dichromate paper from orange to green.	sulfur dioxide from a sulfite
	Colorless gas given off which turns brown where it meets the air.	nitrogen oxide (NO) from a nitrite turning to nitrogen dioxide (NO_2)
Test for halide ions Make a solution of the salt. Acidify with nitric acid, then add silver nitrate solution. Test the solubility of the precipitate in ammonia solution.	White precipitate soluble in dilute ammonia solution.	chloride
	Cream precipitate soluble in concentrated ammonia solution.	bromide
	Yellow precipitate insoluble in excess ammonia.	iodide
Test for sulfate and sulfite ions Make a solution of the salt. Add a solution of barium nitrate or chloride. If a precipitate forms add dilute nitric acid.	White precipitate which redissolves on adding acid.	sulfite
	White precipitate which does not redissolve in acid.	sulfate
Test for nitrates Make a solution of the salt, add sodium hydroxide solution and then a piece of aluminum foil or a little Devarda alloy. Heat.	Alkaline gas evolved which turns red litmus blue. Pungent smell.	ammonia from a nitrate (or nitrite)
Test for chromate ions Make a solution of the salt. Divide into three: • add dilute acid • add a solution of barium nitrate • add a solution of lead nitrate	Yellow solution turns orange.	yellow chromate ions turning to orange dichromate
	Yellow precipitates forms with barium ions and lead ions.	precipitates of insoluble barium and lead chromates chromate

anisotropic solids have properties which depend on the direction in which they are measured. Examples of highly anisotropic substances are fibrous materials such as *asbestos*, and layer structures such as mica. The carbon atoms in graphite are arranged in layers (see *allotropes*). The properties of large single crystals of graphite show the differences between measurements parallel to the planes of atoms and at right angles to the planes of atoms.

annealing: a process of controlled heating and cooling used to modify the properties of a material, for which purpose it is kept for a time in a furnace at a temperature below its melting point. Annealing is used to remove internal stresses from *glass* objects after blowing or casting. Annealing is an important part of the heat treatment used to modify *crystal structures of metals* and hence control their properties.

anode: the *electrode* at which *oxidation* takes place. During *electrolysis* the external power supply removes electrons from the anode so that it becomes the positive electrode, attracting negative ions; the latter lose electrons (oxidation), thus turning into atoms or molecules.

The term anode is also sometimes used in *electrochemical cells,* where the oxidation process at this electrode in one of the *half-cells* forces electrons onto the electrode, which becomes the negative terminal of the cell.

Note that electrons flow out of the anode to the external circuit both during electrolysis and when a current is drawn from a chemical cell.

anodizing: an electrolytic process for thickening the oxide layer on the surface of *aluminum*. It is an example of anodic *oxidation*. In the process, an aluminum sheet or component is the *anode* of an *electrolysis* cell containing sulfuric acid. The newly formed oxide layer can absorb dyes so that the surface of anodized aluminum is easily colored. The thicker oxide film helps to protect the metal, so anodized aluminum is more resistant to *corrosion*.

antacids: ingredients of indigestion tablets taken to neutralize acid in the stomach. After a meal the cells lining the stomach produce gastric juice which contains, among other things, a mixture of hydrochloric acid and enzymes to digest food. Antacids are *bases* which neutralize acids. Examples are sodium hydrogencarbonate, calcium carbonate, aluminum hydroxide and magnesium hydroxide.

antibiotics: chemicals which prevent *bacteria* from growing or kill them. Thus antibiotics can cure bacterial diseases. Most antibiotics are the products of microorganisms. Some of the antibiotics used in medicine are natural antibiotics which have been modified chemically, such as the semisynthetic penicillins. Other antibiotics are now made entirely by chemical synthesis. The misuse of antibiotics in medicine and agriculture has led to the emergence of antibiotic resistant bacteria.

antifreeze: a chemical added to the cooling system of engines to prevent the water from freezing and damaging the engine during cold weather. Antifreeze lowers the *freezing point* of water well below 0°C. Ethylene glycol (1,2-ethanediol) is used as antifreeze because it mixes freely with water but has a higher boiling point (198°C) so it does not evaporate from the coolant when the engine is hot. Ethylene glycol has a high boiling point and mixes freely with water because it has two —OH groups allowing extensive *hydrogen bonding* both between molecules of the diol and between the diol and water molecules. Another advantage of adding ethylene glycol to the water in cooling systems is that it helps to prevent *corrosion* of iron and steel.

antiknock additives prevent knocking in engines by raising the octane number of *gasoline* and preventing the fuel from igniting too early. Since the 1920s the main antiknock additive was a volatile lead compound tetraethyl lead, $Pb(C_2H_5)_4$. Lead compounds in exhaust gases are toxic and poison the metal catalyst in *catalytic converters*. So gasoline companies now produce high-octane fuels by refining the fuel further and adding alcohols and *ethers* such as MTBE.

antiseptics: chemicals which kill microorganisms, but, unlike *disinfectants*, can safely be used on the skin. *Phenol* was the first antiseptic used in surgery, by Joseph Lister in 1857. Lister found that a spray of phenol controlled the growth of bacteria in open wounds but it also made the wounds difficult to heal so he moved on to other methods. Dettol and TCP are improved antiseptics related to phenol.

aqua regia was well known to alchemists for its ability to dissolve noble metals such as gold. Hence the name, which is Latin, meaning "royal water." It is a mixture of three parts concentrated *hydrochloric acid* with one part concentrated *nitric acid*. The nitric acid oxidizes gold atoms to Au^{3+} ions which then form complex ions, $[AuCl_4]^-$, with the high concentration of chloride ions.

aqueous solution: a solution of one or more *solutes* dissolved in water. The state symbol (aq) in *chemical equations* shows that a *species* is in aqueous solution.

aramids: are *polyamides* with benzene rings in the *polymer* chain. The rigid, linear molecules line up parallel to each other, held together by *hydrogen bonding*.

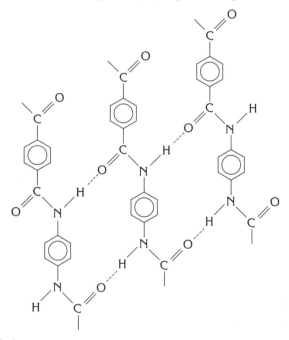

The structure of Kevlar

Trade names for aramids include Kevlar and Nomex. They are very strong, flexible, fire resistant polymers with a low density. They are used to make bullet-proof vests. Aramid ropes are stronger than the same weight of steel cable. The polymer can replace steel in motor vehicle tires.

arenes: *hydrocarbons* such as *benzene* with rings of carbon atoms stabilized by *delocalization of electrons*. Traditionally, chemists have called the arenes "aromatic hydrocarbons," ever since *Kekulé* was struck by the fragrant smell of oils such as benzene. In the modern name, the "ar-" comes from aromatic and the ending "-ene" means that these hydrocarbons are unsaturated compounds like the *alkenes*.

| benzene | methylbenzene (toluene) | 1,3-dimethylbenzene (m-xylene) |

Structures of arenes. The circle in the benzene ring represents six delocalized electrons. This representation explains the shape and stability of benzene. The Kekulé structure is more helpful when describing the mechanism of the reactions of benzene.

As well as arenes related to benzene, there are others with fused ring systems such as naphthalene and anthracene.

Structure of naphthalene with fused rings, showing both the Kekulé structure and delocalization. The delocalized electrons are shared across both the rings.

Arenes are nonpolar. The forces between the molecules are weak *van der Waals forces*. The boiling points of arenes depend on the sizes of the molecules. The bigger the molecules the higher the boiling points. Benzene and methylbenzene (toluene) are liquids at room temperature. Naphthalene is a solid.

The important reactions of benzene and other arenes are very different from *alkenes* because of the delocalized electrons. Instead of *electrophilic addition* reactions, the useful changes are *electrophilic substitution* reactions.

Heating a mixture of an arene bearing a hydrocarbon side chain with alkaline potassium permanganate oxidizes the side chain to a carboxylic acid group.

| methylbenzene | benzoic acid |

Oxidation of the hydrocarbon side chain of methyl benzene

argon (Ar): the third member of the family of *noble gases* coming below neon in the periodic table with the *electron configuration* [Ne]$3s^23p^6$.

The element is a colorless, odorless gas consisting of single atoms.

Argon makes up about 0.9% of dry *air* but it was not identified until the late nineteenth century because of its *inertness*.

Large amounts of argon are produced by the *fractional distillation* of liquid air. Much of the gas is used to provide an inert atmosphere for processes such as welding; some is used to fill filament light bulbs.

aromatic hydrocarbons: the traditional term for *arenes*.

aromatic nitro compounds: see *benzene* and *aniline*.

aromatic substitution: another name for *electrophilic substitution* in arenes.

Arrhenius equation: an equation which describes how the *rate constant* for a reaction varies with temperature and makes it possible to determine the *activation energy* for the reaction. The *collision theory* of reaction rates helps to account for the form of the equation.

Arrhenius equation: $k = Ae^{-E_a/RT}$

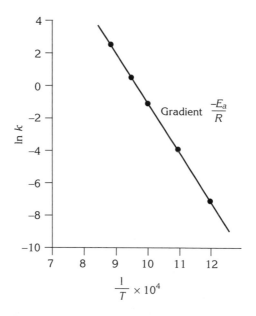

Plot of ln k against $1/T$ for a reaction

k is the rate constant for the reaction and E_a is the activation energy, R is the gas constant and T the temperature in kelvin. A is another constant.

Putting the equation into logarithmic form (using *logarithms* to base e) makes it easier to use.

$$\ln k = \ln A - \frac{E_a}{RT} = \text{constant} - \frac{E_a}{RT}$$

This equation gives a straight line if ln k is plotted against $1/T$. The gradient of the line is $-E_a/R$.

arsenic (As): a greyish-white *metalloid* which forms very toxic compounds. The element comes below phosphorus in group V of the *periodic table* and has the *electron configuration* $[Ar]3d^{10}4s^24p^3$.

aryl group: a group of carbon and hydrogen atoms which forms part of the structure of a molecule. The simplest example is the *phenyl group*, C_6H_5 —, which is *benzene* with one hydrogen atom removed. In general aryl groups are arene molecules minus one hydrogen atom.

asbestos: the name given to various fibrous *silicate* minerals in which SiO_4 groups are linked into long chains. These minerals have very high melting points and they do not burn. The *fibers* are flexible enough to be woven into fabrics which are heat resistant, fireproof and do not conduct electricity. Breathing asbestos fibers can cause lung disease many years later including asbestosis and mesothelioma (a form of cancer). For this reason the use of the asbestos is being phased out in many parts of the world, especially blue asbestos which is particularly hazardous.

ascorbic acid (vitamin C): a white crystalline compound found dissolved in the juices of fresh fruit and vegetables. Lack of ascorbic acid in the diet leads to scurvy. Ascorbic acid is relatively soluble in water and may be washed out of foods during cooking. It is thermally unstable and is progressively destroyed during cooking.

Structure of ascorbic acid

The ascorbic acid molecule does not contain a free *carboxylic acid* group because its carboxyl group reacts with its hydroxyl group, eliminating water to form a ring compound. Ascorbic acid is a good *reducing agent*. It is added to some processed foods as preservative; as an antioxidant, it prevents oxidation of other food components.

aspirin: the common name for the most widely used medical drug. Chemically it is the acetyl *ester* of salicylic acid. (See *acylation*.)

Aspirin is a mild painkiller (*analgesic*); it also helps to lower fevers and reduce inflammation of joints. Large-scale studies have shown that people who have had a heart attack are less likely to have a second attack if they take half an aspirin tablet a day.

assaying: a process of quantitative analysis which determines how much of a given sample is the substance indicated by its name. So a piece of gold is assayed to see what proportion of the metal sample is gold. Gold and silver articles are assayed before being hallmarked. A range of assaying techniques has been developed to determine the purity or biological activity of drugs, enzymes and other substances which affect living things.

astatine (At): the rarest element in the halogen family. It is the element below iodine in group VII of the periodic table. It has the electron configuration $[Xe]4f^{14}5d^{10}6s^26p^5$. There are twenty known isotopes of astatine all of which are highly radioactive. The longest lived is astatine-210 with a half-life of 8.3 hours. The element was discovered in 1940 when it was prepared artificially by bombarding bismuth with alpha particles in a cyclotron.

asymmetric molecules: molecules with no centers, axes or planes of symmetry. Asymmetric molecules are chiral and can exist in distinct mirror image forms giving rise to optical isomerism. Any carbon atom with four different groups or atoms attached to it is asymmetric. See the diagram on page 40.

The mirror image forms of the asymmetric molecule, lactic acid

atactic polymer: a form of addition polymer, such as poly(propene), in which the side groups along the polymer chain are randomly orientated. Atactic poly(propene) is an amorphous, rubbery polymer of little value unlike isotactic poly(propene).

Part of a chain of atactic poly(propene)

atmosphere: the zone around the Earth which contains all the gases in the air. By convention the upper limit of the atmosphere is said to be 1000 km above the Earth's surface. However, because of gravity, most of the air is concentrated in the lower regions. About 50% by mass lies within 5.6 km of the surface and 99% within 40 km. Most of the climate and weather processes take place in the troposphere which extends up to about 17 km.

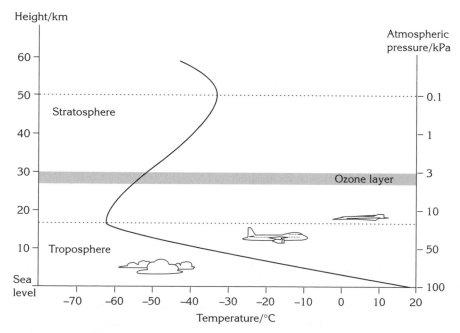

Vertical structure of the Earth's atmosphere showing how temperature and pressure vary with altitude

atmospheric pressure usually means the pressure of the atmosphere at ground level. Standard atmospheric pressure (1 atm) = 101 325 N/m² = 101.325 kPa. The choice of this value dates back to the days when atmospheric pressure was normally measured by a column of mercury in a barometer. A pressure of one atmosphere was defined as the pressure needed to support a column of mercury 760 mm high.

atom: the smallest particles of an element. An atom consists of a tiny nucleus surrounded by a cloud of electrons. The nucleus consists of protons and neutrons. All the atoms of the same element have the same *atomic (proton) number*. The number of neutrons may vary and this accounts for the existence of isotopes. (See also *fundamental particles*.)

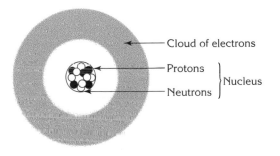

Atomic structure

Chemical reactions involve the electrons in the outer shells of atoms. As a result the electron configuration of an element helps to account for its chemical behavior.

Radioactivity and the changes in nuclear reactors involve the nuclei of atoms. When describing nuclear changes it helps to use symbols which show the mass number and atomic number.

$$\text{Mass number} \longrightarrow {}^{A}_{Z}X \longleftarrow \text{Element symbol}$$
$$\text{Atomic number} \longrightarrow$$

Symbols for atoms. The atomic number gives the number of protons in the nucleus. In a neutral atom the number of electrons is the same. The difference between mass number and atomic number gives the number of neutrons in the nucleus.

atomic absorption spectroscopy: a technique of quantitative analysis for measuring the amount of an element in a sample. The procedure depends on the fact that the atoms absorb radiation at particular frequencies.

The sample is vaporized and split into gaseous atoms by a very hot flame at around 2000°C. Light from a special lamp produces radiation with a wavelength absorbed by the element to be measured.

By measuring the absorbance at a particular wavelength it is possible to determine the amount of certain elements in the sample. The procedure is *calibrated* using samples of known concentration.

Atomic absorption spectroscopy provides a sensitive and reliable method for determining more than 60 elements. Examples include the determination of traces of mercury in the environment, lead in blood, or heavy metals in the effluent from a factory. The technique is widely used in commercial laboratories, for example in the water industry.

atomic emission spectroscopy: a very sensitive method of analysis used to identify and measure the elements in a sample. Atomic emission spectra of pure elements also provide the evidence needed to determine the *energy levels* and *electron configurations* in atoms.

Performing a *flame test* and examining the colored light from the flame with a handheld spectroscope is a simple demonstration of this type of spectroscopy.

The first step in a more sophisticated analysis is to vaporize the sample using a very hot flame or an electric spark to atomize the sample and produce a gas consisting of free atoms. Energy from the flame or spark also excites some of the electrons in the atoms so that they jump to higher energy levels. As the electrons drop back to lower levels they emit radiation. The radiation passes through a prism or diffraction grating to produce the *atomic spectrum* of the elements in the sample.

Important applications of atomic emission spectroscopy include the measurement of sodium, potassium, lithium and calcium ions in biological fluids such as blood and urine.

atomic number: the number of protons in the nucleus of an *atom*. The alternative term is proton number.

atomic orbitals: the subdivisions of the electron shells in atoms. The main shells divide into subshells labelled s, p, d and f. The subshells are further divided into atomic orbitals.

Each orbital is defined by its:

- energy
- shape
- direction in space.

There is one orbital in the first shell, four in the second shell, nine in the third shell and 16 in the fourth shell. Each orbital can contain up to two electrons.

The study of *atomic spectra* makes it possible to determine the energies of atomic orbitals. The terms "energy level" and "orbital" are often used interchangeably.

The energies of atomic orbitals in atoms. The labels s, p, d and f are left over from early studies of atomic spectra which used the words sharp, principal, diffuse and fundamental to describe different series of lines. The pattern of energy levels accounts for the electron configurations of atoms and the arrangement of elements in the periodic table.

The shapes of orbitals are derived from theory. The shapes are determined by solving a mathematical equation (the Schrödinger wave equation) which makes it possible to calculate the probability of finding an electron at any point in an atom.

The probable location of the electrons in the first shell is spherical. It is an example of an s-orbital (1s).

The four orbitals in the second shell are made up of one s-orbital (2s) and three dumbell shaped p-orbitals. The three p-orbitals ($2p_x$, $2p_y$, $2p_z$) are arranged at right angles to each other.

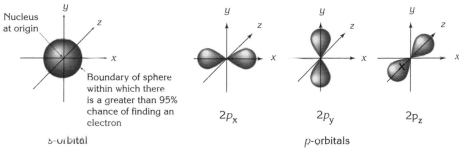

$2p_x$ $2p_y$ $2p_z$

s-orbital   p-orbitals

The shapes of s and p atomic orbitals

atomic radius: a measure of the size of an atom in a crystal or molecule. Chemists use X-ray crystallography and other techniques to measure the distance between the nuclei of atoms. The atomic radius of an atom depends on the type of bonding and on the number of bonds. Usually atomic radii for metals are calculated from the distances between atoms in metal crystals (metallic radii). Atomic radii for nonmetals are calculated from the lengths of covalent bonds in crystals or molecules (covalent radii).

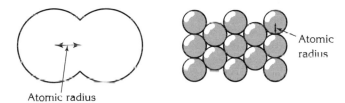

Atomic radii in a molecule and in a crystal

Atomic radii increase down any group in the periodic table as the number of electron shells increases. In group I the metallic radii rise from 157 pm for lithium to 272 pm for cesium.

Trends in atomic radii in the periodic table

Atomic radii decrease across from left to right across a period. Across the period Na to Ar, atomic radii fall from 191 pm for sodium to 99 pm for chlorine. From one element to the next across a period the charge on the nucleus increases by one as the number of electrons in the same outer shell increases by one. *Shielding* by electrons in the same shell is limited so the "effective nuclear charge" increases and the electrons are drawn more tightly to the nucleus.

atomic spectrum: the pattern of lines seen with a spectroscope when the atoms of an element emit radiation. Atoms emit radiation when excited by heat or electricity.

UV   Violet Blue   Green   Red

The line emission spectrum for hydrogen in the visible and ultraviolet (UV) regions

Some elements emit radiation in the visible region of the spectrum and these are the elements which give colored flames in *flame tests*.

Quantum theory explains why atomic emission spectra consist of a series of sharp lines. Each line in an emission spectrum corresponds to an energy jump of a definite size as electrons drop back from a higher to a lower energy level.

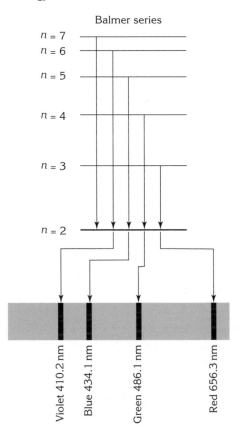

Electron jumps between energy levels in the hydrogen atom giving rise to the visible lines in the emission spectrum

Atoms also absorb radiation. If radiation passes through the vapor of an element and is viewed through a spectroscope, the absorption spectrum appears as dark lines where particular frequencies are absorbed from the continuous spectrum.

atomic theory: the theory that all matter is made up of very tiny particles called atoms. Modern atomic theory was foreshadowed in ancient Greek philosophy but chemists now trace the development of the theory back to the work of John Dalton. It is the ideas developed between 1800 and 1930 that are significant for an advanced chemistry course. Studies of the structure of atomic nuclei continue with the help of huge particle accelerators.

1804 John Dalton's theory of atoms as solid spheres explains the differences between elements and compounds. In this theory all the atoms of the same element weigh the same.

1896 Henri Becquerel discovers *radioactivity* leading to the work of Marie and Pierre Curie, who separate and identify more radioactive elements.

1897 J.J.Thompson discovers the electron showing that atoms are themselves made up of even smaller particles.

1911 Hans Geiger and E. Marsden's *alpha particle* scattering experiment leads to the Rutherford model with a nucleus surrounded by orbiting electrons.

1913 Niels Bohr model explains the *atomic spectrum* of hydrogen by proposing that the electrons orbiting the nucleus could only have definite energies. According to *quantum theory* atoms absorb and emit radiation at particular frequencies depending on the size of the energy jumps between *energy levels*.

1919 Francis Aston uses his mass spectrograph to demonstrate the existence of *isotopes* which had earlier been proposed by Soddy to describe atoms of the same element which have different masses.

1923 Louis-Victor de Broglie puts forward the theory that electrons have wave-like properties as well as particle properties. This theory is confirmed by experiment in 1927 when scientists in the USA and UK obtained diffraction effects with beams of electrons.

1926 Erwin Schrödinger publishes his mathematical theory of wave mechanics The Schrödinger wave equation explains the pattern of energy levels in the hydrogen atoms and gives rise to the picture of electron in *atomic orbitals* with particular energies and shapes.

1932 James Chadwick discovers the neutron, making it possible to explain the existence of isotopes and radioactivity.

atomization: a process in which an element or compound is converted into gaseous atoms. (See *enthalpy change of atomization*.)

ATP (adenosine triphosphate) plays a crucial role in the energy transfer processes of living organisms. *Hydrolysis* of ATP, under the control of *enzymes*, splits off one of the three phosphate groups, releasing 31 kJ/mol to drive biochemical processes and for muscle movement.

(See the structure diagram on page 46.)

adenine

Structure of ATP

ribose

attacking group: a term used to describe the mechanism of chemical reactions. *Nucleophiles* and *electrophiles* are examples of attacking groups.

aufbau principle: the principle that the *electron configurations* of atoms build up according to a set of rules.

The three rules are that:

● electrons go into the *orbital* at the lowest available *energy level*

● each orbital can only contain at most two electrons (with opposite *spins*)

● where there are two or more orbitals at the same energy, they fill singly before the electrons pair up.

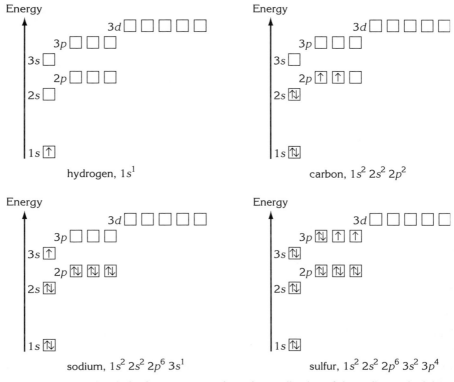

Electrons in energy levels for four atoms to show the application of the aufbau principle

autocatalysis: catalysis of a reaction by one of its products. An autocatalytic reaction starts slowly but then speeds up as the catalytic product starts to form. The oxidation of ethanedioate ions by permanganate ions in acid solution is catalysed by permanganate ions, thus:

$$2MnO_4^-(aq) + 5C_2O_4^{2-}(aq) + 16H^+(aq) \longrightarrow 10CO_2(g) + 2Mn^{2+}(aq) + 8H_2O(l)$$

At first bubbles of carbon dioxide appear very slowly but once the reaction produces some manganese(II) ions it speeds up and gas is evolved more rapidly. This is an example of *homogeneous catalysis*.

autoionization constant of water, K_w: a constant for the equilibrium produced by the ionization of water. There are hydronium and hydroxide ions even in pure water because of a transfer of hydrogen ions between water molecules. This only happens to a very slight extent. At equilibrium in pure water: $[H_3O^+(aq)] = 10^{-7}$ mol/L.

$$H_2O(l) + H_2O(l) \rightleftharpoons H_3O^+(aq) + OH^-(aq)$$

The equilibrium constant $K_c = \dfrac{[H_3O^+(aq)] [OH^-(aq)]}{[H_2O(l)]^2}$

There is such a large excess of water that $[H_2O(l)]$ is a constant, so the relationship simplifies to: $K_w = [H_3O^+(aq)][OH^-(aq)]$

where K_w is the autoionization constant of water.

In pure water $[H_3O^+(aq)] = [OH^-(aq)] = 10^{-7}$ mol/L.

Hence $K_w = 10^{-14}$ mol²/L²

K_w is a constant in all aqueous solutions at 298 K. This makes it possible to calculate the pH of alkalis. (see also *pK*.)

Worked example

What is the pH of a 0.02 mol/L solution of sodium hydroxide?

Notes on the method
Sodium hydroxide is fully ionized in solution. So in this solution

$$[OH^-(aq)] = 0.02 \text{ mol/L}$$
$$pH = -\lg [H_3O^+(aq)]$$

Answer
For this solution:

$$K_w = [H_3O^+(aq)] \times 0.02 \text{ mol/L} = 10^{-14} \text{ mol}^2/\text{L}^2$$

So $[H_3O^+(aq)] = \dfrac{10^{-14} \text{ mol}^2/\text{L}^2}{0.02 \text{ mol/L}} = 5 \times 10^{-13}$ mol/L

Hence pH = $-\lg (5 \times 10^{-13}) = 12.3$

auxochrome: see *chromophore*.

Avogadro constant: the number of atoms, molecules, ions or other chemical *entities* in one mole of a substance. The Avogadro constant (L) = 6.02×10^{23} entities/mol.

amount of substance (mol) × Avogadro constant (entities/mol) = no. of specified entities

So 0.25 mol carbon atoms contains 1.50×10^{23} atoms while 0.5 mol oxygen molecules contains 3.10×10^{23} molecules.

Avogadro's law states that equal *volumes* of gases under the same conditions of temperature and *pressure* contain equal numbers of molecules. This means that one *mole* of any gas occupies the same volume under the same conditions. The volume occupied by a gas at 273 K (0°C) and one atmosphere pressure (101.3 kPa) is about 22.4 L. These are the conditions of standard temperature and pressure (*S.T.P.*). In a warmish laboratory at 298 K (25°C) the volume of one mole of gas is about 24 L.

Amadeo Avogadro (1776–1856) was the Italian scientist who proposed this law to account for *Gay-Lussac's law* of combining volumes.

azeotropic mixture: a mixture of liquids which on *boiling* produces a *vapor* with the same composition as the liquid. This means both that the liquids cannot be separated by fractional distillation and that azeotropes are constant boiling mixtures. A mixture containing 95.6% ethanol with 4.4% water is an azeotrope which boils at 78.2°C. As a result it is not possible to produce pure ethanol (anhydrous or absolute ethanol) by distillation alone.

azo compound: a compound with the group $-N=N-$ in its structure.

azo dyes are made by coupling a *diazonium salt* with one of a variety of coupling agents. Benzene diazonium chloride, for example, couples with naphthalene-2-ol to make a yellow azo compound.

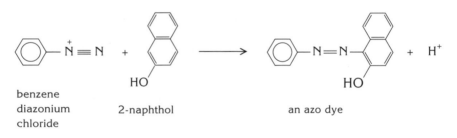

benzene
diazonium   2-naphthol   an azo dye
chloride

Equation for a coupling reaction to make an azo dye

Very many azo *dyes* are manufactured by linking combinations of 50 diazonium salts with 52 coupling agents. Most of the dyes are red, orange or yellow. The *acid–base indicator* methyl orange is an azo dye.

back titration: an analytical technique used when the reaction between the standard solution and the substance to be analyzed is slow. The procedure is to add a measured excess of the standard solution, allow time for the reaction to finish and then to use a titration with a second standard solution to measure how much of the first standard solution remains unused.

This technique is used to measure amounts of ammonium salts. An excess of a standard solution of sodium hydroxide is added to a sample and the mixture heated to drive off ammonia. In the process some of the sodium hydroxide is used up.

$$NH_4^+(aq) + OH^-(aq) \longrightarrow NH_3(g) + H_2O(l)$$

After all the ammonia has been driven off by boiling, a titration with standard hydrochloric acid measures the amount of unused sodium hydroxide remaining. This shows how much of the alkali was used to react with the ammonium salt in the sample and hence the amount of ammonium salt in the sample.

bacteria: a group of microorganisms. Some bacteria are harmful and responsible for diseases such as cholera, tuberculosis and food poisoning. Most bacteria play a vital part in ecological processes. Soil bacteria, for example, help to break down the remains of living things releasing the nutrient elements such as nitrogen, in a form which plants can use for growth.

Bacteria are the basis of much traditional and modern biotechnology. In a traditional process, for example, bacteria are used to make yogurt. In new methods they are altered by genetic engineering to produce substances such as human insulin for diabetics.

bactericide: a chemical which kills bacteria. See *chlorine water treatment* and *water treatment*.

Bakelite: the phenol–methanal plastic discovered by the Flemish-born chemist Leo Baekland in 1905. This was the start of the modern plastics industry. Bakelite is a cross-linked, *thermosetting polymer* formed when phenol and methanal (formaldehyde) are mixed and heated with an acid *catalyst*.

balanced equations show the amounts (in moles) of reactants and products involved in chemical reactions. There is no change in the total number of atoms of each element as reactants turn into products during a chemical reaction. Four steps lead to the balanced equation for a reaction. Take for example the reaction of sodium with water.

Step 1 Identify the reactants and products by name:

sodium reacts with water to form sodium hydroxide and hydrogen.

Step 2 Write down the correct formulas for reactants and products:

$$Na + H_2O \longrightarrow NaOH + H_2$$

49

Step 3 Balance the numbers of atoms of each element by inspection and by writing numbers in front of the formulas as necessary. In this example there has to be an even number of hydrogen atoms on the left (in H_2O molecules) so the number on the right must be even too.

$$2Na + 2H_2O \longrightarrow 2NaOH + H_2$$

Never change any of the formulas to balance an equation. Do not add an extra formula to balance an equation without carefully checking that it is right to do so.

Step 4 Add *state symbols*:

$$2Na(s) + 2H_2O(l) \longrightarrow 2NaOH(aq) + H_2(g)$$

The balanced equation shows the *amounts* (in moles) of reactants and products. In the example, 2 mol sodium atoms react with 2 mol water molecules to give 2 mol sodium hydroxide and 1 mol hydrogen. This is the *stoichiometry* of the reaction. (See also *redox reactions*.)

ball and stick models are used to show in three dimensions the structure and bonding of crystals, molecules and complex ions. They are better than *space-filling models* for showing bond angles, but they do not show the relative size of atoms and ions.

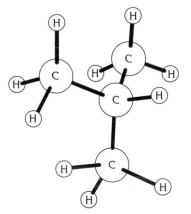

Ball and stick model of 2-methylpropane

Balmer series: a series of lines which appear in the visible region of the *hydrogen emission spectrum*.

barbiturates: drugs which depress the central nervous system. They are used as sedatives and sleeping pills. They are now prescribed much less than before since it was realized that they can produce dependence.

barium (Ba): a silvery-white, soft metal which is one of the alkaline earth metals in *group II* of the periodic table. Barium occurs naturally as barytes ($BaSO_4$) and also as witherite ($BaCO_3$).

Barium sulfate absorbs X-rays strongly so it is the main ingredient in "barium cocktails" used to diagnose disorders of the stomach or intestines. Soluble barium compounds are toxic but barium sulfate is very insoluble and cannot be absorbed into the bloodstream from the gut. X-rays cannot pass through the "barium cocktail" which therefore creates a shadow on the X-ray film.

base: a molecule or ion which can accept a hydrogen ion (proton) from an acid. This is the usual definition based on the *Brønsted–Lowry theory*. A base has a *lone pair of electrons* which can form a *coordinate bond* with a proton. (See also the *Lewis acid/base theory*.)

A hydroxide ion has a lone pair of electrons which can form a dative bond with a hydrogen ion

Common bases are the oxide and hydroxide ions, *ammonia* and *amines* as well as the carbonate and hydrogencarbonate ions. There is a lone pair on the nitrogen atom of ammonia which allows it to act as a base:

$$NH_3(g) \quad + \quad HNO_3(g) \longrightarrow NH_4^+NO_3^-(s)$$

base   acid

In biochemistry the term base often refers to one of the five nitrogenous bases which make up *nucleotides* and *nucleic acids* DNA and RNA. These compounds (adenine, guanine, cytosine, uracil and thymine) are bases in the chemical sense because they have lone pairs on nitrogen atoms which can accept hydrogen ions.

base catalysis: any reaction speeded up by a base *catalyst*. One example is the base catalyzed *hydrolysis* of a nitrile to the salt of a carboxylic acid.

$$CH_3CH_2CH_2-C\equiv N \quad \xrightarrow[\text{heat}]{H_2O/NaOH} \quad CH_3CH_2CH_2-C{\overset{\displaystyle O}{\underset{\displaystyle O^-\,Na^+}{<}}}$$

Hydrolysis of butanenitrile to sodium butanoate with a base catalyst

base dissociation (ionization) constants measure the strength of bases by showing the extent to which they accept hydrogen ions in solution. *Base dissociation constants* are used to compare the strengths of relatively *weak bases*.

For the weak base ammonia: $NH_3(aq) + H_2O(l) \rightleftharpoons NH_4^+(aq) + OH^-(aq)$

The equilibrium constant, $K_c = \dfrac{[NH_4^+(aq)][OH^-(aq)]}{[NH_3(aq)][H_2O(l)]}$

In dilute solution the concentration of water is effectively constant, so the expression can be written in this form where K_b is the base dissociation constant:

$$K_b = \frac{[NH_4^+(aq)][OH^-(aq)]}{[HA(aq)]}$$

The table shows dissociation constants for some weak bases in order of base strength with the strongest of the bases at the top of the list. The figures show that the order of base strength is: primary alkyl *amine* > ammonia > aniline.

Base	K_b, mol/L
ethylamine, $C_2H_5NH_2$	5.4×10^{-4}
methylamine, CH_3NH_2	4.3×10^{-4}
ammonia, NH_3	1.8×10^{-5}
aniline, $C_6H_5NH_2$	3.8×10^{-10}

basic oxide: an oxide of a *metal* which reacts with acids to form salts and water. Copper(II) oxide, for example reacts with acids to produce copper(II) salts.

$$CuO(s) + H_2SO_4(aq) \longrightarrow CuSO_4(aq) + H_2O(l)$$

Note that it is the oxide ion in the basic oxide which acts as a base by taking a hydrogen ion from the acid.

$$O^{2-} + 2H^+ \longrightarrow H_2O$$

Basic oxides which dissolve in water are *alkalis*. As the compound dissolves, the oxide ion, acting as a base, takes a hydrogen ion from water and forms a hydroxide ion.

$$CaO(s) + H_2O(l) \longrightarrow Ca(OH)_2(aq)$$

batch process: a process which produces a specified amount of a product in a single operation. In industry a batch process is used to manufacture chemicals needed in relatively small amounts so that a *continuous process* is not worthwhile. Batch processing is typically used to make fine chemicals on a scale of up to 100 metric tons per year. An industrial batch process is essentially a large scale version of a synthesis carried out in laboratory glassware. After producing a batch of a chemical the apparatus is cleaned and used again.

Some metals are extracted by batch processes (see *titanium extraction*).

battery: two or more *electrochemical cells* connected in series make up a battery of cells. This is the strictly correct meaning of the term battery. It is a term which scientists borrowed from the army where a line of guns is still called a battery. A car battery consists of six rechargeable *lead–acid cells* connected in series to give a 12 volt supply. In everyday conversation, however, it is common to use the word "battery" when referring to a single cell.

bauxite: an ore of aluminum consisting of impure aluminum oxide. Impurities are removed by heating powdered bauxite with sodium hydroxide solution. Aluminum oxide, which is *amphoteric*, dissolves but other oxides such as iron(III) oxide and titanium(IV) oxide do not. After filtering, *seed crystals* are added and hydrated aluminum oxide crystallizes as the solution cools. Heating the hydrate crystals at 1000°C produces anhydrous aluminum oxide ready for *aluminum extraction*.

becquerel (Bq): the *SI unit* for measuring radioactivity. If one atomic nucleus decays per second the activity is one becquerel.

bench reagents: solutions traditionally kept on the laboratory bench for general chemical use such as *qualitative analysis*. The concentrations of bench reagents are only approximate. The concentration of bench acids and alkalis is often around 2 mol/L for hydrochloric acid, nitric acid, sodium hydroxide and ammonia solutions but 1 mol/L for sulfuric acid (which produces two moles of hydrogen ions per mole of acid).

Benedict's solution: a reagent for detecting *reducing sugars*. The deep-blue reagent contains copper(II) ions complexed with citrate ions in alkaline solution. A reducing sugar reduces the reagent to copper(I) oxide. The blue color fades and a reddish-brown precipitate forms. The reagent is similar to Fehling's solution, but safer to use because it is less *corrosive*. Benedict's solution is also more stable and does not have to be stored as two reagents.

benzene (C_6H_6): the simplest *arene* hydrocarbon. It is a nonpolar liquid at room temperature and was once widely used as a solvent until it was found to be *carcinogenic*.

Three ways of representing the structure of benzene. Note that all the atoms in a benzene molecule are in the same plane. The molecule is planar.

At first sight benzene is an unsaturated compound with three double bonds which should be highly reactive like the *alkenes*. Benzene is far more stable and less reactive than the *Kekulé structure* suggests (see *stability of benzene*).

The "three double bonds" in the benzene ring are not isolated. The electron clouds in these double bonds overlap and merge to form a double doughnut-shaped electron cloud of *delocalized electrons* that sandwiches the six-sided carbon skeleton.

Like other arenes, benzene burns with a very smoky, yellow flame. The characteristic reactions of benzene are *electrophilic substitution* reactions.

Summary of the substitution reactions of benzene (see also sulfonation of benzene)

benzene derivates consist of a *benzene* ring with one or more side chains or functional groups substituted for the hydrogen atoms of the ring. Chemists continue to use a mixture of systematic and traditional names for these derivatives. (See diagram on page 54.)

The properties of functional groups such as —OH in *phenol* or —NH_2 in aniline are modified when bonded directly to a benzene ring. These functional groups show their normal behavior, however, when substituted into a side chain.

beryllium: a strong metal with a high melting point which is much less dense than iron. The element makes useful alloys with other metals. Beryllium is the first element in *group II* of the periodic table. Its electron configuration is $[He]2s^2$. The element occurs naturally as beryl, an aluminosilicate mineral, which is colorless if pure but a brilliant green in emerald, where chromium(III) ions replace some of the aluminum(III) ions.

benzoic acid	acetophenone	benzaldehyde
(benzenecarboxylic acid)	(phenylethanone)	(benzenecarbaldehyde)

Benzene derivatives. *In these compounds the functional group is part of a side chain and behaves chemically as it does when attached to an alkyl group. The presence of the benzene ring however affects the physical properties of the compounds. Benzoic acid, for example, is a white crystalline solid which is insoluble in cold water but dissolves on heating (see page 53).*

Beryllium is not a typical group II element. Resemblances between the chemistry of beryllium and aluminum are examples of the *diagonal relationship*.

In the air, beryllium metal is coated with an inert layer of oxide which prevents further reaction with oxygen unless the metal is heated to a high temperature. The oxide layer on beryllium also prevents reaction of the metal with water and dilute acids.

Beryllium oxide has a high melting point and is a useful *refractory*. Chemically it is pretty inert and it does not react with water or dilute acids. Under extreme conditions it can be shown to be an *amphoteric oxide*.

Beryllium hydroxide is insoluble in water. Like the oxide it is amphoteric and this is shown by the fact that it will dissolve both in acids and in alkalis.

Beryllium reacts with chlorine on heating to form a colorless chloride, $BeCl_2$. The bonding in beryllium chloride is covalent because the chloride ion is polarized by the very small, doubly charged beryllium ions (see *polarization of ions*). Beryllium chloride vapor consists of $BeCl_2$ molecules which are linear (see *shapes of molecules*). The beryllium atom in $BeCl_2$ can accept two electron pairs. Solid beryllium chloride has an extended chain-like structure, with chlorine atoms forming *coordinate bonds* with beryllium atoms.

$BeCl_2(g)$
molecules

$BeCl_2(s)$

The structure of beryllium chloride

beta (β) particles: electrons given off from the nuclei of some *radioactive* atoms when they decay. During beta decay a neutron turns into a proton so the *atomic number* of the atom increases by one while the *mass number* is unchanged. An atom of a new element forms. The new element is one place to the right in the *periodic table*.

$$^{228}_{88}Ra \longrightarrow ^{228}_{89}Ac + ^{0}_{-1}\beta$$

radium   actinium   beta
particle

The symbol $_{-1}^{0}\beta$ for a beta particle means that the mass numbers and atomic numbers balance on the two sides of the nuclear equation.

bicarbonate: the traditional name for hydrogencarbonate. Sodium bicarbonate (or "bicarb") can be used as an ingredient in baking powder and as an alkali in antacids.

bidentate ligands form two *coordinate bonds* with metal ions in *complexes*. Examples of bidentate (or "two-toothed") *ligands* are: ethylenediamine, *amino acids* and the oxalate ion. The complexes formed by bidentate ligands are examples of *chelates* which are more stable than complexes with *monodentate ligands*.

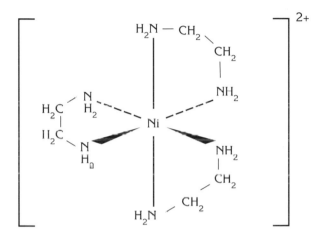

Structure of a complex formed by the bidentate ligand, ethylenediamine

bifunctional compounds: compounds with two *functional groups*. Important examples are the *amino acids*.

biochemistry: the study of chemical changes in living organisms. Metabolism describes the many chemical processes in all living thing. Many of these *enzyme*-catalyzed reactions are involved in the hydrolysis and oxidation of food. Others build up small molecules into the complex molecules which cells need as they grow. The techniques of molecular biology helped to transform biochemistry because they allowed scientists to determine the structures and shapes of large molecules such as *proteins* and *nucleic acids* and hence to begin to discover the way in which genetic material in cells controls metabolism.

biodegradable materials break down in the environment due to the action of microorganisms. Most synthetic polymers are not biodegradable so *polymer waste* does not break down in *landfill* sites.

Biopol was the first totally biodegradable plastic to be manufactured for use as a plastic. Biopol is a trade name for a *polyester* made by bacteria. The bacteria make the natural polyester as an energy store. Manufacturers have developed techniques for growing the bacteria on a large scale, extracting the polymer and converting it to a useful plastic. The plastic is much more expensive than poly(ethene) but microorganisms in the soil can break it down by hydrolyzing the ester links.

biofuels: fuels produced from vegetable matter and organic wastes. Biofuels include *wood* and *biogas*.

Vegetable oils have always been used for heating and lighting on a domestic scale though in industrialized countries they have been largely replaced by natural gas and other fossil fuels.

There is increasing commercial interest in biofuels. Countries such as Italy and Australia are making significant use of diesel fuel manufactured from the oils from crops such as rape. Brazil is the leading country using ethanol as a fuel for motor vehicles. The ethanol is manufactured by *fermentation* of sugars from sugar cane.

Concern about the enhanced *greenhouse effect* is encouraging people to consider biofuels. Crops take in carbon dioxide from the air by *photosynthesis* as they grow, making sugars or oil. The carbon dioxide returns to the air as the biofuels burn. It appears that the use of biofuels would have no overall effect on the level of carbon dioxide in the atmosphere. This, however, ignores the energy from burning *fossil fuels* during the planting, harvesting and processing of the crop. *Fertilizers* may also be needed for growth and these need fossil fuels for their manufacture.

biogas: methane gas produced by the action of bacteria on animal and plant wastes under anaerobic conditions. Many sewage works produce biogas. Biogas also forms in *landfill* sites as the waste breaks down. The gas can be hazardous unless collected and burnt. Small scale biogas digesters are used in many parts of the world to supply nearby homes with gas for cooking.

biological oxygen demand (BOD): a measure of water *pollution* by organic matter from sources such as sewage. BOD is the quantity of dissolved oxygen (measured in mg/L) removed from a sample of water by microorganisms when incubated for five days at 20°C. A sample of water containing much organic matter will support large numbers of bacteria which use dissolved oxygen as they break down the pollutants.

biopolymers: naturally occurring *polymers* in living things. They include *polysaccharides*, *proteins* and *nucleic acids*.

biotechnology makes use of microorganisms or *enzymes* to produce useful products. Traditional biotechnology uses yeast to make beer, bread and wine as well as bacteria to make yogurt. Modern biotechnology, based on genetic engineering, makes it possible to use genetically modified bacteria to produce medically important substances such as insulin, growth hormone and the blood clotting agent, factor VIII.

Biuret test: a test to detect *proteins*. The procedure is to add sodium hydroxide to a test sample followed by a few drops of copper(II) sulfate solution. The solution turns mauve if protein is present. The test actually detects *peptide* bonds.

blank determinations are used in *quantitative analysis* to eliminate errors. The analyst works through the whole procedure with a blank solution containing the solvent and all the reagents but none of the sample. In a *titration* a blank determination makes it possible to allow for the volume of reagent needed to change the color of the indicator at the *end point*. Blank determinations in colorimetry and spectroscopy help to detect, and allow for, errors caused by contamination of reagents or the effects of the apparatus used.

blast furnace: a furnace for extracting metals, which depends on a blast of pre-heated air entering near the bottom of the furnace. *Iron extraction* is a large scale, *continuous process* in blast furnaces.

bleaching: a process for destroying unwanted colors by *oxidation* or *reduction*. Domestic liquid bleach is a solution of sodium hypochlorite formed by the reaction of *chlorine* with cold sodium hydroxide solution. Bleaching powder is calcium hypochlorite, $Ca(OCl)_2$ made by absorbing chlorine gas in calcium hydroxide. These are oxidizing bleaches. Another oxidizing bleach is *hydrogen peroxide*.

Oxidizing bleaches not only remove color but also kill microorganisms.

Sulfur dioxide and sodium sulfite are reducing bleaches. Paper bleached white by a reducing bleach may turn yellow with age as oxygen in the air reverses the process.

Color in organic compounds is caused by delocalized electrons in sequences of alternating double and single bonds (*conjugated systems*). Oxidizing bleaches break some of the double bonds. Reducing bleaches convert double bonds to single bonds.

body-centered cubic structure: a crystal structure with an atom at each corner of a cube surrounding one atom at the center of the cube. This structure is more open than the two *close-packed structures*.

Body-centered cubic structure. Each metal atom is surrounded by eight nearest neighbors.

Metals with the body-centered cubic structure are the *group I* metals lithium, sodium and potassium, also the *d-block* metals chromium, vanadium and tungsten.

boiling: a liquid boils when it is hot enough for bubbles of vapor to form within the body of the liquid. This happens when the *vapor pressure* of the liquid equals the external *pressure*.

The boiling point of a liquid varies with pressure. Raising the external pressure raises the boiling point. Boiling points are usually measured at atmospheric pressure. The normal boiling point is the temperature at which the vapor pressure of the liquid equals one atmosphere (101.3 kPa).

The boiling point of a mixture of two liquids varies with the composition of the mixture. This has to be taken into account when separating mixtures by *fractional distillation*.

If the solute is not volatile, the boiling point of a solution is higher than that of the pure solvent. Boiling-point elevation is a *colligative property*. In dilute solutions the elevation of boiling point is directly proportional to the amount in moles of solute in a given mass of solvent (the *molality*).

Boltzmann distribution: see *Maxwell–Boltzmann distribution*.

bomb calorimeter: an apparatus for measuring energy changes when compounds burn (see diagram on page 58).

A measured amount of a sample burns in oxygen under pressure and the temperature rise of the whole apparatus is measured. The calorimeter is calibrated using benzoic acid for which the standard *enthalpy change of combustion* is known accurately.

A bomb calorimeter operates at constant volume so a correction is needed to convert the results to *enthalpy changes* at constant pressure.

Thermometer

Bomb calorimeter

Oxygen under
pressure

Water

Crucible containing
sample under test

Electrically heated
wire to ignite sample

Stirrer

A bomb calorimeter (see page 57)

bond angle: the angle between two *covalent bonds* in a molecule or *giant structure*. X-ray and electron diffraction studies make it possible to measure bond angles accurately. The results show that *covalent bonds* have a definite direction as well as length.

VSEPR theory makes it possible to predict the *shapes of molecules* and bond angles with surprising accuracy.

Tetrahedral

Trigonal, planar

Linear

Molecules showing the bond angles

bond breaking: see *heterolytic bond breaking* and *homolytic bond breaking*. Note that it takes energy to break bonds. Bond breaking is an *endothermic change*.

bond dissociation energy: the energy change on breaking one mole of a covalent bond in a gaseous molecule. Energy is needed to break covalent bonds. Bond breaking is *endothermic* so the enthalpy change is positive.

Chemists use spectroscopy to measure bond dissociation energies. The bond dissociation energies for the hydrogen halides are as follows:

$$HX(g) \longrightarrow H(g) + X(g)$$

ΔH = +562 kJ/mol for HF

ΔH = +432 kJ/mol for HCl

ΔH = +366 kJ/mol for HBr

ΔH = +299 kJ/mol for HI

For these compounds the bonds get easier to break as the halogen atoms get larger and the bonds get longer down group VII.

In molecules with two or more bonds between similar atoms, the successive bond dissociation energies are not the same. In water, for example, the energy needed to break the first O — H bond in H — O — H(g) is 498 kJ/mol but the energy needed to break the second O — H bond in O — H(g) is only 428 kJ/mol.

bond energies: the average values of *bond dissociation* energies used in approximate calculations to estimate energy changes for reactions. The average values of bond energies take into account the facts that:

- the successive bond dissociation energies are not the same in compounds such as water or methane

- the bond dissociation energy for a specific covalent bond varies slightly from one molecule to another.

Worked example

Use mean bond energies to estimate the enthalpy of formation of *hydrazine*.

Notes on the method
Write out the equation showing all the atoms and bonds in the molecules to make it easier to count the numbers of bonds broken and formed.
Look up the mean bond energies in a reference book. The symbol $E(N — H)$ stands for the bond energy of a covalent bond between a nitrogen atom and a hydrogen atom.

Answer
The equation for the reaction:

$$N\equiv N \ + \ 2H - H \longrightarrow \quad \begin{matrix} H & & H \\ & \diagdown & \diagup & \\ & N - N & \\ & \diagup & \diagdown & \\ H & & H \end{matrix}$$

The energy needed to break the bonds in the reactants

$= E(N\equiv N) \text{ kJ/mol} + 2E(H — H) \text{ kJ/mol}$

$= 945 \text{ kJ/mol} + 2 \times 436 \text{ kJ/mol}$

$= 1817 \text{ kJ/mol}$

The energy given out when the product forms

$= E(N — N) \text{ kJ/mol} + 4E(N — H) \text{ kJ/mol}$

$= 158 \text{ kJ/mol} + 4 \times 391 \text{ kJ/mol}$

$= 1722 \text{ kJ/mol}$

More energy is needed to break bonds than is given out when bonds are formed so the reaction is endothermic and the enthalpy change is positive.

$\Delta H \ = \ + 1817 \text{ kJ/mol} - 1722 \text{ kJ/mol} \ = \ + 95 \text{ kJ/mol}$

bond length: the distance between the nuclei of two atoms linked by one or more *covalent bonds*. For the same two atoms, *triple bonds* are shorter than *double bonds* which in turn are shorter than single bonds.

bond polarity

Bond	Bond length
C — C	154 pm
C = C	134 pm
C ≡ C	120 pm
C ⩵ C in benzene	140 pm

Bond lengths which are intermediate between single and double bonds, as in *benzene*, are a sign of electron *delocalization*.

bond polarity: the extent to which the shared pair of electrons in a *covalent bond* is attracted towards one of the atoms joined by the bond. Bonding electrons are pulled towards the more *electronegative* atom. The more electronegative atom carries a partial negative charge (δ–) and the other atom carries a partial positive charge (δ+). (See also *polar covalent bonds*.)

Examples of molecules with polar covalent bonds

bond rotation is possible around single covalent bonds but is prevented under normal conditions by *double bonds* or *triple bonds*. The lack of rotation around double bonds gives rise to *geometric isomerism*.

The groups at the end of a single C— C bond can rotate freely relative to each other giving rise to different conformations *of molecules*

bonding: see *metallic bonds*, *covalent bonds*, *ionic bonding*, *hydrogen bonding* and *intermolecular forces*.

bonding molecular orbital: a region in space in a molecule where there is a high probability of finding bonding electrons. Molecular orbitals, like *atomic orbitals*, are solutions to the Schrödinger wave equation. At a simplified, descriptive level, a molecular orbital forms by overlap of atomic orbitals of the atoms linked by a covalent bond.

Formation of a sigma orbital   s-orbital   sigma bond

Examples of molecular orbitals are *sigma (σ) bonds* and *pi (π) bonds*.

Sigma and pi bonds in ethene   *p-orbital*

borax ($Na_2B_4O_7 \cdot 10H_2O$): a naturally occurring mineral which has been used as a flux in metallurgy for hundreds of years. Molten borax cleans the surface of hot metals by dissolving metal oxides. This makes for good contact between metal surfaces when metals are welded or soldered together. Most borax is now used to make *borosilicate glass*.

Adding acid converts borax to boric acid (H_3BO_3) which is separated by cooling the solution in ice and filtering off the white crystals. Boric acid is a weak acid and mild antiseptic used in eye solutions and other medicines.

Born–Haber cycle: a thermochemical cycle which can be used to calculate the *lattice energy (enthalpy)* for a compound of a metal with a nonmetal.

Born–Haber cycle for sodium chloride (see the separate entries in this book for the definitions of the terms)

All but one of the terms in the cycle can be measured experimentally. Applying *Hess's law* makes it possible to calculate the one unknown term, which is the *lattice energy (enthalpy)*.

A Born–Haber cycle can help reveal the factors that determine the *stability of compounds*. For a stable compound, such as magnesium chloride, the lattice energy must be bigger than the total energy needed to produce gaseous ions from the elements. Removing two electrons from a magnesium atom is highly endothermic. What accounts for the stability of magnesium chloride ($MgCl_2$) is the large quantity of energy given out when the gaseous ions assemble themselves into a crystal lattice. $MgCl_3$ is not stable because of the very large third ionization energy of magnesium. $MgCl$ might be stable but not as stable as $MgCl_2$. $MgCl_2$ has a much larger lattice energy because an Mg^{2+} ion not only has a larger charge but is also smaller than a Mg^+ ion because it has lost both electrons in the outer shell.

A Born–Haber cycle can also help to determine whether the bonding in a compound is truly ionic. The experimental lattice energy calculated from a Born–Haber cycle can be compared with the theoretical value calculated using the laws of electrostatics and assuming that the only bonding in the crystal is ionic.

Compound	Experimental lattice energy from a Born–Haber cycle, kJ/mol	Theoretical lattice energy calculated assuming that the only bonding is ionic, kJ/mol
NaCl	780	770
KCl	711	702
AgCl	905	833

The values in the table show that ionic bonding can account for the lattice energies of sodium and potassium chlorides. The lattice energy of silver chloride is greater than can be explained by ionic bonding. There must be a contribution from covalent bonding.

(Note that chemists also use a second type of Born–Haber cycle to analyze the energy changes when an ionic crystal dissolves in water – see *enthalpy change of solution*.)

boron (B): a hard, black solid with a high melting point and the only nonmetal in *group III* of the periodic table. Its *electron configuration* is $[He]2s^22p^1$. Boron is in oxidation state +3 in all its compounds. The bonding between boron and other elements is covalent.

Boron has become important as an element since the development of the nuclear industry. The *isotope* boron-10 absorbs neutrons strongly so the element was scattered over the Chernobyl nuclear reactor after the explosion to quench chain reactions.

Boron trichloride (BCl_3) is a colorless liquid and like *aluminum chloride*, it is a strong electron pair acceptor (*Lewis acid*).

Sodium borohydride ($NaBH_4$) is a useful reducing agent.

borosilicate glass is used to make oven glassware and laboratory equipment because it has a higher melting point than cheaper, soda lime glass and is much less likely to crack as it heats up or cools down. Borosilicate glass has a high refractive index making is useful for the production of lenses.

Boyle's law states that the volume of a fixed amount of gas is inversely proportional to its pressure at constant temperature.

$$p \propto \frac{1}{V}$$ at constant temperature for a fixed amount of a gas

Hence $p_1V_1 = p_2V_2 = $ constant

brasses: *alloys* of copper (60–80%) and zinc (20–40%). Brass is easily worked, has an attractive gold color and does not corrode.

Breathalyzer: an instrument for estimating the concentration of alcohol in blood. A Breathalyzer measures the ethanol in a sample of air from the lungs. There is an equilibrium between ethanol dissolved in blood and ethanol vapor in the air in the lungs and the constant ratio at body temperature makes it possible to infer the blood alcohol concentration.

The first successful Breathalyzer was based on the chemical reduction of orange dichromate ions to green chromate(III) ions as they oxidize ethanol to acetaldehyde (ethanal). The driver breathed through a tube containing the orange crystals and the extent to which they turned yellow was *calibrated* to measure the blood alcohol concentration.

Many of the roadside Breathalyzers are now *fuel cells*. At one electrode ethanol is oxidized to acetic acid while at the other electrode oxygen is reduced to water. The higher the concentration of ethanol in the driver's breath, the greater the voltage of the cell. This method is not considered accurate enough for prosecution in court. Suspects are taken to a police station where the alcohol concentration in their breath is measured using an infrared absorption spectrometer. If spectroscopy shows that drivers are above the limit, they may have the right to ask for an analysis of ethanol in a blood sample, determined by *gas-liquid chromatography*.

brine: the term used industrially for a solution of sodium chloride in water. Brine is produced by "solution mining" of underground salt deposits. Water is pumped into the salt down one pipe. The mineral dissolves and flows to the surface through a second pipe. This is the start of the chlor-alkali industry (see *chemical industry*).

Natural brines are solutions of a mixture of salts and can be a useful source of chemicals such as potassium, magnesium and bromine or their compounds.

bromination: a reaction which replaces a hydrogen atom in an organic molecule with a bromine atom. One example is the bromination of *alkanes* – a *radical chain reaction* initiated by ultraviolet radiation.

Another example is the bromination of *benzene* by bromine in the presence of iron bromide as the catalyst. This is an *electrophilic substitution* reaction.

bromine (Br): a corrosive, dark red liquid which evaporates easily to give an orange vapor. It is the only liquid nonmetal at room temperature. Bromine is a halogen – the third element in *group VII* – with the electron configuration $[Ar]3d^{10}4s^24p^5$.

Bromine consists of diatomic molecules with pairs of atoms held together by single covalent bonds. The molecules are nonpolar so the *intermolecular forces* are relatively weak, *van der Waals forces*.

Bromine, like the other halogens, is an oxidizing element. It is a less powerful *oxidizing agent* than chlorine.

Bromine reacts with *s-block* metals to form ionic bromides in which the bromine atoms gain one electron to fill the $4p$ energy levels.

$$\overset{\times\times}{\underset{\times\times}{\overset{\wedge}{\underset{\times}{Br}}}}\!^{-}$$

Outer electron configuration of a bromide ion

Bromine oxidizes *nonmetals* such as sulfur and hydrogen on heating, forming covalent, molecular bromides.

$$H_2(g) + Br_2(g) \longrightarrow 2HBr(g)$$

$$H\!\overset{\times\times}{\underset{\times\times}{\overset{\bullet}{Br}}}\!^{\times}$$

Covalent bonding in hydrogen bromide

Hydrogen bromide is a fuming, acidic gas. Like the other *hydrogen halides* it is very soluble in water and a strong acid.

Bromine only reacts with water to a limited extent but it reacts with sodium hydroxide rapidly *disproportionating* at room temperature to form bromate(v) ions.

oxidized from 0 to +5

$$2Br_2(aq) + 6OH^- \longrightarrow 5Br^-(aq) + BrO_3^-(aq) + 3H_2O(l)$$

reduced from 0 to –1

Bromine oxidizes a range of ions or molecules in solution:

- iodide ions to iodine
- sulfite ions to sulfate ions
- thiosulfate ions to sulfate ions
- hydrogen sulfide to sulfur.

bromine extraction: a process for obtaining bromine from seawater or natural *brine*. The four-stage process concentrates and separates the bromine from seawater which contains about 65 ppm of bromide ions. The process produces one metric ton of bromine from 20,000 metric tons of seawater.

Stage 1   *Oxidation* of bromide ions to bromine.

Seawater is filtered and acidified to pH 3.5 to prevent chlorine and bromine reacting with water, then chlorine displaces bromine:

$$Cl_2 + 2Br^- \longrightarrow 2Cl^- + Br_2$$

Stage 2   Separation of bromine vapor.

A blast of air through the reaction mixture carries away the displaced bromine and helps to concentrate it.

Stage 3   Formation of hydrobromic acid.

The air with bromine vapor meets sulfur dioxide gas and a fine mist of water, producing hydrobromic acid, which is ionized in solution. After this stage the concentration of bromine in the solution is 1500 times greater than in seawater.

$$Br_2(g) + SO_2(g) + 2H_2O(l) \longrightarrow 4H^+(aq) + 2Br^-(aq) + SO_4^{2-}(aq)$$

Stage 4   Displacement and purification of bromine.

The solution from stage 3 now flows down a tower with a flow of chlorine gas and steam passing up it. The chlorine oxidizes bromide ions to bromine which evaporates in the steam. The mixture of steam and bromine is cooled and condensed producing a dense lower bromine layer under a layer of water.

Natural brines contain more bromine than seawater so that only stage 4 is needed to separate the bromine.

Bromine is used to make additives for leaded gasoline, flame retardants, agricultural chemicals, synthetic rubber for the inner lining of tubeless tires, dyes and a range of chemical intermediates. Silver bromide is the photosensitive chemical used in *photography*.

bromine water: the traditional name for a solution of bromine in water used as a laboratory reagent. The reagent is orange because it consists largely of a mixture of bromine molecules and water molecules hence the preferred name "aqueous bromine." Bromine *disproportionates* in water to a very limited extent, unlike *chlorine*. Aqueous bromine is decolorized when shaken with excess *alkene* because the *addition reaction* produces colorless products. Aqueous bromine loses its orange color when mixed with *phenol* and produces a white precipitate of tribromophenol. This is an example of *electrophilic substitution* which happens more readily with phenol than with benzene.

Brønsted–Lowry theory: the theory that *acids* are hydrogen ion (*proton*) donors and that *bases* are hydrogen ion (proton) acceptors. Johannes Brønsted (1879–1947) was a Danish physical chemist. He published his theory in 1923 at the same time as Thomas Lowry (1874–1936) of the University of Cambridge, but the two worked independently.

The Brønsted–Lowry theory is more general than the Arrhenius theory, because it also covers reactions which happen without water. According to the theory, hydrogen chloride molecules give hydrogen ions to water molecules when they dissolve in water producing hydrated hydrogen ions called *hydronium ions*. The water acts as a base:

$$HCl(g) + H_2O(l) \longrightarrow H_3O^+(aq) + Cl^-(aq)$$

An example of an acid–base reaction in the absence of water is the formation of a white smoke of ammonium chloride when *ammonia* gas mixes with hydrogen chloride gas.

$$NH_3(g) + HCl(g) \longrightarrow NH_4^+ Cl^-(s)$$

bronze: an *alloy* of copper with up to 12% tin. Bronze is a strong, hardwearing alloy with good resistance to corrosion. It is used to make gear wheels, bearings, propellers for ships, statues and coins.

Brownian motion: the rapid random motion of minute particles suspended in a liquid or gas. The phenomenon was first seen by the British botanist, Robert Brown, looking at pollen grains in water through a microscope. Brownian motion is evidence for the *kinetic theory*. The erratic random motion of pollen, smoke particles or other *colloid*-sized specks of matter is due to continuous bombardment by fast moving molecules which are too small to be seen through a light microscope.

buckminsterfullerene: a form of carbon consisting of C_{60} particles (see *allotropes*) and named after the architect Richard Buckminster Fuller who designed the geodesic dome for Expo '67 in Montreal. The popular name for C_{60} molecules is "buckyballs."

Buckminsterfullerene was discovered in 1985 by Harry Kroto and his team at the University of Sussex. It was the first of a family of *fullerenes*.

C_{60} molecules are formed as a carbon vapor mixed with helium slowly cools and condenses. C_{60} is special because it is the only structure with pentagons and hexagons which is smooth because every atom has the same curvature. Carbon atoms spontaneously assemble themselves into C_{60} and other fullerenes, including *nanotubes*.

bucky tubes: see *nanotubes*.

buffer solutions: mixtures of molecules and ions in solution which help to keep *pH* more or less constant. A buffer solution cannot prevent pH changes but it evens out the large changes in pH which can happen without a buffer.

Many household products include buffer mixtures, for instance detergents, shampoos, skin creams and a variety of medicines including eye drops. Advertisers sometimes claim that their products are "pH balanced" if the formulation includes a buffer solution.

A B C D E F G H I J K L M N O P Q R S T U V W X Y Z

Buffers are important in living organisms. The pH of blood, for example, is closely controlled by buffers within the narrow range 7.38 to 7.42. Chemists use buffers when they want to investigate chemical reactions at a fixed pH.

Buffers are equilibrium systems which illustrate the practical importance of *Le Chatelier's principle*. A typical buffer mixture consists of a solution of a *weak acid* and one of its salts. For example a mixture of acetic acid and sodium acetate. There must be plenty of the acid and its salt.

By choosing the right weak acid, it is possible to prepare buffers at any pH value throughout the pH scale. If the concentrations of the weak acid and its salt are the same, then the pH of the buffer is equal to pK_a for the acid.

More generally the pH of a buffer mixture can be calculated with the logarithmic form of the equilibrium law (see *Henderson–Hasselbalch equation*).

$$pH = pK_a + \lg \frac{[salt]}{[acid]}$$

Diluting a buffer solution with water does not change the ratio of the concentrations of the salt and acid so the pH does not change (unless the dilution is so great that the assumptions made when deriving the equation no longer apply).

$$CH_3COOH + H_2O \rightleftharpoons H_3O^+ + CH_3COO^-$$

| CH₃COOH acid molecules – a reservoir of H⁺ ions | | H₃O⁺ stays roughly constant – so the pH hardly changes | | CH₃COO⁻ base ions – with the capacity to accept H⁺ ions |

| Plenty of the weak acid to supply more H⁺ ions if alkali is added | Plenty of the ions from the salt to combine with extra H⁺ ions if acid is added |

The action of a buffer solution

Worked example

What is the pH of a buffer solution containing 0.40 mol/L formic acid and 1.00 mol/L sodium formate?

Notes on the answer

Look up the value of pK_a in a book of data. The pK_a of formic acid is 3.8.

Answer

$$pH = 3.8 + \lg \frac{1.00}{0.40}$$

$$pH = 4.20$$

bumping: violent boiling which shakes the apparatus and can throw liquid from the container in which it is being heated. Adding a few fragments of porous pottery or some jagged anti-bumping granules cuts the risk of bumping by helping the bubbles of vapor to form smoothly.

burets: graduated tubes, with taps or valves, used to measure the volumes of liquids or solutions during quantitative investigations such as *titrations*. The *accuracy* of a buret, when clean and properly used, depends on the *precision* with which it is manufactured and cali-brated. (See *uncertainty of measurement*.)

by-products: unwanted products of chemical synthesis or manufacturing. By-products are formed by side reactions which happen at the same time as the main reaction reducing the yield of the product required.

In the laboratory preparation of 1-bromobutane from 1-butanol, for example, the reagent is a mixture of sodium bromide and concentrated sulfuric acid. There are two side reactions which cut the yield. In the presence of acid, some of the 1-butanol dehydrates to an alkene and some reacts to form an ether.

$CH_3CH_2CH_2CH_2OH$

heat with
NaBr/conc H_2SO_4
→
the main reaction

$CH_3CH_2CH_2CH_2Br$

heat with
conc H_2SO_4
→
side reactions
some alkene and ether

The main reaction and the side reactions

caffeine: a *drug* found in tea, coffee, chocolate and cola drinks. Caffeine gives people a "lift" and helps them to be mentally alert. The drug is used medically to stimulate the nervous, respiratory and cardiovascular systems. Caffeine is also added to medicines to counteract sleepiness caused by other ingredients. Caffeine can be isolated from tea or coffee as a white crystalline solid by *solvent extraction*.

Structure of caffeine

calcium (Ca): a reactive, metallic element in *group II* of the periodic table. Its electron configuration is $[Ar]4s^2$. Samples of the silvery metal usually look grey because they are covered with a layer of calcium oxide.

Calcium burns brightly in air with a brilliant red flame forming the white solid calcium oxide (CaO). The metal also reacts with cold water producing hydrogen and a white precipitate of calcium hydroxide.

$$Ca(s) + 2H_2O(l) \longrightarrow Ca(OH)_2(s) + H_2(g)$$

Calcium forms ionic compounds with nonmetals in which the metal is in the +2 oxidation state as Ca^{2+}.

Calcium oxide is a white solid made by heating *calcium carbonate*. It is a *basic oxide*. Its reaction with cold water to make calcium hydroxide is highly *exothermic*. Calcium hydroxide is only sparingly soluble in water forming an alkaline solution often called limewater.

Anhydrous calcium chloride ($CaCl_2$) is a cheap drying agent. The chloride crystallizes from solution as a hydrate, $CaCl_2.6H_2O$.

calcium carbonate occurs naturally as limestone, chalk and marble. Limestone is an important mineral. Some of the rock is quarried for road building and construction. Pure limestone is also used in the chemical industry. Heating limestone in a furnace at 1200 K converts it to calcium oxide (quicklime). The reaction of quicklime with water produces calcium oxide (slaked lime). (See the diagram at the top of page 69.)

calibration: a procedure used in quantitative analysis. Calibration makes it possible to convert the measurement of temperature, volume or light absorption into an accurate *concentration* in moles per liter. When using colorimeters or spectrometers, for example, it is common to plot a calibration curve. Instrument readings are taken for a series of *standard solutions* of known concentration. See the graph opposite.

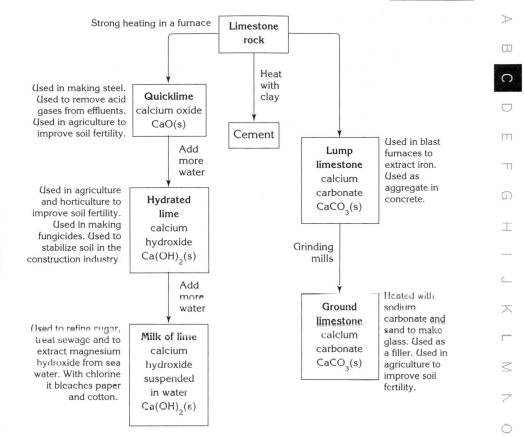

Strong heating in a furnace

Limestone rock

Heat with clay

Quicklime calcium oxide CaO(s)

Used in making steel. Used to remove acid gases from effluents. Used in agriculture to improve soil fertility.

Cement

Lump limestone calcium carbonate CaCO₃(s)

Used in blast furnaces to extract iron. Used as aggregate in concrete.

Add more water

Hydrated lime calcium hydroxide Ca(OH)₂(s)

Used in agriculture and horticulture to improve soil fertility. Used in making fungicides. Used to stabilize soil in the construction industry.

Grinding mills

Add more water

Milk of lime calcium hydroxide suspended in water Ca(OH)₂(s)

Used to refine sugar, treat sewage and to extract magnesium hydroxide from sea water. With chlorine it bleaches paper and cotton.

Ground limestone calcium carbonate CaCO₃(s)

Heated with sodium carbonate and sand to make glass. Used as a filler. Used in agriculture to improve soil fertility.

Products from limestone and their uses

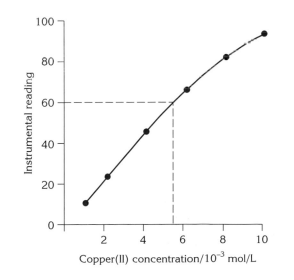

Calibration curve used to determine the concentration of copper(II) ions in aqueous solution obtained by taking instrument readings with a series of standard solutions.
The instrument reading for an unknown sample is 60.0. The concentration of the sample can be read from the graph. The concentration of copper(II) ions in the sample is 5.5×10^{-3} mol/L.

Instrumental reading

Copper(II) concentration/10^{-3} mol/L

calorimeter: an apparatus for measuring the energy change during a chemical reaction. Typically a calorimeter is insulated from its surroundings and contains water. The energy from the reaction heats up the water and rest of the apparatus. An accurate thermometer measures the temperature rise. The apparatus is *calibrated* to determine its overall *specific heat capacity* by measuring the temperature rise for a reaction with a known *enthalpy change*.

Enthalpy changes of combustion are measured using a *bomb calorimeter*.

Enthalpy changes for reactions in solution can be compared quickly using an expanded polystyrene cup with a lid, as the calorimeter. Expanded polystyrene is an excellent insulator and has a negligible specific heat capacity. If solutions are dilute it is sufficiently accurate to assume that their density and specific heat capacity are the same as those of water.

Worked example

When 50 cm^3 of 2.0 mol/L hydrochloric acid mixes with 50 cm^3 2.0 mol/L sodium hydroxide in a polystyrene cup the temperature rise is 13.5°C. What is the enthalpy change for the neutralization reaction?

Notes on the method

Assume that the density of the solutions is the same as water (1 g/cm^3) and that for both, the specific heating capacity, like water, is 4.18 J/g·K.

Note that the total volume of solution is 100 cm^3 so the mass of solution heating up can be taken as 100 g.

Use the formula that the energy change = $mc\Delta T$ where m is the mass, c the specific heat capacity and ΔT the temperature change.

The enthalpy change is the energy exchanged with the surroundings when a reaction proceeds as constant temperature and pressure. Here the energy from the exothermic reaction is kept in the system to heat up the water. The calculation shows the energy which would otherwise be lost to the surroundings during a constant temperature change.

The enthalpy change, ΔH, for a reaction is the energy change for the amounts (in moles) shown in the chemical equation.

Answer

The amount of hydrochloric acid   = the amount of sodium hydroxide

$$= \frac{50}{1000} \text{ L} \times 2.0 \text{ mol/L}$$

$$= 0.1 \text{ mol}$$

Energy given out by the reaction and used to heat the water in the cup

$$= 100 \text{ g} \times 4.18 \text{ J/g·K} \times 13.5 \text{ K}$$

$$= 5643 \text{ J}$$

Energy given out per mole of acid $= \dfrac{5643 \text{ J}}{0.1 \text{ mol}}$

$$= 56\ 430 \text{ J/mol} = 56.4 \text{ kJ/mol}$$

The reaction is exothermic, so the enthalpy change for the system is negative.

$$NaOH(aq) + HCl(aq) \longrightarrow NaCl(aq) + H_2O(l) \qquad \Delta H = -56.4 \text{ kJ/mol}$$

capillary rise: an effect seen when a fine capillary tube is dipped into a liquid. The liquid normally rises up the tube. The larger the surface tension of the liquid and the thinner the tube the greater the effect.

More generally capillary action draws liquids, especially water, through any material which is riddled with fine crevices. Capillary action draws solvent up the paper during *paper chromatography*.

carbocations: *intermediates* formed during organic reactions in which a carbon atom carries a positive charge. Carbocations form, for example, during *nucleophilic substitution* by the S_N1 mechanism and during *electrophilic addition* to alkenes.

Three carbocations showing the order of stability: tertiary > secondary > primary. The inductive effect of electron releasing alkyl groups (shown by arrows on the bonds) helps to spread the positive charge and stabilize the ion. The more alkyl groups the more stable the carbocation.

carbohydrates: compounds of carbon, hydrogen and oxygen. They include sugars, starch and cellulose. The name carbohydrate arises from the formulas in which hydrogen are usually present in the same ratio as in water. The formula for glucose, for example, is $C_6H_{12}O_6$ which could be written $C_6(H_2O)_6$. Neither way of writing the formula gives any idea of how the atoms are arranged in a glucose molecule.

Glucose is a monosaccharide. Sucrose, with two sugar units joined together, is a disaccharide.

glucose

glucose unit

sucrose

fructose unit

Structures of glucose and sucrose (table sugar)

Starch and cellulose are polysaccharides. They both consist of long chains of glucose units. The difference lies in the links between the units.

Plants make glucose by *photosynthesis*. Some of the glucose they store as starch, a reserve of energy food. Some of the glucose builds up the cellulose cell walls as the plants grow. When people eat parts of plants, such as potatoes or rice, they can digest starch but not cellulose. Cows feed on grass but rely on bacteria in their guts to break down the cellulose to glucose. Cows chew the cud to give bacteria time to digest the long chain molecules.

carbon (C): a nonmetal element with three solid *allotropes*: diamond, graphite and *buckminsterfullerene*. *Charcoal* is an impure form of carbon. Carbon is the first element in group IV of the periodic table. Its electron configuration is $[He]2s^22p^2$.

Carbon's remarkable ability to form stable chains and rings of atoms with single, double and triple bonds gives rise to *organic chemistry*.

Carbon forms two oxides both of which are colorless, molecular gases with no smell. *Carbon monoxide* (CO) is a *neutral oxide*.

Carbon dioxide is an acidic oxide. It dissolves in water forming carbonic acid which is a weak acid.

$$CO_2(g) + H_2O(l) \rightleftharpoons H_2CO_3(aq)$$

The gas is strongly absorbed by alkalis such as potassium hydroxide

The simplest chloride of carbon, carbon tetrachloride (CCl_4), is a colorless liquid with tetrahedral molecules. Unlike many other nonmetal chlorides it is not hydrolyzed by water or alkalis (see *hydrolysis of nonmetal chlorides*). Carbon does not react directly with chlorine, so tetrachloromethane is made by the reaction of carbon disulfide (CS_2) with chlorine.

carbon cycle: the cycling of the element carbon in the natural environment as carbon compounds move between the main reservoirs of carbon in the atmosphere, the oceans and on land (see *environmental chemistry*).

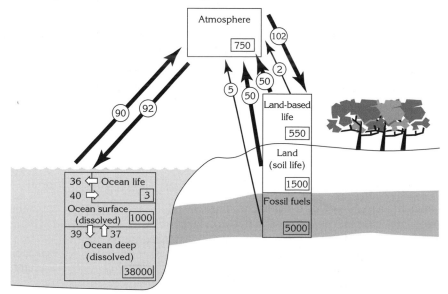

A highly simplified carbon cycle with estimates of the quantities in the main reservoirs where the figures are in gigatonnes of carbon (1 Gt = 1000 million metric tons). Also shown are estimates of the flows of carbon in gigatonnes per year (Gt/yr).

carbon monoxide: a colorless gas with no smell, formed by the incomplete combustion of fuels containing carbon. The gas is highly poisonous because it is held much more strongly by blood *hemoglobin* than oxygen. Death results when most of the hemoglobin is blocked by carbon monoxide. Lower levels of exposure to the gas reduce the ability of blood to carry oxygen and this puts a strain on the heart.

Carbon monoxide is also an important industrial chemical, both as a reducing agent and as a starting point for the synthesis of organic chemicals. Carbon monoxide is used as a reducing agent in metallurgy, for example during *iron extraction* in a blast furnace.

Steam reforming is the source of synthesis gas which is a mixture of carbon monoxide and hydrogen.

Carbon monoxide is used on a large scale to manufacture *methanol*. A mixture of carbon monoxide and hydrogen gases combines to make methanol at about 250°C over a catalyst made from copper and zinc oxide. This is an example of *heterogeneous catalysis*.

$$CO(g) + 3H_2(g) \rightleftharpoons CH_3OH(l)$$

carbonium ions: ions in which a carbon atom carries a positive charge (see *carbocations*).

carbonyl compounds contain the carbonyl group ($C=O$). The two main classes of carbonyl compounds are the *aldehydes* and the *ketones*. The $C=O$ double bond is polar with the electrons drawn towards the more electronegative oxygen atom. The characteristic reactions of carbonyl compounds are *nucleophilic addition reactions*

$$\overset{\delta+}{\underset{}{C}}=\overset{\delta-}{O}$$

The carbonyl group in aldehydes and ketones

carborundum: the common name for silicon carbide (SiC), a shiny, black solid made by heating silicon dioxide with carbon at 2000°C. It has the same crystal structure as diamond but with every other carbon atom replaced by a silicon atom. Carborundum is a useful abrasive being harder than corundum (*aluminum oxide*) but not as hard as diamond. Like diamond, carborundum has a very high melting point making it a useful *refractory*.

carboxylic acids: compounds with the formula R—COOH where R represents an *alkyl group*, *aryl group* or a hydrogen atom. The carboxylic acid group (—COOH) is the *functional group* which gives the acids their characteristic properties.

Names and structures of carboxylic acids

Carboxylic acids are named by changing the ending of the corresponding *alkane* to "-oic" acid. So ethane becomes ethanoic acid.

Reactions which produce carboxylic acids

Even the simplest acids, such as formic acid and acetic acid, are liquids at room temperature because of *hydrogen bonding* between the carboxylic acid groups. Also, because of hydrogen bonding, these acids mix freely with water. Hydrogen bonding leads to *dimerization* of carboxylic acids in nonpolar solvents.

Benzoic acid is a solid at room temperature. It is only very slightly soluble in cold water but more soluble in hot water.

Carboxylic acids are *weak acids*. The —OH group in the carboxylic acid group is more acidic than the —OH group in *alcohols* because the carboxylate ion formed is stabilized by *delocalization of electrons*. Carboxylic acids are sufficiently acidic to produce carbon dioxide when added to a solution of sodium carbonate (or sodium hydrogencarbonate). This reaction distinguishes carboxylic acids from weaker acids such as *phenols*.

Ionization of acetic acid. Delocalization of electrons stabilizes the carboxylate ion and favors ionization of the acid.

Carboxylic acids form a number of related compounds (derivatives): *acyl chlorides, acid anhydrides, esters* and *amides*. The reactions of acetic acid are typical of carboxylic acids.

Reactions of acetic acid

74

carcinogens: compounds which can cause cancer. In a healthy body, the rate at which cells divide to produce new cells is strictly controlled. Cancer is a disorder which happens when some cells somehow escape from the normal control and multiply to produce a growth, or tumor. Chemicals which cause cancer probably produce changes (mutations) in the genes that control cell division.

catalysts speed up the rates of chemical reactions without themselves changing permanently. Catalysts can be recovered at the end of the reaction. Often a small amount of catalyst is effective.

Catalysts work by providing for the reaction an alternative pathway with a lower *activation energy*. A lower activation energy means that there is an increased proportion of molecules with enough energy to react at a particular temperature.

Reaction profile showing the effect of a catalyst on the activation energy of a reaction

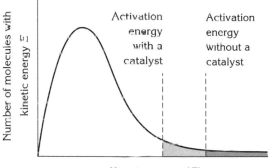

Distribution of molecular energies showing how the proportion of molecules able to react increases when a catalyst lowers the activation energy

Often a catalyst changes the mechanism of the reaction and makes it more productive by increasing the yield of the desired product and reducing waste. The alternative route for the reaction often involves the formation of *intermediates* which are combinations of reacting molecules and the catalyst. These intermediates have a short life. The catalyst regenerates as the product forms. So the catalyst can go on to catalyze the reaction of more molecules. This gives rise to a continuous catalytic cycle.

Many catalysts are specific to a particular reaction. This is especially true of *enzymes*. If the catalyst is in the same phase as the reactants it is a *homogeneous catalyst*. If the catalyst is in a different phase it is a *heterogeneous catalyst*.

Ideally a catalyst should be highly selective. This means that the reaction with the catalyst produces a high yield of the required product with a minimum of by-products. Developing more selective catalysts is one of the aims of *green chemistry*.

Catalysts speed up reactions but they do not change the position of equilibrium for a reversible reaction.

Here are some highlights in the development of industrial catalysts:

*   1908 Fritz Haber discovered how to make *ammonia* from nitrogen and hydrogen with a modified iron catalyst.

*   1912 Paul Sabatier first used a nickel catalyst to *hydrogenate* unsaturated *vegetable oils* and turn them to solid *fats* for margarine.

*   1930 Eugene Houdry developed *catalytic cracking* of oil fractions to make *gasoline*.

*   1942 Vladimir Ipatieff and Herman Pines found a catalytic method of alkylating *hydrocarbons* to produce branched hydrocarbons with high octane numbers to prevent *knocking* in gasoline engines.

*   1976 General Motors and the Ford Motor Corporation developed *catalytic converters* to cut pollution from motor vehicles.

*   1984 The company Takasago International introduced a *chiral* catalyst to manufacture menthol (one of the most important flavoring chemicals).

*   1996 The oil company BP announced a new method for making acetic acid from methanol and carbon dioxide with an iridium iodide catalyst. The new catalyst has helped to increase yields, cut costs and reduced harmful impacts on the environment.

*   1998 After decades of development work, two new, large, advanced ammonia plants brought on stream using ruthenium in place of iron as the catalyst.

catalyst poisons inactivate *catalysts*, especially heterogeneous catalysts. Lead compounds in *gasoline*, for example, poison the catalyst in catalytic converters and stop it from working. Carbon monoxide poisons the iron catalyst used for *ammonia manufacture*.

catalytic converter: a device in the exhaust system of a car which contains a *catalyst* to covert pollutants in the exhaust gases to less harmful substances. Car exhausts pollute the air because the engine does not burn all the fuel and because the temperature and pressure in the cylinders are high enough for nitrogen from the air to react with oxygen.

Thin coating of catalyst on the ceramic

From the engine

CO_2
NO_x
C_xH_y

Out to the air

N_2
H_2O
CO_2

Diagram of a catalytic converter

Ceramic block with many fine tubes

Unleaded gasoline must be used in cars fitted with a catalytic converter because lead would poison the catalyst and stop it working.

The presence of sulfur in fuels lowers the efficiency of catalytic converters. For this reason, lowering the sulfur concentration in gasoline helps to cut down on harmful pollutants from car exhausts.

Pollutant	Origin of the pollutant
carbon dioxide, CO_2	complete combustion of hydrocarbons in gasoline
carbon monoxide, CO	incomplete combustion of fuel
hydrocarbons, C_xH_y	unburnt fuel
nitrogen oxides, NO_x	reaction of nitrogen and oxygen from the air in the hot engine
lead compounds	from anti-knock additives in leaded gasoline

The catalyst is a finely divided alloy of platinum and rhodium supported on an inert ceramic, pierced with many fine tubes to give a very large surface area. Once the catalyst is hot enough it converts the pollutants to steam, carbon dioxide and nitrogen.

catalytic cracking: a process in an oil refinery for converting fractions from the *fractional distillation of oil* into more useful products by breaking up larger molecules into smaller ones.

The catalyst powder flows to the vertical reactor where cracking takes place. The cracked vapors pass to a fractionating column while the catalyst flows to the regenerator.

The problem in refining is to produce the oil products in the amounts needed by customers. Generally crude oil contains too much of the high-boiling fractions (bigger molecules) and not enough of the low-boiling fractions (smaller molecules) needed for fuels such as *gasoline*.

The *heterogeneous catalyst* is a *zeolite*. Cracking takes place on the surface of the catalyst. The mechanism involves *heterolytic bond* breaking and the intermediates are *carbocations*.

Catalytic cracking is a *continuous process*. The finely powdered catalyst made of silicon and aluminum oxides gradually gets coated with carbon so it circulates through a regenerator where the carbon burns away in a stream of air.

catenation: the ability of the atoms of an element to join together in long chains or rings. *Carbon* atoms have an exceptional ability to catenate – hence the wide range of organic compounds. *Silicon*, the second element in group IV, does not have the same ability to catenate. Silicon forms *hydrides*, Si_nH_{2n+2}, analogous to the *alkanes* but the largest known molecule has n equal to eight, and only silane (SiH_4) is stable for any length of time.

Sulfur atoms can catenate. Both the rhombic and monoclinic *allotropes* consist of S_8 rings. Plastic sulfur is a tangled mass of long chains of sulfur atoms but it rapidly reverts to the rhombic form at room temperature.

cathode: the *electrode* at which *reduction* takes place. During *electrolysis* the external power supply adds *electrons* to the cathode so that it becomes the negative electrode attracting positive ions which gain electrons (reduction) turning into atoms or molecules.

The term cathode is also sometimes used in *electrochemical cell* where the reduction process at this electrode in one of the *half-cells* takes electrons from the electrode which becomes the positive terminal of the cell.

Note that electrons flow into the cathode from the external circuit both during electrolysis and when a current is drawn from a chemical cell.

cathodic protection: an electrochemical method of preventing *corrosion* used with pipelines, oil rigs and the hulls of ships. Steel corrodes where it is oxidized. This happens wherever the metal is an *anode*. In anodic regions the iron atoms give up electrons turning into ions. Attaching a more reactive metal, such as zinc or magnesium, to iron creates an electrochemical cell in which the iron is the *cathode* so the more reactive metal corrodes. The zinc or magnesium acts as a sacrificial anode.

cations: *positive ions* attracted to the *cathode* during *electrolysis*. (See *cation tests* for examples of common cations.)

cation tests are used in qualitative analysis to identify the positive ions in salts. (See the table at the top of page 79.) Adding sodium hydroxide produces a precipitate if the metal hydroxide is insoluble. The precipitate dissolves in excess of the alkali if the hydroxide is *amphoteric*.

Adding *ammonia* solution also precipitates insoluble hydroxides. These redissolve in excess if the metal ion forms stable *complex ions* with ammonia molecules.

These tests can be supplemented with *flame tests* to identify metal ions such as sodium and potassium which do not give precipitates.

Cation tests (see page 78)

Positive ion in solution	Observations on adding sodium hydroxide solution drop by drop and then in excess	Observations on adding ammonia solution drop by drop and then in excess
calcium, Ca^{2+}	white precipitate but only if the calcium ion concentration is high	no precipitate
magnesium, Mg^{2+}	white precipitate insoluble in excess reagent	white precipitate insoluble in excess reagent
barium, Ba^{2+}	no precipitate	no precipitate
aluminum, Al^{3+}	white precipitate which dissolves in excess reagent	white precipitate insoluble in excess reagent
chromium(III), Cr^{3+}	green precipitate which dissolves in excess to form a dark green solution	green precipitate insoluble in excess reagent
manganese(II), Mn^{2+}	off-white precipitate insoluble in excess reagent	off-white precipitate insoluble in excess reagent
iron(II), Fe^{2+}	green precipitate insoluble in excess reagent	green precipitate insoluble in excess reagent
iron(III), Fe^{3+}	brownish-red precipitate insoluble in excess reagent	brownish-red precipitate insoluble in excess reagent
copper(II), Cu^{2+}	pale blue precipitate insoluble in excess	pale blue precipitate dissolving in excess to form a dark blue solution
zinc, Zn^{2+}	white precipitate which dissolves in excess reagent	white precipitate which dissolves in excess reagent
lead, Pb^{2+}	white precipitate which dissolves in excess reagent	white precipitate insoluble in excess reagent
ammonium, NH_4^+	alkaline gas (ammonia) given off on heating	no visible change

cell: see *electrochemical cell.*

Celsius temperature scale: a scale in which the melting point of ice is set at 0°C and the boiling point of water at 100°C. The scale is based on the absolute or *Kelvin* temperature scale.

Temperature in °C = absolute temperature (K) − 273 K

cement is manufactured by heating clay with powdered limestone in a rotating kiln and then grinding the lumps of product to a fine powder. Cement is a complex mixture of calcium silicates and calcium aluminosilicates which becomes hydrated and sets when mixed with water. Stirring cement with sand, gravel and water makes concrete.

ceramics: materials such as pottery, glasses, cement, concrete and graphite. Ceramics also include a wide range of crystalline materials such as *carborundum*, silicon nitride, *aluminum oxide* and magnetic *ferrites*. What all these materials have in common is that they are nonmetallic, inorganic materials which are heated to a high temperature in a furnace at some stage during their manufacture.

Typical advantages of ceramics are that they are:

- very hard
- strong in compression
- chemically *inert*
- *refractories*
- electrical insulators.

Typical disadvantages of many ceramics are that they are:

- weak in tension
- brittle
- liable to crack if there is a sudden temperature change.

cesium chloride structure: the cubic crystal structure of the ionic compound cesium chloride (CsCl). In general it is a structure for a compound M⁺X⁻ in which each positive ion is surrounded by eight nearest neighbors at the corners of a cube and each negative ion is similarly surrounded by eight positive ions. So the *coordination numbers* for both elements is 8.

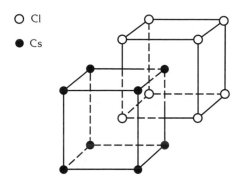

Structure of cesium chloride showing 8:8 coordination. The structure consists of a simple cubic array of positive ions interpenetrating a cubic array of negative ions.

Other compounds with this structure include CsBr, CsI and NH_4Cl.

This structure is only possible in compounds with a positive ion which is relatively large so it can be in contact with eight neighboring negative ions. Cesium, in period VI of the periodic table, forms larger ions than, for example, sodium. (See also *unit cell*.)

CFCs (chlorofluorocarbons) such as CCl_3F, CCl_2F_2 and CCl_2FCClF_2 are compounds containing carbon, chlorine and fluorine. They have advantages: they are unreactive, do not burn and are not *toxic*. Also it is possible to make CFCs with different boiling points to suit different applications. These properties made CFCs ideal as the working fluid in refrigerators and air conditioning units. They are also used to make the bubbles in expanded plastics and insulating foams. CFCs can be good solvents for dry cleaning and removing grease from electronic equipment.

The problem with CFCs is that they escape into the *atmosphere* where they are so stable that they last for many years, long enough for them to diffuse up to the stratosphere. In the stratosphere the intense ultraviolet light from the Sun splits CFCs into *radicals* including chlorine atoms. Chlorine atoms react with *ozone*:

$$Cl\bullet + O_3 \longrightarrow ClO\bullet + O_2$$
$$ClO\bullet + O\bullet \longrightarrow Cl\bullet + O_2$$

The first reaction is much faster than other reactions in the stratosphere. The second reaction involves oxygen atoms which are common in the stratosphere and this recreates the chlorine atom. This means in effect that one chlorine atom can rapidly destroy many ozone molecules.

Now that the damaging effects of CFCs are known they are being phased out where possible. The chemical industry has developed HFCs (hydrofluorocarbons) instead.

changes of state: changes from one state of matter to another. The following are all examples of changes of state:

• a solid melting to a liquid
• a liquid freezing to a solid
• a liquid evaporating and becoming a gas
• a gas condensing to a liquid.

Energy is taken in from the surroundings during melting and evaporating (they are *endothermic* processes). The energy is needed to break the bonds between atoms, molecules or ions.

Energy is given out to the surroundings during freezing or condensing (they are *exothermic* processes). Energy is given out as the bonds between particles reform.

A few solids turn directly to gas when heated at atmospheric pressure. On cooling the vapor turns directly back to a solid. This is *sublimation*.

charcoal: a form of *carbon* made by heating wood or bones in the absence of air. Charcoal consists of minute graphite crystals. Heating charcoal in steam at about 1000°C produces activated charcoal by driving out volatile compounds from the pores of the solid. Activated charcoal is an excellent absorbent used to filter out impurities from gases and solutions.

Charles' law states that the volume of a fixed amount of gas at constant pressure is proportional to its absolute temperature.

$$V \propto T \text{ or } \frac{V}{T} = \text{constant}$$

The law follows from the *ideal gas* equation $pV = nRT$ if p and n are constant. Real gases deviate from ideal gas behavior. Note that the law only applies if there is no chemical change altering the number of gas molecules as the gas gets hotter or colder.

chelates: *complex ions* in which each *ligand* molecule or ion forms more than one *coordinate bond* with the central metal ion. Chelates are formed by *bidentate* and *polydentate* ligands such as *edta*. The term chelate comes from the Greek word for a crab's claw, reflecting the claw-like way in which chelating ligands grab hold of metal ions. Powerful chelating agents trap metal ions and effectively isolate them in solution.

Chelate complexes are generally more stable than complexes formed by monodentate ligands (see *formation constants*). The chelate effect can be explained by considering the entropy change for the ligand exchange reactions involved. When a bidentate ligand replaces monodentate ligands in a complex there is an increase in the number of molecules and ions. The value of ΔS_{system} is positive, increasing the tendency for the reaction to happen.

$$[Ni(NH_3)_6]^{2+}(aq) + 3H_2NCH_2CH_2NH_2(aq) \longrightarrow [Ni(H_2NCH_2CH_2NH_2)_3]^{2+}(aq) + 6NH_3(aq)$$

chemical equations describe what happens during reactions by identifying the reactants and products. State symbols show the *states of matter* of the chemicals involved. Equations may also show the reaction conditions by including information about temperature, pressure and catalysts above or below the arrow leading from reactants to products. There are various types of chemical equation.

- Full, *balanced equations* with formulas are used to calculate the amounts of reactants and products:

$$2H_2(g) + O_2(g) \longrightarrow 2H_2O(l)$$

- *Ionic equations* are balanced equations leaving out any *spectator ions* which do not change:

$$H^+(aq) + OH^-(aq) \longrightarrow H_2O(l)$$

- *Half equations* are balanced equations which show part of a *redox* reaction:

$$2H^+(aq) + 2e^- \longrightarrow H_2(g)$$

- Unbalanced symbol equations show just the main starting chemical and the main product with the conditions for reaction written by the arrow:

$$C_2H_5OH \xrightarrow[\text{heat}]{Cr_2O_7^{2-}/H^+} CH_3CO_2H$$

- Mechanistic equations show intermediate steps in the *mechanism* for a reaction:

$$H_2 + Br\bullet \longrightarrow HBr + H\bullet$$

chemical industry: the industry which converts raw materials such as crude oil, natural gas and minerals into useful products such as pharmaceuticals, *fertilizers*, *detergents*, paints and *dyes*.

A chemical plant consists not only of the reaction vessels and equipment for separating and purifying products, but also the storage vessels, pumps and pipes, sources of energy and heat exchangers together with the control room.

Bulk chemicals are manufactured on a scale of thousands or even millions of metric tons per year. Examples are ethene, sulfuric acid, ammonia and chlorine. They are mainly used as the starting point for making other substances. Fine chemicals such as pesticides and pharmaceuticals are made on a much smaller scale – a few metric tons or hundreds of metric tons per year.

Speciality chemicals are manufactured for their particular properties as thickeners, stabilizers, flame retardants and so on.

Main raw materials	Large scale processes and products	Uses of the products
Crude oil and *natural* gas for the petrochemical industry	*Fractional distillation, thermal cracking, steam reforming* and *isomerization* and many other processes to produce the building blocks for the industry such as *ethene*, propene and benzene	Manufacture of *polymers,* solvents, pharmaceuticals, *agrochemicals, dyes,* pigments and *detergents*
Salt (sodium chloride) and limestone for the chlor-alkali industry	*Electrolysis of brine* for making chlorine, sodium hydroxide and hydrogen. Solvay process to make sodium carbonate	Manufacture of *bleaches, disinfectants,* solvents, some polymers and in the *glass* and paper industries
Sulfur from underground deposits of the element or from the purification of oil and gas plus oxygen from the air	Contact process for *sulfuric acid manufacture*	Manufacture of *paints,* pigments, *fertilizers,* detergents, *plastics* and many uses in the chemical, metallurgical and petrochemical industries
Nitrogen from the air, natural gas and oil fractions	Haber process for *ammonia manufacture* and catalytic oxidation of ammonia for *nitric acid manufacture*	Manufacture of fertilizers, dyes, pigments, detergents, *explosives,* plastics and *fibers*
Calcium phosphate rock	Treatment of phosphate rock with concentrated sulfuric acid to make *phosphoric acid* and phosphates	Fertilizer Industry, manufacture of washing powders, toothpaste, food industry, enamels and glazes
Fluorite (calcium fluoride)	Action of concentrated sulfuric acid on fluorite to make hydrogen fluoride; electrolysis of fluorides in hydrogen fluoride to make fluorine	Etching and polishing glass and integrated circuits, manufacture of fluorocarbons and hydrofluorocarbons (to replace *CFCs*); pharmaceuticals and the polymer ptfe.

chemotherapy: the use of chemicals to treat disease. Chemotherapy began with the work of Paul Ehrlich (1854–1915) who had the idea that it might be possible to find chemicals which kill the microorganisms which cause disease without harming healthy living cells. Paul Ehrlich worked as an assistant to Robert Koch (1843–1910) who pioneered the use of *dyes* to stain and identify *bacteria*. Ehrlich was particularly interested in selective dyes. Some dyes, for example, take well on cotton but not on wool. Ehrlich found that methylene blue would dye nerve cells well but not other parts of the body. This inspired him to search for chemical "magic bullets" to target microorganisms and diseased cells. After a long series of experiments Ehrlich and his Japanese colleague, Sahachiro Hata, finally discovered an arsenic compound which cured syphilis. For the first time a synthetic chemical was used to cure a bacterial disease.

The high hopes raised by Ehrlich's work led to many disappointments until 1932 when Gerhard Domagk was involved in testing the medical effects of new dyes produced by a German chemical company. He found that the dye Prontosil red was remarkably effective against streptococcal infections in mice. Domagk saved his daughter's life with the new drug when she accidentally picked up a serious infection by pricking her finger in his laboratory. This led to the development of sulfonamide drugs which were used to treat bacterial diseases until the discovery of *antibiotics*.

Today chemotherapy is widely used to treat cancer but few drugs are "magic bullets." If they are effective in destroying cancerous cells they are generally toxic and damage other parts of the body too (see *cisplatin*).

chiral compounds are *asymmetric* so that they have mirror-image forms which are not identical. The commonest chiral compounds have a carbon atoms attached to four different atoms or groups. The two mirror image forms are known as enantiomers.

Mirror-image forms of lactic acid

Enantiomers behave identically in ordinary chemical reactions and their main physical properties are the same. They differ in their effect on *polarized light* – they are optically active. One mirror-image form rotates the plane of polarized light in one direction. The other form has the opposite effect. They are *optical isomers*.

There are three common ways of naming compounds to distinguish between pairs of enantiomers. The symbols (+) and (–) indicate the direction in which the molecules rotate the plane of polarized light. The other two systems (D/L and R/S) distinguish the enantiomers according to the arrangement of atoms in three dimensions. The D/L system is commonly used for *amino acids*. The R/S system is based on a set of rules related to the *atomic numbers* of the atoms bonded to the chiral carbon atom.

Chirality is very important in living organisms. Living cells are full of messenger and carrier molecules which interact selectively with the active sites and receptors in other molecules such as *enzymes*. The messenger and carrier molecules are all chiral and the body works with only one of the mirror-image forms. In most living things all the *amino acids*, for example, are L-amino acids.

The importance of chirality in living things was brought home to people forcibly by the impact of the thalidomide tragedy. Thalidomide was a sedative used in the early 1960s. Doctors believed that it was very safe and prescribed it widely. Sadly it was soon discovered that thalidomide could harm babies if taken by mothers during the early months of pregnancy.

Thalidomide is chiral. One isomer is the effective sedative. The other isomer causes malformations in babies (it is *teratogenic*). Ever since the thalidomide affair, the pharmaceutical and agrochemical industries have to test the mirror image forms of chiral chemicals separately before they can be used as drugs or agrochemicals.

Some complex ions are chiral.

Mirror

Mirror-image forms of a complex between nickel(II) ions and ethylenediamine, en, (see bidentate ligand). Three-dimensional models make it easier to see that they are non superimposable.

chlorination: a reaction which replaces a hydrogen atom in an organic molecule with a chlorine atom. One example is the chlorination of *alkanes* – a *free-radical chain reaction* initiated by ultraviolet radiation.

Another example is the chlorination of *benzene* by chlorine in the presence of an iron chloride catalyst. This is an *electrophilic substitution reaction*.

chlorine (Cl) occurs as a greenish, toxic gas consisting of Cl_2 molecules. It is the second element in the family of nonmetals called the halogens (*group VII*). Its electron configuration is $[Ne]3s^23p^5$.

Chlorine is manufactured on a large scale by the *electrolysis of brine*.

Chlorine is a powerful *oxidizing agent* which reacts directly with most elements. In its compounds chlorine is usually present in the -1 oxidation states but chlorine can be oxidized to positive oxidation states by oxygen and fluorine.

$+7$	ClO_4^-	$KClO_4$
$+5$	ClO_3^-	$KClO_3$
$+3$	ClO_2^-	$KClO_2$
$+1$	ClO^-	$HOCl$
0	Cl_2	
-1	Cl^-	HCl

Oxidation states of chlorine

Chlorine forms ionic chlorides with metals.

Formation of ions in magnesium chloride when chlorine reacts with magnesium

Chlorine forms covalent, molecular chlorides with most nonmetals but it does not react directly with carbon, oxygen or nitrogen. Hydrogen burns in chlorine to produce the colorless, acidic gas hydrogen chloride, HCl. Hydrogen chloride is very soluble in water forming *hydrochloric acid*.

$$H \overset{\circ}{\underset{\circ}{}} H \quad + \quad \overset{\times\times}{\underset{\times\times}{}}Cl\overset{\times\times}{\underset{\times\times}{}}Cl\overset{\times\times}{\underset{\times\times}{}} \quad \longrightarrow \quad 2\ H\overset{\circ}{\underset{\circ}{}}\overset{\times\times}{Cl}\overset{}{\underset{\times\times}{}}$$

Reaction of hydrogen with chlorine to produce hydrogen chloride molecules

Chlorine dissolves in water. It reacts reversibly with water forming a mixture of weak hypochlorous acid and strong hydrochloric acid. This is an example of a *disproportionation reaction*.

$$Cl_2(aq) + H_2O(l) \rightleftharpoons HOCl(aq) + Cl^-(aq) + H^+(aq)$$

Chlorine oxidizes a range of ions or molecules in solution:

- iron(II) ions to iron(III) ions

- bromide ions to bromine

- iodide ions to iodine

- sulfite ions to sulfate ions

- thiosulfate ions to sulfate ions

- hydrogen sulfide to sulfur.

chlorine oxoanions form when chlorine reacts with water and alkalis. In these compounds chlorine is oxidized to positive oxidation states through a series of *disproportionation reactions*.

When chlorine dissolves in potassium (or sodium) hydroxide solution at room temperature it produces hypochlorite and chloride ions.

$$Cl_2(aq) + 2OH^-(aq) \longrightarrow ClO^-(aq) + Cl^-(aq) + H_2O(l)$$

On heating, the hypochlorite ions disproportionate to chlorate and chloride ions:

$$3ClO^-(aq) \longrightarrow ClO_3^-(aq) + 2Cl^-(aq)$$

Potassium chlorate can be crystallized from the solution. Careful heating just above the melting point converts potassium chlorate to potassium perchlorate and potassium chloride.

$$KClO_3(s) \longrightarrow 3\ KClO_4(s) + KCl(s)$$

chlorine water treatment: a lifesaving application of *chlorine* chemistry to kill microorganisms in drinking water and swimming pools. Since the nineteenth century, the treatment of drinking water with chlorine has helped to control the spread of diseases such as typhoid and cholera.

Chlorine disinfects tap water by forming hypochlorous acid, HOCl.

$$Cl_2(aq) + H_2O(l) \rightleftharpoons HOCl(aq) + H^+(aq) + Cl^-(aq)$$

Hypochlorous acid is a powerful oxidizing agent and a weak acid. It is an effective disinfectant because the molecule can pass through the cell walls of bacteria, unlike the negatively charged ClO^- ions. Once inside the cell the HOCl molecules break it open and kill the organism by oxidizing and chlorinating within the cell.

Chlorine gas is very hazardous, so for household cleaning it is dissolved in sodium hydroxide to make domestic bleach, sodium hypochlorite. Sodium hypochlorite ionizes fully in water.

$$NaOCl(aq) \longrightarrow Na^+(aq) + OCl^-(aq)$$

Hypochlorous acid is a *weak acid* so in bleach solution some of the hypochlorite ions take hydrogen ions from water molecules and turn into the un-ionized acid.

$$OCl^-(aq) + H^+(aq) \rightleftharpoons HOCl(aq)$$

cholesterol: a *steroid* which plays an important part in metabolism. It is a part of cell membranes and the precursor of *steroid* hormones such as testosterone and progesterone.

Structure of cholesterol

High levels of cholesterol in the blood may lead to deposits building up in arteries resulting in heart disease. Cholesterol levels are monitored in some older people and in those thought likely to suffer from heart disease.

chromatography: a method for separating and identifying the chemicals in a mixture. Chromatography can be used to:

- separate and identify the chemicals in a mixture
- check the purity of a chemical product
- identify impurities in a product
- purify a chemical product (on a laboratory or industrial scale).

All types of chromatography have a stationary phase (a solid or a liquid held by a solid) and a mobile phase (a liquid or gas). The components of a mixture separate as the mobile phase moves through the stationary phase. Components which tend to mix with the mobile phase move faster. Components which tend to be held by the stationary phase move slower.

The basic principle underlying the separation is:

- *adsorption* when the stationary phase is a solid
- *partition* when the stationary phase is a liquid held as a thin layer on the surface of a solid.

There are several types of chromatography including: *liquid chromatography, high-performance liquid chromatography* (HPLC), *paper chromatography, thin-layer chromatography* (TLC) and *gas–liquid chromatography* (GLC).

chromium (Cr): a hard, silvery *d-block* metal with the electron configuration $[Ar]3d^54s^1$. This electron configuration is an exception to the normal $[Ar]3d^x4s^2$ pattern for the first series of *d*-block elements. Energetically it is more favorable to have one electron in each *d-orbital* and thus to half-fill the *d*-subshell.

In solution chromium forms ions in the +2, +3 and +6 *oxidation states*.

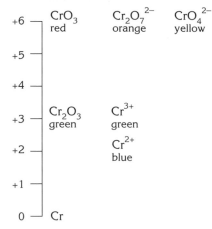

Oxidation states of chromium

There is an equilibrium between yellow chromate ions and orange dichromate ions in aqueous solution. The position of equilibrium depends on the pH. In acid the hydrogen ion concentration is high so the solution is orange because dichromate ions predominate. Adding alkali removes hydrogen ions and turns the solution yellow as chromate ions form. These shifts in the position of equilibrium are as predicted by *Le Chatelier's principle*. Note that this is not a redox reaction. Chromium is in the +6 state on both sides of the equation.

$$2CrO_4^{2-}(aq) + 2H^+(aq) \rightleftharpoons Cr_2O_7^{2-}(aq) + H_2O(l)$$

Dichromate ions in acid solution are used to oxidize *alcohols* and *aldehydes*. Paper moistened with dichromate solution is used in the *gas test* for sulfur dioxide. The paper turns from orange to green as dichromate ions are reduced to chromium(III) ions. Potassium dichromate is used as a *primary standard* in *redox titrations*.

$$Cr_2O_7^{2-}(aq) + 14H^+(aq) + 6e^- \longrightarrow 2Cr^{3+}(aq) + 7H_2O(l)$$

Chromium(III) ions in solution exist as aquo complexes: $[Cr(H_2O)_6]^{3+}$. The hydrated ions are acidic and behave in a very similar way to aluminum(III) ions.

The removal of successive protons from the hydrated chromium(III) ion, $[Cr(H_2O)_6]^{3+}$, produces a neutral complex, $[Cr(H_2O)_3(OH)_3]$, which is sparingly soluble and precipitates. The removal of further protons produces a negative ion, $[Cr(OH)_6]^{3-}$, and the ionic compound redissolves. This is possible because of the high polarizing power of the small, highly charged chromium(III) ions.

Under alkaline conditions *hydrogen peroxide* oxidizes chromium(III) to chromate.

Zinc reduces a green solution of chromium(III) to a blue solution of chromium(II) ions. Chromium(II) is a powerful *reducing agent* and is quickly converted to chromium(II) by oxygen in the air.

chromophore: the part of a molecule which gives rise to its color. Chromophores in carbon compounds have an extended system of *delocalized electrons*. Families of dyes with the same chromophore are made by attaching different *functional groups* to modify the properties of the dyes.

Functional groups linked to a chromophore may act as:

- auxochromes which change the color of the dye
- solubilizing groups which make the dye more soluble in water and help it stick to the fibers of a textile.

chrysoidine

alizarin yellow

Two dyes based on the same chromophore

cisplatin: an anticancer drug consisting of a complex ion of platinum, $Pt(NH_3)_2Cl_2$. The ion is planar and can exists as *cis* and *trans* isomers (see *geometrical isomerism.*). The *cis* isomer inhibits cell division but does not prevent cell growth. This makes it a useful treatment for cancer. Unfortunately cisplatin is toxic and has unpleasant side effects.

cis trans isomerism: see geometrical isomerism

Structure of the cis and trans isomers of $Pt(NH_3)_2Cl_2$

cis isomer

trans isomer

clay minerals: hydrated aluminosilicates formed by the weathering of feldspars in igneous rocks. Kaolinite is the main component of china clay. Much china clay is used as a coating agent and filler. Kaolinite is also the main constituent of ball clay used to make *ceramic* tableware, porcelain and wall tiles.

Kaolinite has a layer structure. Each layer consists of a sheet of *silicate* (SiO_4) tetrahedra interlocked with an aluminate sheet.

close packed structures are found in metal crystals. In a layer of close packed spheres each atom has six other spheres touching it. In three dimensions, layers of close packed atoms stack up in two possible ways. In hexagonal close packing the third layer is directly over the first layer (aba). In cubic close packing it is the fourth layer which corresponds with the first layer (abca). In the two structures each atom touches 12 nearest neighbors so for both the *coordination number* is 12.

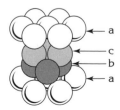

Hexagonal and cubic close packing of metal atoms

The *unit cell* of the cubic close packed structure is a face-centered cube.

A B C D E F G H I J K L M N O P Q R S T U V W X Y Z

coal: a fossil fuel formed by the action of heat and pressure on the remains of plant buried under sediments. Heating coal in the absence of air at 1000°C drives off coal tar, ammonia and coal gas leaving a residue of coke. Coke is required for a range of industrial processes including *iron extraction* in blast furnaces. Coal tar was a major source of organic chemicals until crude oil took over in the 1940s. Coal tar is a rich source of arenes and was the main source of *benzene* when William Perkin was pioneering the production of synthetic dyes.

cobalt (Co): a hard, silvery *d-block* metal which is less reactive than iron. It has the electron configuration $[Ar]3d^7 4s^2$.

Cobalt is an ingredient of alloys *steels* such as the ferromagnetic alloy, Alnico which makes excellent permanent magnets.

In solution cobalt forms ions in the +2 and +3 oxidation states. Cobalt(II) is the more stable state. Anhydrous cobalt(II) chloride is blue but it turns pink on adding water as the cobalt ions are hydrated to the $[Co(H_2O)_6]^{2+}$ ion. This is used as a test to detect water. The granules of the drying agent "self-indicating *silica gel*" are blue because they contain anhydrous cobalt ions. When the gel's drying action is exhausted the granules turn pink.

A dilute solution of cobalt chloride is pink because the cobalt(II) ions are hydrated. A concentrated solution is blue. A dilute solution also turns blue on adding concentrated hydrochloric acid. The color change is due to a *ligand* exchange reaction as chloride ions replace the water molecules. The reaction is reversible:

$$[Co(H_2O)_6]^{2+} + 4Cl^-(aq) \rightleftharpoons [CoCl_4]^{2-} + 6H_2O(aq)$$

It is normally very difficult to oxidize aqueous cobalt(II) to cobalt(III) but the reaction works easily if the cobalt(II) ions are complexed with ammonia molecules. The Co(III) complex with ammonia is more stable than the Co(II) complex. The value for the *standard electrode potential* shows that aqueous Co(III) is a stronger oxidizing agent than potassium permanganate in acid solution:

$$[Co(H_2O)_6]^{3+} + e^- \rightleftharpoons [Co(H_2O)_6]^{2+} \qquad E° = +1.82 \text{ V}$$

When the two states are complexed with ammonia the standard electrode potential shifts to a value that shows that the Co(III) state is much more stable. Cobalt(II) is now a reducing state and can be oxidized to Co(III) by oxygen or hydrogen peroxide.

$$[Co(NH_3)_6]^{3+} + e^- \rightleftharpoons [Co(NH_3)_6]^{2+} \qquad E° = +0.10 \text{ V}$$

colligative properties: the properties of solutions which depend on the concentration of solute particles but not on the nature of the particles. Dissolving salt, sugar or any other solute in water lowers its *vapor pressure*, raises its *boiling* point, lowers its *freezing point* and increases its *osmotic pressure*. The extent of these changes depends only on the number of solute particles per solvent molecule and not on the nature of the dissolved molecules or ions.

collision theory accounts for the effects of concentration, temperature and *catalysts* on *reaction rates*. The idea is that a chemical reaction happens when the molecules or ions of reactant collide, making some bonds break and allowing new bonds to form.

It is not enough for the molecules to collide. In soft collisions the molecules simply bounce off each other. Molecules are in rapid random motion and if every collision led to reaction all reactions would be explosively fast. Only pairs of molecules which collide with enough energy to stretch and break chemical bonds can lead to new products. Reactant molecules have to overcome the *activation energy*.

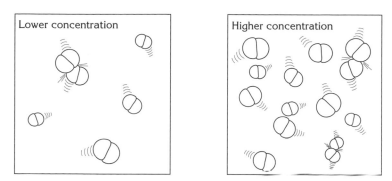

Raising the concentration means that the reacting particles are closer together. There are more collisions and reactions are faster.

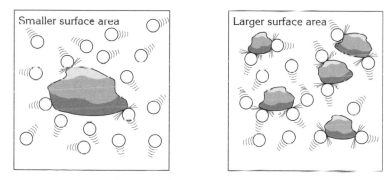

In a heterogeneous reaction of a solid with either a liquid or gas the reaction is faster if the solid is broken up into smaller pieces. Crushing the solid increases its surface area, collisions can be more frequent and the rate of reaction is bigger.

Collision theory refers to the *Maxwell–Boltzmann distribution* of energies of molecules to explain the effects of temperature and catalysis on reaction rates.

colloids consist of fine particles (the disperse *phase*) of one substance finely dispersed in another (the continuous phase). The dispersed particles in liquids and gases are large enough to scatter light but they do not settle out and they show *Brownian motion*. The diameter of colloid particles is typically around 10 to 1000 nm (much larger than atomic diameters which are around 0.2 nm).

See the table on the next page.

Continuous phase	Disperse phase	Type	Example
gas	liquid	liquid aerosol	mist
gas	solid	solid aerosol	smoke
liquid	gas	foam	whipped cream
liquid	liquid	emulsion	hand cream, mayonnaise
liquid	solid	sol	paint, muddy river water, sewage
solid	gas	solid foam	pumice
solid	liquid	gel or solid emulsion	jelly, butter
solid	solid	solid sol	pearl, pigmented plastics

colorimetry: a method for measuring the *concentration* of compounds in solution which can be used with chemicals which are themselves colored or which give a color when mixed with a suitable reagent.

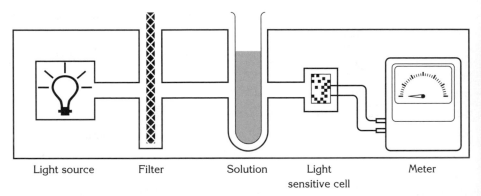

| Light source | Filter | Solution | Light sensitive cell | Meter |

Diagram of a colorimeter

The filter lets through light which is strongly absorbed by the solution. The extent of absorption depends on the path length of the light through the solution (which is a constant for the sample chamber used) and on the concentration of the solution. The instrument is *calibrated* with a series of *standard solutions*. It can then be used to measure the concentration of unknown samples.

A colorimeter can be used to determine the formula of a complex ion. The graph on the opposite page shows the results of measuring the absorbance of a series of mixtures of 0.01 mol/L Cu^{2+}(aq) ions and 0.01 mol/L EDTA(aq).

The peak of absorption corresponds to mixing equal volumes of the two solutions which both have the same molar concentration. This shows that 1 mol Cu^{2+}(aq) forms a complex with 1 mol EDTA(aq). (For a diagram see *EDTA*.)

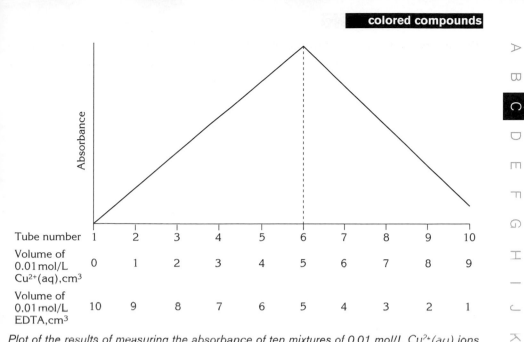

Plot of the results of measuring the absorbance of ten mixtures of 0.01 mol/L $Cu^{2+}(aq)$ ions and 0.01 mol/L EDTA(aq)

colored compounds in many instances get their color by absorbing some of the radiation in the visible region of the *electromagnetic spectrum* with wavelengths between 400 nm and 700 nm.

Color of compound	Wavelength absorbed/nm	Color of light absorbed
greenish yellow	400–430	violet
yellow to orange	430–490	blue
red	490–510	blue-green
purple	510–530	green
violet	530–560	yellow-green
blue	560–590	yellow
greenish blue	590–610	orange
blue-green to green	610–700	red

It is the electrons in colored compounds that absorb radiation as they jump from their normal state to a higher excited state. According to *quantum theory* there is a fixed relationship between the size of the energy jump and the wavelength of the radiation absorbed. In many compounds the jumps are so big that they absorb in the ultraviolet part of the spectrum. Such compounds are colorless.

Color of transition metal complex ions – the color of a transition metal complex ion depends on:

- the metal
- the *oxidation state* of the metal
- the *ligand*
- the *coordination number*.

Color in transition metal ions arises from electronic transitions between *d-orbitals*. In a free atom all of the five *d-orbitals* have the same energy. When a transition metal ion forms complex ions the *d-orbitals* split into two groups with different energies. The size of the split depends on the number and nature of the ligands in the complex. This helps to account for color changes.

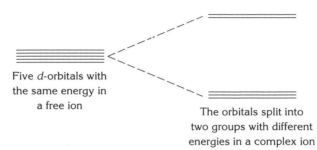

Five *d*-orbitals with
the same energy in
a free ion

The orbitals split into
two groups with different
energies in a complex ion

Energy difference between d-orbitals in a transition metal complex ion

If all the *d-orbitals* are full there is no possibility of electronics transitions between them. This explains why Zn^{2+} and Cu^+ ions are colorless. Sc^{3+} ions are also colorless because the ion has no *d*-electrons.

Color in organic compounds – color in organic compounds is often associated with *delocalized electrons*. The energy jumps between molecular orbitals absorb visible radiation in molecules with extended systems of alternating double and single bonds (see *conjugated systems*).

Stick structure of β-carotene, the main orange color in carrots, used as a colorant in foods, drugs and cosmetics

Other sources of color – some colors are caused by physical effects rather than electronic transitions. Examples of these effects are:

- light scattering – moonstones, blue skies
- interference – soap bubbles and oil films on water
- diffraction – opal, liquid crystals
- refraction of some wavelengths more than others – "fire" in gemstones.

combinatorial chemistry: a technique of chemical synthesis which speeds up the search for new drugs. The method takes advantage of automation to prepare very large numbers of compounds very rapidly. In this way chemists build up "libraries" of closely related compounds.

The starting materials for synthesis are attached to polymer beads. After each step of a synthesis the beads with the product attached can be filtered off from the other reagents by filtering and washing. The final step removes the polymer beads to leave the products.

At first the compounds are tested for their biological activity as mixtures. This is much quicker than the traditional method of making, purifying and testing individual compounds. It then only becomes necessary to synthesize the compounds individually if the tests show that a particular mixture shows promising biological activity.

combustion: the reaction of a fuel with oxygen to release energy. If there is plenty of air or oxygen, the carbon in the fuel turns to carbon dioxide and hydrogen turns to steam (water).

$$CH_4(g) + 2O_2(g) \longrightarrow CO_2(g) + 2H_2O(l)$$

methane in natural gas

Incomplete combustion in a limited supply of oxygen is hazardous because the products are then soot and the toxic gas *carbon monoxide*.

During combustion, any sulfur in a fuel is oxidized to sulfur dioxide. Nitrogen in a fuel may end up as the element after combustion or, if the temperature is high enough, combined with oxygen to form a mixture of *nitrogen oxides*. The formation of these acidic oxides contributes to *acid rain*.

combustion analysis: a method for determining the *empirical formulas* of organic compounds. A weighed sample of the compound is burnt in excess oxygen mixed with helium. This converts the carbon to carbon dioxide and the hydrogen to water. A catalyst ensures that combustion is complete. The inert helium carries the products of combustion and the excess oxygen through a tube which contains chemicals to remove any volatile halogen, sulfur or phosphorus compounds. Oxides of nitrogen are converted to nitrogen gas and the excess oxygen combines with copper.

The water vapor is absorbed in magnesium perchlorate. Carbon dioxide is absorbed in soda lime. In a modern instrument, measurements of the thermal conductivity of the helium before and after absorption make it possible to determine the masses of water, carbon dioxide and nitrogen formed by burning the sample.

From the results it is possible to calculate the *percentage composition* of the compound. Any mass of the sample not accounted for is assumed to be due to oxygen.

Worked example

Complete combustion of 0.15 g of a liquid compound produced 0.22 g of carbon dioxide and 0.09 g water. What is the empirical formula of the compound?

Notes on the method

The *molar mass* of carbon dioxide, CO_2 = 44 g/mol of which carbon is 12 g/mol

The molar mass of water, H_2O = 18 g/mol of which hydrogen is 2 g/mol

Answer

The mass of carbon in the sample = $\dfrac{12}{44} \times 0.22$ g = 0.06 g

The mass of hydrogen in the sample = $\dfrac{2}{18} \times 0.09$ g = 0.01 g

The total mass of carbon and hydrogen = 0.07 g in a sample with mass 0.15 g

So the difference gives the mass of oxygen in the sample which is 0.08 g

These are the *amounts* of the elements in the sample:

A B C D E F G H I J K L M N O P Q R S T U V W X Y Z

carbon: 0.06 g ÷ 12 g/mol = 0.005 mol

hydrogen: 0.01 g ÷ 1 g/mol = 0.01 mol

oxygen: 0.08 g ÷ 16 g/mol = 0.005 mol

The ratio C:H:O is 1:2:1

The empirical formula of the compound is CH_2O.

---

complex ions consist of a central metal ion linked to a number of molecules or ions with lone pairs of electrons. The surrounding molecules or ions are *ligands* which use lone pairs of electrons to form a *coordinate bond* with the metal ion. The number of ligands in a complex ion is typically two, four or six.

Transition metal ions form aquo complexes when dissolved in water. Examples are $[Fe(H_2O)_6]^{2+}$ and $[Cr(H_2O)_6]^{3+}(aq)$.

The overall charge of a complex ion is the sum of the charges on the metal ion and its ligands. (See also *names of complex ions, shapes of complex ions, colored compounds, coordination compounds* and *coordination number.*)

complexes with ammonia molecules are useful in *qualitative analysis* in both anion tests and cation tests. Example are the diamminesilver(I) ion and the tetraammine copper(II) ion. The *lone pair* on the nitrogen atom of each ammonia molecule forms a *coordinate bond* with the central metal ion.

$$[Ag(NH_3)_2]^+ \qquad\qquad [Co(NH_3)_6]^{2+}$$

diamminesilver(I) ion   hexaamminecobalt(II) ion

Complexes between metal ions and ammonia

Adding ammonia solution to a precipitate of a *silver halide* can distinguish the chloride from the bromide and the iodide. Only silver chloride, the least insoluble, will freely dissolve in dilute ammonia solution to form the diammine complex.

Formation of a complex may produce a color change or redissolve a precipitate. Adding ammonia solution to a solution with copper(II) ions first produces a pale blue precipitate of the hydroxide, but with excess ammonia the precipitate dissolves to give a deep blue solution as the tetraammine complex forms.

complex-forming titration: a practical technique used to determine the *concentration* of metal ions. A *titration* measures the volume of a *standard solution* of a complex-forming reagent needed to react exactly with a measured volume of the unknown solution of metal ions, such as zinc ions.

Complex-forming titrations often use a standard solution of *EDTA* which forms very stable *complex ions* with metal ions. The procedure is the same as for any other titration.

To find the end point the analyst adds an indicator and a *buffer solution* which forms a colored but unstable complex with the metal ion in the flask. A suitable indicator is Eriochrome black T which is blue in solution at pH 10. At the start of the titration it produces a wine-red complex. EDTA from a buret forms a more stable complex with zinc ions and so takes the metal ions from the indicator. At the end point all the zinc ions have been complexed by titration with EDTA. The last drop of EDTA leaves no Zn^{2+} ions to form the red complex with the indicator so the indicator turns blue again.

Worked example

An alkaline buffer and a few drops of Eriochrome black T indicator were added to 25.0 cm³ of a solution of zinc sulfate. In the titration 23.2 cm³ of 0.010 mol/L EDTA were run in from a buret until the indicator changed from red to blue. What was the concentration of the zinc ions?

Notes on the method

Always start by writing the equation for the reaction. See *titration* for a general method for the calculations.

Remember to convert volumes in cm³ to volumes in L by dividing by 1000.

In any titration there is one unknown – in this case the concentration of the zinc ions, c_A.

Answer

The equation for the reaction is:

$$Zn^{2+}(aq) + EDTA^{4-}(aq) \longrightarrow [Zn(EDTA)]^{2-}(aq)$$

The volume of zinc sulfate solution in the flask, $V_A = \dfrac{25.0}{1000}$ L

Let the concentration of zinc ions be c_A.

The volume of hydrochloric acid added from the buret, V_B $\dfrac{23.2}{1000}$ L

The concentration of hydrochloric acid, c_B = 0.010 mol/L

$$\frac{V_A \times c_A}{V_B \times c_B} = \frac{n_A}{n_B}$$

$$\frac{^{25}\!/_{1000} \times c_A}{^{23.2}\!/_{1000} \times 0.010} = \frac{1}{1}$$

Therefore $c_A = \dfrac{23.2 \times 0.010}{25.0}$ – 0.0928 mol/L

The concentration of the zinc ions was 0.0928 mol/L.

composites are made by combining two or more materials to create a new material. A composite combines the desirable properties of its constituents and compensates for their disadvantages. Steel reinforced concrete is a composite material as is galvanized steel. Kitchen countertops are composites consisting of chipboard covered with paper impregnated with a polymer such as a *thermosetting* melamine–methanal resin.

Many important composites consist of fibers of one material, such as *glass*, aramids, or graphite, embedded in a *polymer*, *metal* or *ceramic* matrix. The combinations are designed to give new materials with better properties especially high stiffness per unit weight which is important in road and rail transport, aircraft, sporting goods and many other applications.

Glass fibers in a *polyester* matrix (so-called fiberglass) are used to make boat hulls and car bodies. Parts of aircraft and some sports gear are made from carbon fibers in an epoxy matrix.

computational chemistry takes advantage of powerful computers and ingenious software to investigate chemical behavior without using test tubes. Chemists use computer models to explore reaction mechanisms, to predict the behavior of novel catalysts and polymers

A B C D E F G H I J K L M N O P Q R S T U V W X Y Z

and to visualize the interaction between drugs and their target molecules in living organisms. Central to all the software is a theory (such as quantum mechanics) for calculating the energy of each possible state of a set of interacting atoms or molecules. The software uses its calculations to predict the arrangement of atoms or molecules that gives rise to the lowest energy for the system.

concentrations of solutions are usually measured in moles per liter of solution (mol/L). There are small volume changes when chemicals dissolve in water so it is important to note that concentrations normally refer to liters of solution not to liters of the solvent.

$$\text{concentration (mol/L)} = \frac{\text{amount of solute (mol)}}{\text{volume of solution (L)}}$$

Writing the formula of a chemical in square brackets is the usual shorthand for "concentration in mol/L." (For example: $[CaCl_2] = 0.1$ mol/L.)

When ionic crystals dissolve the ions separate and become independent.

$$CaCl_2(s) + aq \longrightarrow Ca^{2+}(aq) + 2Cl^-(aq)$$

So if $[CaCl_2] = 0.1$ mol/L, then $[Ca^{2+}] = 0.1$ mol/L but $[Cl^-] = 0.2$ mol/L.

Other ways of measuring the concentrations are to use *mole fractions, molality* or *parts per million, ppm*.

Worked example

What is the concentration of a solution of potassium permanganate made by dissolving 3.95 g of the solid in water and making the solution up to 500 cm³?

Answer

The molar mass of potassium permanganate, $(KMnO_4) = 158.0$ g/mol

$$\text{Amount of } KMnO_4 \text{ in solution } = \frac{3.95 \text{ g}}{158.0 \text{ g/mol}} = 0.025 \text{ mol}$$

$$\text{Volume of the solution } = \frac{500}{1000} \text{ L} = 0.5 \text{ L}$$

$$\text{Concentration of the solution } = \frac{0.025 \text{ mol}}{0.5 \text{ L}} = 0.05 \text{ mol/L}$$

condensation polymers are produced by a series of condensation reactions splitting off water between the functional groups of the *monomers*. Examples of condensation polymers are *polyamides* and *polyesters*. Where each monomer has two function groups this type of polymerization produces chains. Condensation polymers make good fibers because they form long, straight-chain molecules with few side chains and relatively strong intermolecular forces between polar bonds in neighboring chains. *Cross linking* is possible if one of the monomers has three functional groups.

condensation reaction: a reaction in which molecules join together by splitting off a small molecule such as water. The *addition–elimination reactions* of carbonyl compounds are examples of condensation reactions. The formation of an ester from an acid and an alcohol is also a condensation reaction.

Condensation reaction to form the ester ethyl ethanoate (ethyl acetate)

conditions for organic reactions specify concisely but accurately the reagents, catalysts, temperatures and pressures needed for good yields. Factors to consider include:

- whether a solvent is needed – if so, whether the solvent is water or a non-aqueous solvent such as ethanol

- the concentration of the reagents – whether acids, for example, are dilute or concentrated

- the temperature – whether the mixture has to be cooled, heated or kept at room temperature

- the pressure – many industrial process take place at several times the pressure of the atmosphere

- whether a catalyst is required.

conductiometric titrations use a conductivity cell and meter to measure the conductivity of the reaction mixture during a titration and hence to determine the end point.

The conductivity can change during a titration because there is a change in:

- the ability of the ions present to move through the solution and conduct electricity

- the proportion of electrolytes which are only slightly ionized relative to electrolytes which are fully ionized.

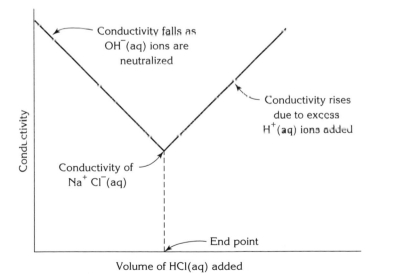

Changes in conductivity during a titration of NaOH(aq) with HCl(aq). In water H^+ (aq) and OH^- (aq) ions are very mobile so they conduct electricity very well.

99

confidence limits: the limits around an experimental mean value within which there is a high probability that the true mean lies.

Statistical analysis shows that there is a 95% probability that the true mean lies within $\pm 1.96\sigma$ of the experimental mean.

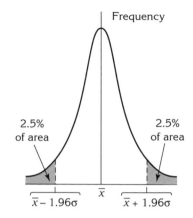

Normal distribution of a large number of experimental results spread randomly about a mean value $\bar{x}$ where σ is the standard deviation

conformation of molecules: the possible shapes of molecules which can arise because of *rotation about covalent bonds*. Cyclohexane molecules, for example, shift to and fro between "chair" and "boat" forms. In both conformations the four single bonds around each carbon atom are arranged tetrahedrally as normal. The chair form is preferred because it is a little more stable. The energy difference between conformations is not large enough for them to be separated as distinct compounds.

Chair form

Boat form

Chair and boat forms of cyclohexane

conjugate acid–base pairs: see *acid–base equilibria*.

conjugated system: a system of alternate double and single bonds in a molecule. Organic molecules with extended conjugated systems absorb radiation in the visible part of the spectrum and are therefore colored. Examples are the orange β-carotene molecule in carrots and the indicator phenolphthalein in alkali (see *colored compounds*). (See the equation on page 101.)

<table>
<tr><td>colorless</td><td>red</td></tr>
</table>

colorless   red

Structures of phenolphthalein in acid and alkali (see page 100)

conservation of mass (law of): the law which chemists now take for granted every time they write a balanced chemical equation. The law states that matter is neither created nor destroyed during a chemical reaction.

constant-boiling mixture: see *azeotropic mixture.*

contact process: see *sulfuric acid manufacture.*

continuous processes manufacture chemicals on a large scale in industrial plants which operate for 24 hours a day. Raw materials are constantly fed into the plant and products continuously removed. Examples of continuous processes are the *fractional distillation of oil*, the Haber process for *ammonia manufacture, iron extraction* in a blast furnace and the contact process for *sulfuric acid manufacture.*

conversion of units is often needed to make sure that the units are consistent before carrying out calculations. Typically the units of measurement (such as cm³) have to be converted to different units for calculation (in this case liters).

Worked example

In a titration, the volume of acid added from the buret was 23.5 cm³. What is the volume in liters (L)?

Notes on the method

Multiply by an appropriate factor to convert from the unit used for measurement to the unit needed for the calculation. Find the factor by:

- writing down the relationship between the two units, then by
- writing the relationship as a ratio, so that the units cancel when it multiplies the measurement, such that the original unit is replaced by the unit required.

Answer

$1 \text{ L} = 1000 \text{ cm}^3$

Hence $23.5 \text{ cm}^3 \times \dfrac{1 \text{ L}}{1000 \text{ cm}^3} = 0.0235 \text{ L}$

coordinate bond: a covalent bond formed when one atom contributes both of the shared pair of electrons. Once formed there is no difference between a coordinate bond and any other covalent bond.

Ammonia forms a coordinate bond with a hydrogen ion when it is acting as a base.

A B C D E F G H I J K L M N O P Q R S T U V W X Y Z

Formation of an ammonium ion

Coordinate bonding also accounts for *the structures of carbon monoxide*, the Al_2Cl_6 molecules in *aluminum chloride* vapor and the solid chloride of *beryllium*.

hydronium ion	nitric acid	carbon monoxide

Examples of coordinate bonds

Coordinate bonding is important in *complex ions* and other *coordination compounds*.

coordination compounds contain complexes which may be cations, anions or neutral molecules. In a coordination compound, *ligand* molecules or negative ions form *coordinate bonds* with a metal ion.

Examples of coordination compounds:

- $K_3[Fe(CN)_6]$ containing the negatively charged complex ion $[Fe(CN)_6]^{3-}$
- $[Ni(NH_3)_6]Cl_2$ containing the positively charged complex ion $[Ni(NH_3)_6]^{2+}$
- $Ni(CO)_4$ a neutral complex between nickel atoms and carbon monoxide molecules.

coordination number: the number of nearest neighbors to an atom or ion in a *crystal structure* or the number of *coordinate bonds* formed between the ligands and the metal ion in a *complex ion*.

copolymerization is used to modify the properties of *polymers* by producing polymer chains from a mixture of monomers. ABS, for example, is a rigid, tough plastic widely used for the casing of domestic equipment and for parts of cars. It is a co-polymer of **A**crylonitrile ($CH_2 = CH — CN$), **B**utadiene ($CH_2 = CH — CH = CH_2$) and **S**tyrene ($C_6H_5 — CH = CH_2$).

copper (Cu): a ductile metal with a familiar reddish color. It has the electron configuration $[Ar]3d^{10}4s^1$. This electron configuration is an exception to the normal $[Ar]3d^x4s^2$ pattern for the first series of *d-block elements*. Energetically it is more favorable to have fill the *d*-subshell and leave only one electron in the 4s.

Copper is relatively unreactive. It corrodes very slowly in moist air and is not attacked by dilute non-oxidizing acids. Copper is a good conductor of electricity; it is widely used in electricity cables and for domestic water pipes.

Copper's mechanical properties are enhanced by making alloys such as *brass* and *bronze*.

Copper forms compounds in the +1 and +2 states. Under normal conditions copper(ıı) is the stable state in aqueous solution. Copper(ı) *disproportionates* in aqueous solution. So when Cu_2O dissolves in dilute sulfuric acid the products are copper(ıı) sulfate and copper.

$$2Cu^+(aq) \rightleftharpoons Cu^{2+}(aq) + Cu(s)$$

The equilibrium lies well to the right. Copper(I) in the presence of water can exist as very insoluble compounds such as Cu_2O, CuI or $CuCl$. Iodide ions, for example, reduce copper(II) to copper(I) ions which immediately precipitate with more iodide ions as white copper(I) iodide.

$$2Cu^{2+}(aq) + 4I^-(aq) \longrightarrow 2CuI(s) + I_2(s)$$

Copper(I) can exist in aqueous solution as stable complexes such as $[Cu(NH_3)_2]^+$ or $[Cu(CN)_4]^{3-}$.

Fehling's and *Benedict's* reagents contain deep blue copper(II) complexes in alkali. The reagents are used to detect *reducing sugars* and to distinguish *aldehydes* from *ketones*. A reducing sugar or aldehyde reduces the reagent on heating to copper(I) oxide. The blue color goes and a reddish-brown precipitate forms.

(See *names of complex ions* and *ligand substitution reactions* for more examples of copper complexes. See *carbon monoxide* for an example of copper acting as a catalyst. See *cation tests* and *flame tests* for methods used to detect the presence of copper(II) ions.)

copper refining uses electrolysis to turn copper which is 99.5% pure into 99.99% pure metal. High purity is important especially when copper is to be used as an electrical conductor.

Impure copper is cast into anodes while the cathodes are thin sheets of pure copper. The electrolyte is a mixture of copper(II) sulfate and sulfuric acid.

At the anodes: $Cu(s) \longrightarrow Cu^{2+}(aq) + 2e^-$

At the cathodes: $Cu^{2+}(aq) + 2e^- \longrightarrow Cu(s)$

Valuable impurities such as gold and silver are recovered from the bottom of the electrolysis cell because they do not dissolve in the electrolyte.

corrosion of a metal: a redox process in which oxygen, water and acids attack metals. Most metals are extracted from oxides. Corrosion turns them back into oxides.

The most familiar and economically serious example of corrosion is the rusting of iron. Rusting is an electrochemical reaction.

An electrochemical cell on the surface of iron in contact with water and the air.

When iron is in contact with air and water, the regions rich in oxygen are cathodic. In these regions oxygen is reduced to hydroxide ions. Other parts of the metal surface are anodic. In these regions iron is oxidized to iron(II) ions. Iron(II) ions and hydroxide ions diffuse together and form a precipitate of iron(II) hydroxide which is then oxidized to rust, hydrated iron(III) oxide.

coulomb (symbol C) is the *SI unit* of electric charge.

electric charge (C) = current (C/s) × time (s)

A coulomb is the amount of electric charge flowing each second past a point in a circuit when the current is one ampere (1 A = 1 C/s).

covalent bonds form when atoms share electrons. The atoms are held together by the attraction between the positive charges on their nuclei and the negative charge on the shared electrons.

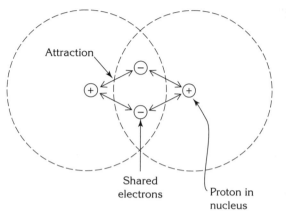

Forces of attraction giving rise to a covalent bond between hydrogen atoms in H₂

Attraction / Shared electrons / Proton in nucleus

Covalent bonding holds together the atoms of nonmetals in molecules and in network solids. Molecules have a definite shape because covalent bonds have a definite length and direction.

Electron sharing in methane and the shape of a methane molecule

One shared pair of electrons gives rise to a single bond. *Double bonds* and *triple bonds* are also possible with two or three shared pairs.

Lewis electron dot formulas *to show single covalent bonding in molecules. Also shown is a simpler way of showing the bonding in molecules. A line between two symbols represents a covalent bond.*

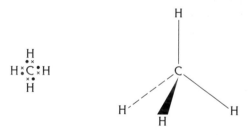

chlorine   water   ammonia

The nonmetals common in organic chemistry generally form a fixed number of covalent bonds. This helps us to determine the structures of molecules.

Element	Number of covalent bonds	Examples		
carbon, C	4	H—C—H (with H top and bottom)	H₂C=O	O=C=O
hydrogen, H	1	H₂O (bent)	H—N—H (with H top)	H—Cl
oxygen, O	2	O=O	H₂O (bent)	H—C(=O)O—H
nitrogen, N	3	N≡N	Cl—N—Cl (with Cl top)	CH₃—N with H and two lines
halogens, F, Cl, Br, I	1	H—Br	Cl—C—Cl (with Cl top and bottom)	H—C—I (with H top and bottom)

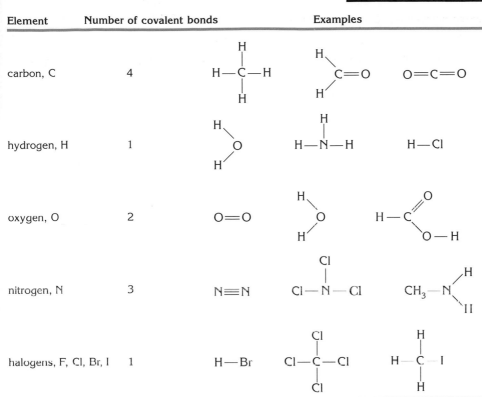

covalent radius: the covalent radius of an element is half the *bond length* when two atoms of the element are linked by a single covalent bond.

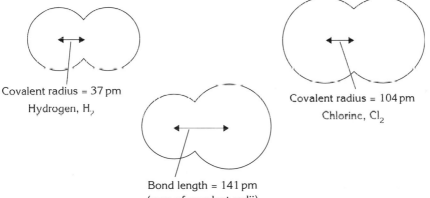

Covalent radius = 37 pm
Hydrogen, H₂

Covalent radius = 104 pm
Chlorine, Cl₂

Bond length = 141 pm
(sum of covalent radii)
Hydrogen chloride, HCl

The length of a single covalent bond between atoms of different elements can be estimated fairly accurately by adding the covalent radii for the two atoms

cracking: see *catalytic cracking* and *thermal cracking*.

critical temperature: the temperature above which it is impossible to liquefy a gas however high the pressure (see *vapors*).

cross-linking: the formation of chemical bonds between polymer chains to modify the properties of polymers such as *thermosetting* plastics as well as natural *rubber* during vulcanizing.

Cross-linking determines the three-dimensional shape of *proteins*. Cross-links in protein molecules take various forms including:

- *hydrogen bonding* between amino acid side-chains
- disulfide bridges formed by *covalent bonding* between cystine side-chains.

crude oil: a complex mixture of *hydrocarbon* molecules formed over millions of years when the remains of microscopic sea creatures trapped in sediments were converted by heat and pressure to petroleum. Crude oil is now the main source of *fuels* and organic compounds. The composition of crude oil varies from one oilfield to another. Some crudes contain significant quantities of sulfur and nitrogen compounds as well as traces of various metals. *Fractional distillation of oil* is the first step in refining to produce fuels and *lubricants* as well as feedstocks for the *petrochemical industry*.

crystallinity of polymers arises when the long chain molecules are regular enough to lie more or less parallel to each other in some regions of the solid. The rest of the solid where the chains are tangled together is *amorphous*.

The crystallinity of *polymers* with regular, unbranched chains is generally higher than in polymers with branched or irregular molecules. Relatively strong *intermolecular forces* between chains also favor crystallinity. Highly crystalline polymers are stronger and less flexible than more amorphous polymers.

crystal structures of ionic compounds include the *sodium chloride, cesium chloride* and *fluorite* structures. In an ionic crystal the ions behave like charged spheres in contact. The structures are only stable if each ion is in contact with its nearest neighbors. Cesium ion is large enough to have eight chloride ions around it as in the cesium chloride structure. Sodium ions are smaller and only big enough to touch six neighboring chloride ions as in the sodium chloride structure.

Compounds with ionic structures are hard, melt at high temperatures and only conduct electricity when molten or dissolved in water.

The *Born–Haber cycle* makes it possible to decide whether or not the bonding in a crystal is purely ionic due to *electrostatic forces* between ions.

crystal structures of metals: the important metal structures are the two *close packed structures* and the *body-centered cubic* structure. All *metals* consist of atoms held together by *metallic bonding*. Metal crystals are good conductors of electricity because of the delocalized bonding electrons. The layers of atoms in a metal can slide over each other so that metals are *malleable* and *ductile*.

crystal structures of nonmetals: most solid *nonmetals*, such as *iodine*, *sulfur* and white *phosphorus* are molecular so the forces between the particles in the crystals are weak *intermolecular forces*. The crystals easily melt or turn to vapor on gentle heating.

Some nonmetals are network solids with the atoms held together by *covalent bonds*. Examples are *carbon* (as graphite or diamond) and *silicon*. Covalent bonds are strong and point in a definite direction so these structures have very high melting points.

Diamond is very hard. Graphite, the other *allotrope* of carbon with a giant structure, has a layer lattice. It too has a high melting point but it can be used as a *lubricant* because the forces between the layers are weak and they can slide over each other.

An iodine molecule

The crystal structure of iodine. Molecules of I_2 held in a regular lattice by intermolecular forces

current: a flow of electric charge, measured in *amperes*. In a metal the flow of charge is carried by electrons. In an electrolyte the charge carriers are negative ions moving towards the anode and positive ions moving towards the cathode.

curved arrows describe the movement of electrons as bonds break and form in the steps of the mechanism of a reaction. A curved arrow with both halves of the arrow head shows the movement of a pair of electrons. The tail of the arrow starts where the electron pair begins. The head of the arrow points to where the electron pair ends up.

$$CH_3\!-\!\underset{\underset{CH_3}{|}}{\overset{\overset{CH_3}{|}}{C}}\!\diagup\!Br \longrightarrow CH_3\!-\!\underset{\underset{CH_3}{|}}{\overset{\overset{CH_3}{|}}{C}}\!+ \quad + \quad Br^-$$

Movement of an electron pair during heterolytic bond breaking

A curved arrow with only half an arrow head indicates the movement of a single electron.

$$Cl\!:\!Cl \longrightarrow Cl^\bullet + Cl^\bullet$$

Movement of single electrons during homolytic bond breaking

Dalton's law of partial pressures states that the total pressure of a mixture of gases is the sum of the *partial pressures*. This means that the gases in a mixture do not interfere with each other. Each gas makes its own independent contribution to the total pressure.

***d*-block elements:** the elements in the three horizontal rows of elements in *periods* 4, 5 and 6 for which the last electron added to the atomic structure goes into a *d-orbital*. In period 4, the *d*-block elements run from scandium ($1s^2 2s^2 2p^6 3s^2 3p^6 3d^1 4s^2$) to zinc ($1s^2 2s^2 2p^6 3s^2 3p^6 3d^{10} 4s^2$).

The changes in the properties across a series of *d*-block elements are much less marked than the big changes across a *p-block* series. This is because from one element to the next, as the *atomic number* of the nucleus increases by one, the extra electron goes into the inner *d*-subshell. In period 4, the outer shell is always the 4s *orbital* which is filled before the 3*d* starts to fill.

Electron configurations of d-block elements as free atoms. Note that orbitals fill singly before the electrons start to pair up. Note that the configurations chromium *and* copper *do not fit the general pattern.*

The chemistry of an atom is to a large extent determined by its outer electrons because they are the first to get involved in reactions. So the elements Sc to Zn in period 4 are similar in many ways.

All the *d*-block elements are *metals* with useful properties for engineering and construction. Most have high melting points. A plot of *physical properties* against atomic number often has two peaks corresponding to the half-filling and then filling of the *d*-shell. (See the graph on the next page.)

decantation involves gently pouring off most of a liquid or solution from a solid after centrifuging or simply allowing it to settle to the bottom of a container.

Filtering is often quicker if the solid is first allowed to settle and then most of the liquid decanted through the filter paper before pouring in the bulk of the solid. As a result most of the liquid passes through the filter paper before its pores are clogged by solid particles.

d-*block elements: plot of melting point against atomic number for the elements Ca to Zn*

decomposition: a reaction in which compounds break down into simpler substances which may be compounds or elements. Heating is often necessary for decomposition (see *thermal decomposition*).

dehydration removes the elements of water from a compound to form a new compound. Concentrated sulfuric acid is a powerful dehydrating agent. It dehydrates blue copper sulfate crystals, sucrose and ethanol.

$$CuSO_4 \cdot 5H_2O(s) \xrightarrow{-5H_2O} CuSO_4(s)$$
$$\text{blue} \qquad\qquad\qquad \text{white}$$

$$C_{12}H_{22}O_{11}(s) \xrightarrow{-11H_2O} 12C(s)$$
$$\text{sucrose, white} \qquad\qquad \text{black}$$

$$C_2H_5OH(l) \xrightarrow{-H_2O} CH_2{=}CH_2(g)$$
$$\text{ethanol} \qquad\qquad\qquad \text{ethene}$$

deliquescent substances take up water vapor from the air and dissolve in it. Examples are calcium chloride, potassium hydroxide and sodium hydroxide. Deliquescent substances make good drying agents in *desiccators*.

Deliquescent substances cannot be used as *primary standards* in volumetric analysis because they cannot be weighed accurately.

delocalized electrons: bonding electrons which are not fixed between two atoms in a bond but are shared between three or more atoms. Electron delocalization accounts for the shape, stability and properties of *benzene* rings, *nitrate* ions and other oxoanions, the acidity of *carboxylic acids* and *phenols*, and the *color* of some organic compounds.

Electron delocalization takes places in molecules where the conventional structure shows alternating double and single bonds. Delocalization also affects ions where an atom with a lone pair of electrons and a negative charge is separated by a single bond from a double bond.

The extreme example of delocalization is *metallic bonding* where electrons are shared between all the atoms in a crystal. Extended delocalization over the planes of carbon atoms explains the electrical conductivity of graphite.

109

Examples of delocalization in molecules and ions. The bond lengths *of the bonds affected by delocalization in each of these examples are the same – shorter than single bonds but longer than double bonds.*

denaturation of proteins happens when the three-dimensional shape of a *protein* is disrupted by heating, by extremes of *pH*, or by the presence of *heavy metal* ions. The chemical activity of protein molecules is linked to their shape. Changing the structure of proteins means that they lose their normal activity. *Enzymes*, for example, cease to act as *catalysts* when denatured.

density is the mass per unit volume of a material.

$$\text{density} = \frac{\text{mass}}{\text{volume}}$$

The symbol for density is ρ and the SI unit is kg/m^3. In chemistry densities are generally given in g/cm^3. The densities for most solids and liquids are in the range 0.5 to 10 g/cm^3.

The density of water at 0°C is 1.00 g/cm^3 and in approximate laboratory work it is common to assume that dilute aqueous solutions have the same density as water. Solids and immiscible liquids float if they are less dense than water. The density of *ice* at the same temperature is 0.917 g/cm^3 so ice floats on water.

At room temperature and pressure the densities of gases are about a thousand times smaller than those of solids and liquids. Measurement of gas densities can be used to calculate the *molar masses* of gases.

deprotonation: the removal of a *proton* from a molecule or ion. Adding aqueous hydroxide ions to a solution of *aluminum(III) ions* leads to a series of deprotonation reactions.

derivatives are used by chemists to identify unknown organic compounds. Converting a compound to a crystalline derivative produces a substance which can be purified by recrystallization and then identified by measuring its melting point.

desalination: a process for obtaining pure water from seawater or other sources of water containing dissolved salts. Methods of desalination include *distillation* at low pressure, freezing, reverse *osmosis*, electrodialysis and *ion exchange*.

desiccator: a container with a *drying agent* used to remove the moisture (or other liquids) from chemical products which decompose if warmed in an oven. Chemicals which can be oven dried may also be stored in a desiccator as they cool to prevent them from picking up moisture from the air. A desiccator is a useful place to store chemicals which must be kept dry.

Vacuum desiccators are fitted with a tap so that air can be pumped out. The partial vacuum speeds up evaporation and diffusion of the vapor to the drying agent.

detergents clean things by removing dirt from surfaces. The chemicals which act as detergents are surface active agents or *surfactants*. Water is a polar solvent with a high *surface tension*. This means that water alone is not good at removing dirt and grease. Detergents help to clean by:

- lowering the surface tension of water so that it spreads out and wets the surface
- separating grease and particles from the surface
- suspending the dirt in water so that it can be rinsed away.

There are two main types of detergent:

- soap detergents (usually just called *soaps*) made from animal *fats* or *vegetable oils*
- soapless detergents (usually just called *detergents*) which are made using chemicals from oil.

Powder or liquid laundry detergents for clothes are complex formulations which may include:

- detergents to increase wetting power, separate grease from fabrics and keep dirt in suspension,
- *sequestering agents* such as citrates or *zeolites* to soften *hard water*
- *enzymes* to break down protein stains such as blood stains
- optical brighteners to keep white fabrics looking bright and white
- oxygen *bleach* such as sodium peroxoborate(III) which only acts above 60°C
- *perfume* and color to make the product distinctive and attractive to customers.

deuterium: the *isotope* of hydrogen with atomic number 1 but mass number 2. The symbols used for deuterium are $_1^2H$ or D. About 0.015% of natural hydrogen is the deuterium isotope. Deuterium oxide (heavy water) is D_2O.

The relative mass of D is twice that of H and this difference means that deuterium compounds react more slowly than their normal hydrogen equivalents. During *electrolysis* of acidified water H_2 is formed at the cathode more readily than D_2 so the concentration of deuterium increases as electrolysis continues. Eventually it is possible to make almost pure heavy water; this is used as a moderator in some nuclear power stations.

diagonal relationship: the similarities between the first member of one group in the periodic table with the second element in the next group, to the right, found particularly with these three pairs of elements: Li–Mg, Be–Al, and B–Si.

The diagonal relationship is a consequence of the relatively small size of the ions of the elements in the second short period. The *polarizing power* of the positive ion of an element determines to a large extent the type of bonding between the element with nonmetals such as oxygen and chlorine and hence the chemical characteristics of the compounds (see *polarization of ions*).

Decreasing
polarizing power:
same charge,
larger size

Increasing polarizing power:
increasing charge, smaller size

Diagonal relationships related to polarizing power

These are some of the similarities between *beryllium* and *aluminum* which are not shared by the other group II elements:

● their bonding in compounds is mainly covalent

● their oxides and hydroxides are amphoteric

● their anhydrous chlorides vaporize easily and form dimers in the vapor (Be_2Cl_4 and Al_2Cl_6).

dialysis is used to separate dissolved ions and small molecules from *colloid* particles. Dialysis uses a *selectively-permeable membrane* which lets through the small ions and molecules but traps larger colloid particles. Dialysis is used in medicine to treat the blood of patients with kidney failure. Dialysis removes the waste products of metabolism from blood and helps to adjust the concentration of ions. Blood cells and colloidal sized protein molecules cannot pass through the membrane.

diamond: one of the *allotropes* of carbon; it is a *network solid*. Larger diamonds are valued as gemstones because of the mineral's high *refractive index* and dispersive power. Diamond is also a valuable industrial material because it is the hardest natural substance and the best solid conductor of heat energy. Its thermal conductivity is five times as great as that of copper. This makes diamonds very useful for cutting, grinding and polishing other hard materials. Diamond tipped tools are durable and they do not overheat. Industrial diamonds are manufactured by heating graphite to 1800°C at very high pressure in the presence of a metal such as nickel or iron.

diatomic molecule: strictly a molecule with two atoms such as N_2 or HCl but the term is generally used for the molecules of elements with two identical atoms. Examples of diatomic elements are oxygen, O_2, nitrogen, N_2 and the halogens, F_2, Cl_2, Br_2 and I_2.

diazonium salts: salts formed when aryl amines, such as *aniline*, react with *nitrous acid* (HNO_2) below about 10°C.

Diazonium salts are unstable so they are made as needed and kept cold. Above 10°C, benzene diazonium chloride decomposes to phenol and nitrogen.

The commercial importance of diazonium salts is based on their coupling reactions to form *azo dyes*.

Diazonium salts are also useful intermediates which make it possible to form derivatives of arenes.

Diazotization – the formation of benzene diazonium chloride

dienes are hydrocarbons with two double bonds. 1,3-butadiene is a product of *thermal cracking* which is a useful intermediate for making other organic compounds. Synthetic *rubbers* are copolymers of butadiene, or one of its derivatives, with other unsaturated compounds such as styrene.

diesel fuel is heavier and oilier than *gasoline*. The hydrocarbon molecules in diesel typically contain 14 carbon atoms ($C_{14}H_{30}$), while the molecules in gasoline are generally smaller. Less *refining* is needed to make diesel and so it is cheaper. Diesel *fuel* also has a higher energy density than gasoline.

Diesel engines emit air pollutants that include oxides of nitrogen (NO_x), particulate matter, unburnt hydrocarbons as well as carbon monoxide from their exhausts. The emission of *particulates* is much more of a problem with diesel than with gasoline engines. There are technologies based on catalysts to trap particulates and adsorb oxides of nitrogen which can cut emissions by up to 90% but they only work with very low sulfur fuel.

diffusion: a spreading out and mixing process in gases or solutions as molecules or ions mingle with each other. Molecules diffuse from a region where they are more concentrated to a region where their concentration is lower. Eventually diffusion evens out differences in concentration. The smaller the molecules or ions the faster they diffuse. Diffusion is a consequence of the rapid random motion of molecules in gases and liquids. Diffusion through membranes is important in *dialysis* and *osmosis*.

dilution: the process of adding more solvent to a solution to lower the concentration.

Quantitative dilution is an important procedure in analysis. The purpose is to make a solution with known concentration by accurately diluting a *standard solution*. Successive dilutions provide a series of solutions which can be used to *calibrate* instruments such as *colorimeters*.

The procedure is to take a measured volume of the more concentrated solution with a pipet and run it into a graduated flask. The flask is then carefully filled to the mark with purified water.

The key to calculating the volumes to use when diluting a solution is to remember that the amount in moles of the reagent in the final solution must equal the amount in moles of the sample taken from the concentrated solution. If c is the concentration in mol/L and V is the volume in L:

- the amount in moles of the reagent in the concentrated solution = $c_A V_A$

- the amount in moles of the reagent in the diluted solution = $c_B V_B$

  So: $c_A V_A = c_B V_B$

Worked example

What volume of a 1.00 mol/L solution of copper(II) sulfate is needed to prepare 100 cm^3 of a 0.1 mol/L solution?

Notes on the method

Start by a *conversion* to give the volume of the diluted solution in L.

V_A can be calculated because all the other terms in the relationship $c_A V_A = c_B V_B$ are known.

Answer

Final volume of the diluted solution is to be: $100 \text{ cm}^3 \times \dfrac{1 \text{ L}}{1000 \text{ cm}^3} = 0.1 \text{ L}$

$1.0 \text{ mol/L} \times V_A = 0.1 \text{ mol/L} \times 0.1 \text{ L}$

$V_A = \dfrac{0.1 \text{ mol/L} \times 0.1 \text{ L}}{1.0 \text{ mol/L}} = 0.01 \text{ L} = 10 \text{ cm}^3$

Pipetting 10 cm^3 of the concentrated solution into a 100 cm^3 graduated flask and making it up to the mark gives the required dilution.

---

dimerization: two molecules linking together. Dimers may be held together by *covalent bonds* (see *aluminium chloride*) or by *hydrogen bonding*.

An acetic acid dimer in a non-polar solvent. *In aqueous solution, acetic acid does not dimerize because there are so many water molecules with which they can form hydrogen bonds.*

diols: alcohols with two —OH groups. An example is 1,2-ethanediol (ethylene glycol) used as *antifreeze*.

Structure of 1,2-ethanediol

dioxins: a family of stable compounds which persist in the environment. Chemically they contain chlorine atoms bonded to benzene rings. Some dioxins are toxic and can cause skin disease, cancer and birth defects. Traces of dioxins may form during the manufacture of chlorinated phenols, the bleaching of wood pulp for paper and during the *incineration of wastes* containing chlorine compounds such as *PCBs* and the *addition polymer*, PVC.

dipole moments: see *polar molecules*.

dipole–dipole interactions: *intermolecular forces* involving *polar molecules* with permanent dipoles. The cohesive forces are stronger than *intermolecular forces* and account for the fact that compounds such as acetone are liquids at room temperature while comparable non-polar hydrocarbons are gases.

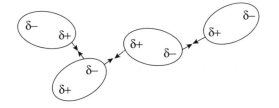

Attractions between molecules with permanent dipoles

diprotic acid: an *acid* such as sulfuric acid (H_2SO_4), which can give away two *protons* (H^+).

disaccharides: see *carbohydrates.*

disinfectants: chemicals which destroy microorganisms. Chlorine and oxygen *bleaches* are used as disinfectants. Unlike *antiseptics*, disinfectants cannot be used on skin and other living tissues.

displacement reactions: redox reactions which can be used to compare the relative strengths of metals as *reducing agents* and nonmetals as *oxidizing agents*.

A more reactive metal displaces a less reactive metal from one of its salts. Zinc, for example, displaces copper from a copper(II) sulfate. The zinc atoms reduce the copper ions.

$$Zn(s) + Cu^{2+}(aq) \longrightarrow Zn^{2+}(aq) + Cu(s)$$

In *group VII*, a more reactive halogen displaces a less reaction halogen. The order of reactivity for the halogens is $Cl>Br>I$. The more reactive halogen oxidizes the ions of a less reactive halogen.

$$Br_2(aq) + 2I^-(aq) \longrightarrow 2Br^-(aq) + I_2(s)$$

Standard electrode potentials make it possible to predict the direction of change in a displacement reaction (see *electrochemical series*).

disproportionation reaction: a reaction in which the same element both increases and decreases its oxidation number. Copper(I) ions disproportionate in aqueous solution to a mixture of $Cu^{2+}(aq)$ ions and Cu atoms.

A series of disproportionation reactions is used to make *chlorine oxoanions*. The oxygen in *hydrogen peroxide* disproportionates when it decomposes. Note that disproportionation refers to a particular element in a compound not to the compound as a whole.

(See the *electrochemical series* for the use of standard electrode potential values to predict whether or not disproportionation will take place.)

dissociation: a reaction in which a compound splits into two or more smaller products which can recombine to form the original compound if the conditions change.

Some compounds dissociate on heating. Examples are ammonium chloride and dinitrogen tetroxide. These *thermal dissociation* reactions are *reversible*. On cooling the original compounds reform.

$$NH_4Cl(s) \underset{cool}{\overset{heat}{\rightleftharpoons}} NH_3(g) + HCl(g)$$

$$N_2O_4(g) \underset{cool}{\overset{heat}{\rightleftharpoons}} 2NO_2(g)$$

Acids dissociate into ions on solution in water. The extent to which they ionize distinguishes strong and weak acids and this is measured by the *acid dissociation constant*.

distillation: a technique for separating and purifying a liquid by heating to vaporize the liquid and then cooling to condense it. Simple distillation is used to purify water, to recover pure solvents and to separate liquid product from reaction mixtures. Substances which do not evaporate are left behind in the distillation flask.

Apparatus for simple distillation

One of the easiest ways to determine the boiling point of a liquid is to distil a pure sample of the liquid in an apparatus with a thermometer.

(See also *fractional distillation, steam distillation* and *vacuum distillation*.)

disulfide bridges: see *cross-linking*.

DNA (deoxyribonucleic acid): a double helix made up of two poly*nucleotide* chains. Both chains have a sugar phosphate backbone. Every sugar unit has one of four bases linked to it. The four bases are adenine (A), cytosine (C), guanine (G) and thymine (T). The links between the chains are formed by hydrogen bonding between pairs of bases: A pairs with T and C pairs with G.

adenine (A)   thymine (T)

Hydrogen bonding between adenine and thymine in a DNA double helix

Genes are sections of DNA molecules found in the chromosomes of cells.

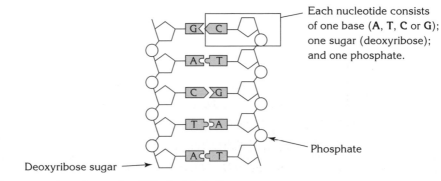

Each nucleotide consists of one base (**A**, **T**, **C** or **G**); one sugar (deoxyribose); and one phosphate.

Phosphate

Deoxyribose sugar

Representation of part of a DNA double helix

dolomite: a carbonate mineral with the composition $MgCO_3.CaCO_3$. It is a raw material for the production of magnesium and its salts.

***d*-orbitals:** the five atomic orbitals in a *d*-subshell. In a free atom or ion the five *d*-orbitals are at the same energy level. The five orbitals are not the same shape and they split into two groups with different energies when a *d-block element* ion is surrounded by molecules or ions in a complex ion. This helps to account for the *color* of complex ions.

double bond: two covalent bonds between two atoms as in *oxygen, alkenes* and *ketones*. With two electron pairs involved in the bonding, there is a region of high electron density between the two atoms joined by a double bond.

Examples of molecules with double bonds

The molecular orbital model for a double bond shows that one of the bonds is a normal sigma(σ) bond while the second bond is a pi(π) bond.

double indicator titration: a procedure for using two *indicators* successively to determine the two end points during a *titration*. This procedure can be used to find the concentrations in a solution containing a mixture of sodium carbonate and sodium hydrogencarbonate by titration with a standard solution of an acid such as hydrochloric acid.

The solution starts alkaline because carbonate and hydrogencarbonate ions are bases. Phenolphthalein is added at the start and it turns red. Acid is run in from the buret. The phenolphthalein becomes colorless at the first end point. This corresponds to converting all $CO_3^{2-}(aq)$ ions in the mixture to $HCO_3^-(aq)$.

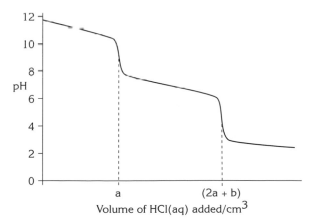

The pH curve during a titration of a mixture of $Na_2CO_3(aq)$ and $NaHCO_3(aq)$ with dilute 0.10 mol/L hydrochloric acid

Next, methyl orange indicator is added and it starts yellow. More acid is run in from the buret and the indicator turns red at the second end point. The second end point corresponds to the complete neutralization of all the $HCO_3^-(aq)$ ions.

$$Na_2CO_3 \xrightarrow{\text{a}} NaHCO_3 \xrightarrow{\text{a}} NaCl$$

The chemical changes and the titers during a double indicator titration

$$NaHCO_3 \xrightarrow{\text{b}} NaCl$$

present in the
original mixture

double salts: solid ionic salts with two different metal cations in a lattice of negative ions. Crystalline double salts are not mixtures but distinct compounds with a definite structure.

When dissolved in water double salts behave just like a mixture of two simple *salts*. The usual way of making a double salt is to mix equal amounts (in moles) of the two single salts in solution and then to crystallize the double salt. Examples of double salts are the *alums*.

drugs: substances which affect natural chemical processes in the body.

In medicine, drugs are substances used for the treatment, relief, diagnosis or prevention of disease. A drug is what a doctor prescribes. A *medicine* is the whole formulation.

There are many classes of medical drug which include:

- drugs which act on the central nervous system, such as hypnotics, sedatives, tranquilizers, antidepressants, *narcotics, anesthetics and analgesics*
- drugs which kill bacteria, such as *antibiotics* and sulfonamides (see *chemotherapy*)
- antiviral drugs, such as those used to suppress HIV
- cardiovascular drugs, which help to lower blood pressure, stimulate the heart or control the heartbeat
- drugs such as *cisplatin* used in cancer treatment
- drugs used for the digestive system, such as *antacids*, laxatives and drugs to treat ulcers
- *hormones* such as *insulin*, growth hormone and *steroids*.

People also take drugs for pleasure, stimulation and relaxation. Some recreational drugs such as alcohol and nicotine are legal. Others such as ecstasy, marijuana, cocaine and opiates are illegal. Drugs used to treat disease, such as steroids, can also be exploited to enhance performance in sports. Sensitive methods of chemical analysis are widely used to detect traces of drugs in blood, breath and urine (see *Breathalyzer*).

drug development: the research and development program which leads to the launch of a new drug. Scientists in the *pharmaceutical industry* use the term "new chemical entity" (NCE) to describe a newly-discovered chemical for treating disease. Discovering an NCE can be hugely profitable, so pharmaceutical companies spend large sums on research and development.

Before a new compound can become a marketable drug the manufacturer has to show that:

- the drug meets a definite need of patients and their doctors
- it is technically possible to make it on a large scale
- there is a big enough market to make it commercial
- it is safe.

The pharmaceutical chemist Gertrude Elion (born 1918) developed a rational process of drug design based on detailed studies of the differences between the *biochemistry* and molecular biology of human cells and the cells of microorganisms which cause disease. Since the 1940s, the team she worked with has developed drugs to treat a range of diseases including leukemia, malaria, gout and HIV infection.

dry ice: solid carbon dioxide. Above −79°C it *sublimes* to carbon dioxide gas. This makes it a useful coolant.

drying agents remove water from moist liquids and gases. Liquid products of organic preparations are often dried with small amounts of anhydrous sodium sulfate or anhydrous magnesium sulfate before the final distillation step. Gases made with laboratory reagents are often moist and can be dried by bubbling them through concentrated sulfuric acid, or more safely by passing them though a U-tube containing *silica gel*. Anhydrous calcium chloride is frequently used in *desiccators* to absorb water, alcohols or amines.

drying oils: oils such as linseed oil which polymerize and set when exposed to air. As a drying oil linseed oil is used in the manufacture of *paints* and varnishes.

ductility: a property of materials, especially metals, which can be drawn out under tension without breaking. *Copper* is a ductile metal which can be pulled through a die (narrow hole) to make thin wire.

dyes: colored materials used to dye fabrics and other materials such as paper, hair, leather and food. A good dye is a fast dye which means that it is not only colored but can also attach itself strongly to a material so that it does not wash out nor should it fade in the light.

Plants were the main source of natural dyes until William Perkin chanced on the first synthetic dye, mauveine, in 1856. Alizarin and indigo were plant dyes which were the basis of large scale industries. The discovery of synthetic routes to these dyes in the late nineteenth century were early triumphs for *organic chemistry*.

Examples of synthetic dyes are *azo dyes*, *vat dyes*, and *fiber reactive dyes*.

dynamic equilibrium: the state of balance in a *reversible* process when neither the forward change nor the backward change is complete; both changes are still going on at equal rates so that they cancel each other out and there is no overall change.

Examples of systems which can be in dynamic equilibrium include:

• water and ice at 0°C, or more generally the liquid and solid states of a substance at its melting point

• a liquid with its saturated *vapor* in a closed container

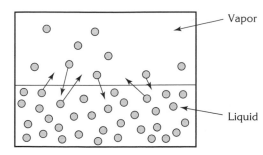

Dynamic equilibrium between a liquid and its vapor

- a solid with a *saturated solution* of the solid
- a solute distributed between two immiscible solvents
- a *weak acid* in aqueous solution, or any other reversible reaction at constant temperature and pressure.

EDTA: the common abbreviation for a particular ion which binds so firmly with metal ions that it holds them in solution and effectively makes them chemically inactive. A very little EDTA is added to salad dressing to trap traces of metal ions which otherwise would catalyze the oxidation of oils. Another practical application is where EDTA is included in bathroom cleaners to help remove scale by dissolving deposits of calcium carbonate left by hard water. Thanks to this ability to act as a *chelating agent*, EDTA can be used to treat lead poisoning.

In laboratories EDTA is the disodium salt of **e**thylene **d**iaminete**t**ra**a**cetic **a**cid. The ion can fold itself around metal ions so that four oxygen atoms and two nitrogen atoms can present lone pairs to form coordinate bonds with the metal ion. It is a hexadentate *ligand*.

(See also *complex forming titration*.)

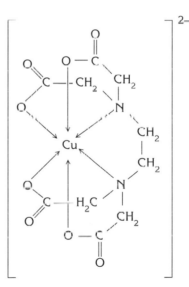

Complex ion formed by EDTA with a metal ion.

efflorescence: the loss of *water of crystallization* from a hydrated salt kept in an open container. Hydrated crystals of sodium carbonate (washing soda, $Na_2CO_3.10H_2O$) turn powdery in an unsealed container as they gradually lose most of their water.

effusion: the escape of a gas through a small hole into a vacuum. Effusion happens whenever a porous barrier separates a gas from a vacuum. The rate is governed by *Graham's law of effusion*.

elastomers: *polymers* which can be stretched to several times their normal length and recover their original size and shape when released. They are elastic materials. Natural rubber is an elastomer. So are synthetic rubbers such as the *copolymer* of butadiene and styrene.

Spandex (trade name Lycra) is an elastic fiber with an exceptional ability to stretch and recover. It is a condensation polymer combining two polymer segments – *polyurethane* $(-NH-CO-O-)$ and polyether $(-CH_2-CH_2O-)$. The polyether sections of the chains are the soft stretchy parts, while the urethane sections make sure that the fibers snap back into shape after stretching. Many sporting fabrics now include spandex fibers but a garment is never made entirely from spandex.

electric arc furnace: a furnace for making *steel* by *recycling* scrap iron and steel. The heat of an arc from carbon electrodes melts the metal. Added lime combines with impurities to form a *slag*. Other ingredients are added as necessary. The process can be precisely controlled to give small batches of steel matched to a specified composition.

electrical conductors: solids, liquids or gases which conduct electricity because they contain charged particles which are free to move. They include:

- metals and graphite in which the charge carriers are *delocalized electrons*
- electrolytes in which the charge carriers are positive and negative ions.

electrical insulators: solids, liquids or gases which do not conduct electricity because they contain no free moving charged particles. Increasing the applied voltage may turn an insulator into a conductor if the voltage becomes high enough to ionize the atoms or molecules.

electrochemical cells produce an electric potential difference from a redox reaction. Some electrochemical cells are designed for practical use. Examples include *alkaline manganese cells* and rechargeable *lithium*, *nicad* (nickel–cadmium) and *lead–acid cells*.

The reaction of zinc metal with aqueous copper(II) ions is a redox reaction which can be described by two half equations. Zinc is oxidized to zinc ions as copper(II) ions are reduced to copper metal.

$$Zn(s) \longrightarrow Zn^{2+}(aq) + 2e^-$$
$$Cu^{2+}(aq) + 2e^- \longrightarrow Cu(s)$$

In an electrochemical cell the two half-reactions happen in separate *half-cells*. The electrons flow from one cell to the other through a wire connecting the electrodes. The electric circuit is completed by a *salt bridge* connecting the two solutions.

The tendency for the current to flow is measured with a high resistance voltmeter which measures the cell's *emf* when no current is flowing. In the example illustrated on page 123 electrons tend to flow out of the zinc electrode (negative) around the circuit and into the copper electrode (positive). The emf of the cell is 1.14 V under standard conditions (298 K and concentrations of 1.0 mol/L).

There is a convenient shorthand for describing cells. The standard emf of the cell, E°_{cell} is written alongside the cell diagram. The agreed convention is that the sign of E°_{cell} is the charge on the right-hand electrode.

$$Zn(s)| \ Zn^{2+}(aq) \ \| \ Cu^{2+}(aq)|Cu(s) \qquad E^\circ_{cell} = +1.14 \text{ V}$$

If the cell emf is positive the reaction tends to go according to the cell diagram read from left to right. As a current flows in a circuit connecting the two electrodes, zinc atoms turn into zinc ions and dissolve, as copper ions turn into copper atoms and deposit on the copper electrode.

$$Zn(s) + Cu^{2+}(aq) \longrightarrow Zn^{2+}(aq) + Cu(s)$$

$E°_{cell}$ values can be determined for any redox reaction with the help of a table of *standard electrode potentials*.

$$E°_{cell} = E°_{\text{right-hand electrode}} - E°_{\text{left-hand electrode}}$$

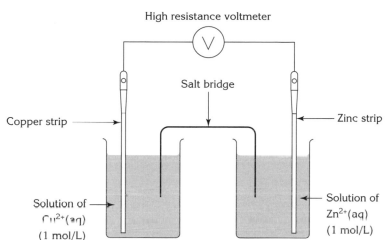

Diagram of an electrochemical cell

electrochemical series: a series showing half-reactions for oxidation and reduction, with the most powerful reducing agent at the top (or bottom) of the list and the most powerful oxidizing agent at the bottom (or top) of the list.

The order of the half-reactions is determined by their *standard electrode potentials*. In the list below, the *half-cell* with the most negative electrode potential is at the top of the list. The most negative electrode has the greatest tendency to give up electrons, so it is powerfully reducing.

The half-cell with the most positive electrode potential is shown at the bottom of the list. The most positive electrode has the greatest tendency to gain electrons so it is the most powerfully oxidizing.

Half cell	Half reaction	E°/V
[Cl₂(aq),2Cl⁻(aq)]│Pt	$Cl_2(aq) + 2e^- \longrightarrow 2Cl^-(aq)$	+1.51
[Br₂(aq),2Br⁻(aq)]│Pt	$Br_2(aq) + 2e^- \longrightarrow 2Br^-(aq)$	+1.09
[I₂(aq),2I⁻(aq)]│Pt	$I_2(aq) + 2e^- \longrightarrow 2I^-(aq)$	+0.54
Cu²⁺(aq)│Cu(s)	$Cu^{2+}(aq) + 2e^- \longrightarrow Cu(s)$	+0.34
2H⁺(aq)│H₂(g)	$2H^+(aq) + 2e^- \longrightarrow H_2(g)$	0.00 (by definition)
Zn²⁺(aq)│Zn(s)	$Zn^{2+}(aq) + 2e^- \longrightarrow Zn(s)$	–0.76
Mg²⁺(aq)│Mg(s)	$Mg^{2+}(aq) + 2e^- \longrightarrow Mg(s)$	–2.37
Na⁺(aq)│Na(s)	$Na^+(aq) + e^- \longrightarrow Na(s)$	–2.71

The bottom part of the above electrochemical series shows half-reactions for metal ions and metals. Sodium is the most reactive of the metals when it reacts as a reducing agent forming metal ions; copper is the least reactive. This broadly corresponds to the reactivity series for metals and the order of reaction shown by metal/metal ion *displacement reactions.*

The top part of the above electrochemical series shows half-reactions for halogens and halide ions. Chlorine is the most reactive of these halogens when it acts as an oxidizing agent forming chloride ions. Iodine is the least reactive. This corresponds to the order of reactivity of the halogens as shown by their displacement reactions.

An electrochemical series based on electrode potentials can be used to predict the direction of change in redox reactions. First identify the two half-equations. Write them down one above the other. The more positive half-reaction tends to go from left to right taking in electrons, while the more negative half-reaction goes from right to left.

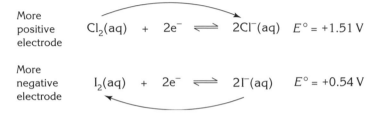

More positive electrode

$$Cl_2(aq) + 2e^- \rightleftharpoons 2Cl^-(aq) \quad E^\circ = +1.51\ V$$

More negative electrode

$$I_2(aq) + 2e^- \rightleftharpoons 2I^-(aq) \quad E^\circ = +0.54\ V$$

Using half-reactions and E° values to predict the direction of change

In this way we can predict from electrode potentials whether or not a *disproportionation* reaction is likely to occur. Copper(ı) ions disproportionate in aqueous solution, but iron(ıı) ions tend not to disproportionate.

More positive electrode

$$Cu^+(aq) + e^- \rightleftharpoons Cu(s) \quad E^\circ = +0.52\ V$$

Copper(I) disproportionates

More negative electrode

$$Cu^{2+}(aq) + e^- \rightleftharpoons Cu^+(aq) \quad E^\circ = +0.15\ V$$

More positive electrode

$$Fe^{3+} + e^- \rightleftharpoons Fe^{2+} \quad E^\circ = +0.77\ V$$

Iron(III) reacts with iron metal to form iron(II)

More negative electrode

$$Fe^{2+}(aq) + 2e^- \rightleftharpoons Fe(s) \quad E^\circ = -0.44\ V$$

Using standard electrode potential values to decide whether or not ions are likely to disproportionate

Standard electrode potentials predict the direction of change but say nothing about the rate of change. A reaction which is *feasible* may not in fact happen because it is so slow. Also, the predictions apply only under standard conditions. Changing the concentrations can alter the direction of change.

electrode: see *anode* and *cathode*.

electrode potential: see *standard electrode potential*.

electrolysis: a process which uses an electric current to decompose a molten ionic compound, or a solution of ions, into elements. Ions can conduct electricity only when they are free to move, which is why electrolytes are molten compounds or solutions.

Large scale manufacturing processes which use electrolysis include *aluminum extraction* and the *electrolysis of brine*. *Electroplating, copper refining* and *anodizing* are also electrolytic processes.

During the electrolysis of molten salts, such as lead bromide:

- metals deposits appear at the *cathode*, e.g. $Pb^{2+} + 2e^- \longrightarrow Pb$ (reduction)
- nonmetals appear at the *anode*, e.g. $2Br^- \longrightarrow Br_2 + 2e^-$ (oxidation).

During the electrolysis of a solution of a salt in water the change at the cathode depends on the type of metal ion in the salt. If the metal is:

- unreactive it forms at the cathode,
  e.g. $Cu^{2+}(aq) + 2e^- \longrightarrow Cu(s)$
- reactive the product at the cathode is hydrogen, from hydrogen ions produced by the ionization of water,
  $2H^+(aq) + 2e^- \longrightarrow H_2(g)$

During the electrolysis of a solution of a salt in water the change at the anode may depend on the type of electrode. If the anode is made of carbon or platinum the products are:

- halogen molecules if the ions in the solution are chloride, bromide or iodide ions,
  e.g. $2Cl^-(aq) \longrightarrow Cl_2(g) + 2e^-$
- otherwise, oxygen from the water, $4OH^-(aq) \longrightarrow O_2(g) + 2H_2O(l) + 4e^-$

The concentration of the solution can affect the product at the anode. Chlorine forms at the anode during electrolysis of a concentrated solution of sodium chloride, but as the solution gets more dilute increasing amounts of oxygen form.

If the anode is a metal such as copper or silver, the atoms in the electrode turn into ions,
  e.g. $Cu(s) \longrightarrow Cu^{2+}(aq) + 2e^-$

The amount of change at an electrode depends on:

- the charge on the ions
- the amount of electric charge that flows.

electrolysis of brine: the basis of the chlor-alkali industry. Electrolysis of brine is used to manufacture chlorine, hydrogen and sodium hydroxide. The process is also the source of chlorine bleach, made by allowing the chlorine and sodium hydroxide to mix and react forming sodium hypochlorite.

Brine is a solution of sodium chloride in water. During electrolysis chlorine forms at the positive electrode (anode). Hydrogen bubbles off the negative electrode (cathode) while the solution turns into sodium hydroxide.

The cell used for electrolysis of brine has to be carefully designed because chorine reacts with sodium hydroxide. The cell has to keep the chlorine and sodium hydroxide apart.

Two of the types of cell used for the process are the flowing mercury cell and the membrane cell. The mercury cell is being phased out because mercury is expensive and hazardous. Traces of mercury escaping into the environment cause pollution.

Main inputs and outputs of a membrane cell. The membrane allows the solution to pass through but stops the chlorine mixing with the alkali. The membrane has ion exchange properties; it lets positive ions through but not negative ions.

electromagnetic radiation: radiation such a X-rays, ultraviolet radiation, visible light and infrared rays which can travel through space as oscillating electric and magnetic waves.

There is a spectrum of types of electromagnetic radiation which differ only by their *wavelength*, λ, and *frequency*, ν. In space all radiation travels at the same speed, that of light, $c = 2.99 \times 10^8$ m/s.

$$\lambda = \frac{c}{\nu} = \frac{2.99 \times 10^8 \text{ m/s}}{\nu}$$

	radio frequency	microwave	infrared	visible	ultraviolet	X-rays	γ-rays

Frequency ν/Hz 10^5 10^6 10^7 10^8 10^9 10^{10} 10^{11} 10^{12} 10^{13} 10^{14} 10^{15} 10^{16} 10^{17} 10^{18} 10^{19} 10^{20}

Wavelength λ/m $\quad 10^3 \qquad\qquad 1 \qquad\qquad 10^{-3} \qquad\qquad 10^{-6} \qquad\qquad 10^{-9}$

The electromagnetic spectrum

When electromagnetic radiation interacts with matter it behaves as if made up of packets of energy. These energy quanta are also called photons. The higher the frequency of the radiation the higher the energy of the quanta where $E = h\nu$. The higher energy photons of ultraviolet radiation and X-rays can have a much greater chemical effect than lower energy photons of IR and radio waves.

electron affinity: the enthalpy change when gaseous atoms of an element gain electrons to become negative ions. Electron affinities are precisely defined and only used in thermo-chemical cycles such as the *Born–Haber cycle*. Chemical reactions in laboratories do not normally involve free gaseous atoms and ions.

The first electron affinity of an element is the enthalpy change when one mole of gaseous atoms gains electrons to form one mole of gaseous ions. These two equations define the first and second electron affinities for oxygen:

$$O(g) + e^-(g) \longrightarrow O^-(g) \qquad \Delta H^\circ = -141 \text{ kJ/mol}$$

$$O^-(g) + e^-(g) \longrightarrow O^{2-}(g) \qquad \Delta H^\circ = +798 \text{ kJ/mol}$$

The gain of the first electron is exothermic but adding the second electron to a negatively charged particle is an endothermic process. Overall adding two electrons to a gaseous oxygen atom is endothermic.

(Note that chemists use *electronegativity* values and **not** electron affinities to account for polar covalent bonds.)

electron configuration: the number and arrangement of electrons in an atom of an element. Electrons fill energy levels according to the *aufbau principle*.

The electron configuration helps to make sense of the chemistry of an element. It is the electrons in the outer shell which largely determine the chemical properties of an element. Elements in the same *group* of the periodic table have similar properties because they have the same outer electron configuration. There are trends in properties down a group because of the *shielding* effect of the increasing number of inner full shells.

There are several common conventions for writing electron configurations.

A shortened form of electron configuration uses the symbol of the previous *noble* gas to stand for the inner full shells. According to this convention the electron configuration of sodium is [Ne]$3s^1$.

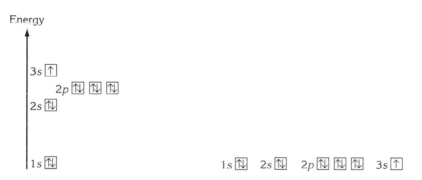

Representations of the electron configuration of sodium, $1s^2 2s^2 2p^6 3s^1$

electron density maps show the whereabouts of electrons in crystals. The lines in the maps connect points with equal electron densities. Each map shows the pattern of electron density for a two-dimensional slice through the crystal. The maps are an interpretation of

Sodium chloride   Urea

Electron density maps for an ionic crystal and for a molecule

the information from X-ray diffraction studies. This is possible because it is the electrons in a crystal which scatter X-rays (see *X-ray crystallography*).

Electron density maps for an ionic crystal provide evidence for electron transfer from the metal to the nonmetal. The ions show up as distinct particles. Electron density maps for molecules show electrons shared between atoms in *covalent bonds*.

electron transfer takes place during *redox reactions*. An atom, molecule or ion loses electrons when oxidized (**o**xidation **is l**oss). The electrons are transferred to the atom, molecule or ion being reduced (**r**eduction **is g**ain). Hence the memory aide: OIL RIG. *Oxidation numbers* and *half-equations* for redox reactions help to keep track of electron transfer reactions.

electronegativity measures the pull of an atom of an element on the electrons in a chemical bond. The stronger the pulling power of an atom the higher its electronegativity. There are two quantitative scales of electronegativity; one devized by Linus Pauling and the other by Robert Mulliken. However the term electronegativity is generally used to compare one element with another qualitatively so it is enough to know the trends in values across and down the periodic table.

Electronegativity
increases

Trends in electronegativity for s *and* p *block elements. Values are low to the left of each period and rise towards the right, giving rise to a repeating (periodic) pattern.*

The highly electronegative elements, such as fluorine and oxygen, are at the top right of the periodic table. The least electronegative elements, such as cesium, are at the bottom left.

The bigger the difference in the electronegativity of the elements forming a bond, the more polar the bond. Oxygen is more electronegative than hydrogen so an O — H bond is *polar* with a slight negative charge on the oxygen atoms and a slight positive charge on the hydrogen atom.

Examples of polar covalent bonds *between atoms with different electronegativities*

The bonding in a compound becomes ionic if the difference in electronegativity is large enough for the more electronegative element to remove completely electrons from the other element. This happens in compounds such as sodium chloride, magnesium oxide or calcium fluoride.

electronic structure: the arrangement of electrons in an atom in its main energy levels and sublevels. (See *electron configuration*.)

electrophiles: reactive ions and molecules which attack parts of molecules which are rich in electrons. They are "electron-loving" reagents. Electrophiles form a new bond by accepting a pair of electrons from the molecule attacked during a reaction.

Examples of electrophiles are the H^+ and the NO_2^+ ions. Other electrophiles are the atoms at the δ^+ end of a *polar covalent* bond. See *electrophilic addition* and *electrophilic substitution*.

electrophilic addition takes place when *alkenes* undergo *addition reactions*. The *electrophile* attacks the electron-rich region of the double bond between two carbon atoms. Electrophiles which add to alkenes include hydrogen bromide, bromine, sulfuric acid and water (in the presence of an acid catalyst).

Hydrogen bromide molecules are *polar*. The hydrogen atom, with its δ^+ charge is the electrophilic end of the molecule.

Electrophilic addition of hydrogen bromide to ethene. The reaction takes place in two steps. The intermediate has a positive charge on a carbon atom. It is a carbocation. Curly arrows show the movement of a pair of electrons. Bond breaking is heterolytic and the intermediates ions.

Bromine molecules are not polar but they become polarized as they approach the electron-rich double bond. Electrons in the double bond repel electrons in the bromine molecule. The δ^+ end of the molecule is electrophilic.

Electrophilic addition of bromine to ethene

(See also *Markovnikov's rule*.)

electrophilic substitution: the characteristic reaction of *arenes* such as benzene. The electrophile attacks the electron-rich benzene ring with its six *delocalized electrons*. When describing the mechanism of this type of reaction it is easier to keep track of what is happening to the electrons using the *Kekulé structure* for *benzene*.

Electrophilic substitution of bromine in benzene is rapid in the presence of iron. Iron reacts with some bromine to form iron(III) bromide, which is the catalyst. Iron(III) bromide is an electron pair acceptor (*Lewis acid*) which produces the electrophile, Br^+. An alternative catalyst is *aluminum chloride* (another Lewis acid). The catalysts for this reaction are sometimes described as "halogen carriers."

The first step of electrophilic substitution in benzene is very similar to the first step of *electrophilic addition* to alkenes. The intermediate could then complete the reaction by adding a bromide ion but in fact what happens is the loss of H⁺ leading to substitution. Loss of H⁺ is preferred because it keeps the stable benzene ring stabilized by six delocalized electrons.

Formation of the electrophile

The mechanism of electrophilic substitution. Note that an alternative second step (addition of a bromide ion) is not favored because it gives a product which does not have a stable benzene ring.

Other examples of electrophilic substitution are the *nitration of benzene* and the *Friedel–Crafts reaction.*

electrophoresis: a technique for separating and identifying organic compounds. It can be used to separate compounds such as *amino acids* which become electrically charged when dissolved in water at a particular *pH*. Electrophoresis is used in hospitals to diagnose disease. It is important in genetic profiling to separate out the bands which form the "finger-print." It is also used in research to study the structures of *proteins* and *nucleic acids.*

Electrophoresis takes place on strips of a gel soaked in a buffer solution to keep the pH constant. Small spots of several samples can be put on each gel. The charged molecules move when an electric voltage is connected between the ends of the gel and they separate into bands according to their size and charge. Most samples are colorless so, once the separation is complete, the gel is dipped into a stain to reveal the bands.

electroplating uses *electrolysis* to coat one metal with another. The process is often used to coat metals with thin layers of copper, nickel, chromium, silver and zinc. The object to be plated forms the cathode in an electrolysis cell containing a solution of ions of the metal to be plated. Good results depend on careful control of the conditions, including the composition and concentration of the electrolyte, the temperature and the size of the current. Often the metal ion is present as a *complex ion*. During silver plating, for example, the silver ions are complexed with cyanide ions.

electrostatic forces: the forces between charged particles. Unlike charges attract each other. Positive ions, for example, attract negative ions in ionic compounds. Like charges repel each other. Positive ions in an electrolyte are repelled by the positive anode while being attracted to the negative cathode.

The size of the electrostatic force between two charges varies according to Coulomb's law. The bigger the charges the stronger the force – the force is proportional to the size of the charges, Q. The greater the distance, *d*, between the two charges the smaller the force – the force is inversely proportional to the distance squared.

$$\text{electrostatic force} \propto \frac{Q_1 \times Q_2}{d^2}$$

elements: chemically the simplest substances. All the atoms of an element have the same number of protons in the nucleus. The atomic number identifies an element and fixes its position in the *periodic table*. When an atom turns into an ion by gaining or losing electrons it becomes a charged atom of the same element. The *isotopes* of an element have different numbers of neutrons in the nucleus but they have the same chemical properties because they have the same number of protons and therefore the same *electron configuration*.

elimination reaction: a reaction which splits off a simple compound from a molecule while forming a double bond.

Elimination of a *hydrogen halide* from a *haloalkane* produces an *alkene*. The conditions are to heat the haloalkane with a solution of potassium hydroxide in ethanol. Using ethanol as the solvent instead of water makes the alternative *nucleophilic substitution reaction* less likely.

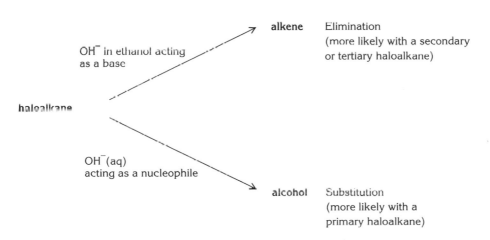

Competition between elimination and substitution

The elimination reaction is favored if the halogen atom is in the middle of the carbon chain or at a branch in the chain.

Elimination of hydrogen bromide from 2-bromopropane

Another example of an elimination reaction is the removal of water from an *alcohol* to form an alkene. This is a useful reaction in synthesis, for introducing double bonds into molecules. The conditions for reaction are either to pass the vapor of the alcohol over a hot, solid catalyst such as aluminum oxide or to dehydrate the alcohol by heating it with concentrated sulfuric acid.

Formation of an alkene from an alcohol. The acid catalyst protonates the — OH group so that the leaving group is a water molecule – a better leaving group than a hydroxide ion. The proton is reformed in the second step of the reaction, so overall the acid catalyst is unchanged.

Ellingham diagrams explain the conditions for metal extraction from oxide ores. The diagrams are graphs showing how the *free energy changes* of formation (ΔG°) of oxides vary with temperature.

EMF (electromotive force) measures the "voltage" of an electrochemical cell. The symbol for EMF is *E* and the SI unit is the volt (V). The EMF is the energy transferred in joules per coulomb of charge flowing through a circuit connected to a cell. The definition of the volt as a "joule per coulomb" makes it possible to calculate the *free energy change*, ΔG, for reactions from the values of the EMF for *electrochemical cells*, E°_{cell}.

$$\Delta G^{\circ} = - nFE^{\circ}_{cell}.$$

where *n* is the number of electrons transferred according to the equation for the reaction and *F* is the *Faraday constant*.

empirical formula: the formula of a compound found by calculation from the combining masses of elements. The empirical formula shows the simplest ratio of the amounts of elements in the compound; it therefore gives the ratio of the numbers of atoms.

For molecular compounds, the *relative molecular mass* shows whether or not the empirical formula is the same as the *molecular formula*.

Worked example

Analysis of a sample of a salt shows that it consists of 1.30 g zinc combined with 0.24 g carbon and 0.96 g oxygen. What is the formula of the salt?

Notes on the method
Look up the molar masses of the elements in a reference book.

Recall that: amount of substance (mol) $= \dfrac{\text{mass of substance (g)}}{\text{molar mass (g/mol)}}$

Answer

	zinc	carbon	oxygen
Combining masses	1.30 g	0.24 g	0.96 g
Molar masses of elements	65 g/mol	12 g/mol	16 g/mol
Amounts combined	$\dfrac{1.30 \text{ g}}{65 \text{ g/mol}}$	$\dfrac{0.24 \text{ g}}{12 \text{ g/mol}}$	$\dfrac{0.96 \text{ g}}{16 \text{ g/mol}}$
	= 0.02 mol	= 0.02 mol	= 0.06 mol
Simplest ratio of amounts	1	1	3

The formula is $ZnCO_3$.

emulsions consist of droplets of one liquid finely dispersed in another liquid. Emulsions are *colloids*. Milk is an emulsion of fat droplets in water. Salad dressing is an emulsion of vegetable oil in vinegar.

An emulsion is often "thicker" (more *viscous*) than either liquid on its own. Many cosmetic and medical creams are emulsions. Foundation creams and shaving creams are oil-in-water emulsions. Cold creams and cleansing creams are water-in-oil emulsions.

Just shaking two liquids together generally produces droplets which quickly separate back into two layers on standing. It is usually necessary to add an emulsifying agent to stabilize an emulsion. Salad dressing is an emulsion stabilized using mustard as the emulsifying agent. Mayonnaise is stabilized with egg yolk. Many creamy foods are emulsions and the lists of their ingredients include emulsifying agents such as lecithin and alginates.

Emulsifying agents are *surfactants*.

enantiomers: see *chiral compounds.*

endothermic changes take in energy from their surroundings. Melting and evaporation are endothermic *changes of state.*

For an endothermic reaction the enthalpy change, ΔH, is positive. During an endothermic reaction more energy is taken up in breaking bonds in the reactants than is given out as new bonds form in the products.

The fact that there are *spontaneous endothermic reactions* shows that it is not safe to predict that a reaction will not happen just because the *enthalpy change,* ΔH, is positive. Even if ΔH is positive, the *free energy change* (ΔG) can be negative if the *entropy change* (ΔS_{system},) is large enough and positive.

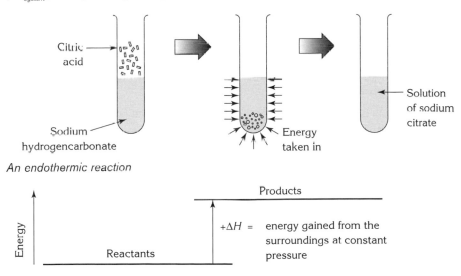

An endothermic reaction

Energy level diagram *for an endothermic reaction*

end point: the point during a titration when a color change shows that enough of the solution in the buret has been added to react exactly with the amount of chemical in the flask. Ideally the endpoint corresponds with the *equivalence point.* In *acid–base titrations* it is common to use a colored *acid–base indicator* to detect the endpoint.

In some *redox titrations* the endpoint is shown by a permanent change in the color of the solution in the flask once a drop of excess reagent has been added. In *potassium permanganate* titrations the solution in the flask becomes permanently pink with only a small excess of the purple-colored solution of permanganate ions.

energy conservation: a consequence of the first law of *thermodynamics*. In the sciences, energy conservation helps to keep account of *energy transfers* because the total quantity of energy before and after any change is the same.

Public campaigns to "save energy" have a different meaning. They refer to the importance of conserving fuels and other useful sources of energy. Energy is conserved in the scientific sense when fuels burn but the energy is spread around at a relatively low temperature in the environment so it can no longer be used for heating and to drive engines.

energy levels in atoms are the energies of the electrons in the available *atomic orbitals*. According to *quantum theory* each electron in an atom has a definite energy. When atoms gain or lose energy the electrons jump from one energy level to another.

energy level diagrams are a convenient way of displaying and comparing energy changes. The diagrams show the difference in energy between one state and another. Chemists use these diagrams to show the *energy levels* in atoms, and to summarize *exothermic* and *endothermic* processes.

energy transfers from one *system* to another system, or between a system and its surroundings, happen in chemistry by heating, by the emission and absorption of radiation (such as light) or by the flow of electric charges.

enthalpy change: the exchange of energy between a reaction mixture (*system*) and its surroundings when the reaction takes place at constant *pressure*. The enthalpy change is calculated when the starting temperature of the reactants and final temperature of the products are the same. The symbol for an enthalpy change is ΔH and the units are kJ/mol.

The sign of ΔH is decided by noting what happens to the reaction mixture.

- If the reaction is *exothermic* the system loses energy to its surroundings and ΔH is negative

- If the reaction is *endothermic* the system gains energy from its surroundings and ΔH is positive.

Standard enthalpy changes for reactions are calculated for carefully specified conditions and for precise amounts of chemicals (in moles). The standard conditions for enthalpy changes are 298 K and 1 bar (= 100 kPa). The symbol for a standard enthalpy change is ΔH°_{298}.

Books of data list standard enthalpy changes but it is often impossible to measure them directly under the stated conditions. Many values are determined indirectly using *Hess's Law*. Values measured under nonstandard conditions are corrected to give the value under standard conditions.

Enthalpy changes for changes of state (*melting*, vaporizing or *subliming*) are usually measured at the melting or boiling point.

enthalpy change of atomization: the *enthalpy change* to produce one mole of gaseous atoms from an element. The element at the start has to be in its *standard state* (normal, stable state) at 298 K and 1 bar *pressure*.

$$\text{element in its standard state} \xrightarrow{\Delta H^{\circ}_{at}} \underset{\text{1 mol}}{\text{gaseous atoms}}$$

Enthalpy change of atomization

Values for the standard enthalpy of atomization of elements are used in the *Born–Haber cycle* to determine *lattice energies*. They are also used to calculate *bond energies*.

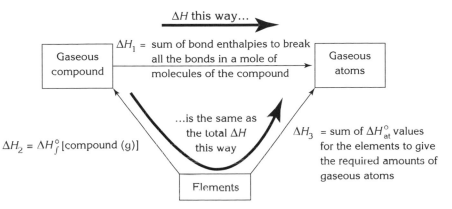

Thermochemical cycle to determine average bond energies in methane,

$$\Delta H_1 = -\Delta H_2 + \Delta H_3$$

Like all thermochemical quantities the precise definition is important. The standard enthalpy change of atomization of an element, $\Delta H^{\circ}_{at,\ 298}$, is the enthalpy change when one mole of gaseous atoms forms from the element in its *standard state* under *standard conditions*.

enthalpy change of combustion: the *enthalpy change* when one mole of a compound burns completely in oxygen. The compound and the products of burning must be in their normal stable states. For a carbon compound complete combustion means that all the carbon burns to carbon dioxide and that there is no soot or carbon monoxide. When burning a compound containing hydrogen the water formed must end up as a liquid, not gas.

Values of enthalpies of combustion are much easier to measure than many other enthalpy changes. They can be calculated from measurements taken with a *bomb calorimeter*. The importance of these values is that, with the help of *Hess's law*, they can be used to calculate *enthalpy changes of formation*.

Like all thermochemical quantities the precise definition of standard enthalpy of combustion is important.

The standard enthalpy change of combustion of a substance, $\Delta H^{\circ}_{c,298}$, is the enthalpy change when one mole of the substance completely burns in oxygen under *standard conditions* with the reactants and products in their *standard states*.

enthalpy change of formation: the *enthalpy change* when one mole of a compound forms from the elements in their normal stable states. The more stable state of an element is chosen where there are *allotropes*. For carbon, perhaps surprisingly, graphite turns out to be more stable thermochemically than diamond.

Tables of standard enthalpies of formation are very useful because, with a *Hess's law* cycle, they can be used to calculate the enthalpy changes for reactions.

Outline of a thermochemical cycle for calculating standard enthalpies of formation from standard enthalpies of combustion, $\Delta H_1 = \Delta H_2 - \Delta H_3$

Like all thermochemical quantities the precise definition of standard enthalpy of formation is important.

The standard enthalpy change of formation of a compound, $\Delta H^{\circ}_{f,298}$, is the enthalpy change when one mole of the compound forms from its elements under *standard conditions* with the elements and the compound in their *standard states*.

Worked example

Calculate the enthalpy of formation of propane, C_3H_8, at 298 K given these standard enthalpies of combustion:

- propane, $\Delta H^{\circ}_c [C_3H_8] = -2220$ kJ/mol
- carbon, $\Delta H^{\circ}_c [C] = -393$ kJ/mol
- hydrogen, $\Delta H^{\circ}_c [H_2] = -286$ kJ/mol

Notes on the method

Draw up a thermochemical cycle. Use Hess's law to produce an equation for the enthalpy changes. All the enthalpy changes are given except $\Delta H^{\circ}_f [C_3H_8]$.

Pay careful attention to the signs. Put the value and sign for a quantity in brackets when multiplying, adding or subtracting enthalpy values.

Answer

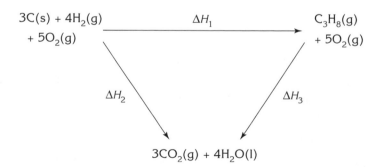

According to Hess's law $\Delta H_1 = \Delta H_2 - \Delta H_3$

$$\Delta H_1 = \Delta H_f^\circ[C_3H_8]$$

$$\Delta H_2 = 3 \times \Delta H_c^\circ[C] + 4 \times \Delta H_c^\circ[H_2] = 3 \times (-393 \text{ kJ/mol}) + 4 \times (-286 \text{ kJ/mol})$$

$$= -2323 \text{ kJ/mol}$$

$$\Delta H_3 = \Delta H_c^\circ[C_3H_8] = -2220 \text{ kJ/mol}$$

$$\text{Hence } \Delta H_f^\circ[C_3H_8] = (-2323 \text{ kJ/mol}) - (-2220 \text{ kJ/mol})$$

$$= -2323 \text{ kJ/mol} + 2220 \text{ kJ/mol}$$

$$= -103 \text{ kJ/mol}$$

enthalpy change of hydration: the *enthalpy change* when one mole of gaseous ions is hydrated under standard conditions to produce a solution in which the concentration of ions is 1 mol/L.

$$Na^+(g) + aq \longrightarrow Na^+(aq) \qquad \Delta H^\circ_{\text{hydration}} = -390 \text{ kJ/mol}$$

In this equation "+ aq" is short for adding water.

enthalpy change of melting (fusion): the *enthalpy change* when one mole of a solid turns to a liquid at its melting point.

Physicists general use "specific latent heat" values when studying the energy changes when solids melt. Specific latent heats of melting are measured in joules per kilogram rather than joules per mole.

enthalpy change of neutralization: the *enthalpy change of reaction* when an acid and an alkali neutralize each other.

$$HCl(aq) + NaOH(aq) \longrightarrow NaCl(aq) + H_2O(l) \quad \Delta H_{\text{neutralization}} = -57.1 \text{ kJ/mol}$$

The size of enthalpy of neutralization for dilute solutions of *strong acid* with *strong base* is always close to 57.1 kJ/mol. The reason is that these acids and alkalis are fully ionized, so in every instance the reaction is the same:

$$H^+(aq) + OH^-(aq) \longrightarrow H_2O(l) \qquad \Delta H = -57.1 \text{ kJ/mol}$$

The enthalpy changes of neutralization for reactions involving weak acids and weak bases differ from this value because there are other energy changes, such as the energy needed to remove a proton from a weak acid.

Enthalpies of neutralization can be measured approximately by mixing solutions of acids and alkalis in a *calorimeter*.

enthalpy change of reaction: the enthalpy change when the amounts shown in the chemical equation react. The standard enthalpy change for reaction is defined at 298 K, 1 bar pressure and with the reactants and products in their normal, stable states under these conditions. The concentration of any solutions is 1 mol/L.

The enthalpy change for a reaction can easily be calculated from tabulated values for standard *enthalpy changes of formation* using *Hess's law*.

According to Hess's law:

$$\Delta H^\circ_{\text{reaction}} = -\Delta H_1 + \Delta H_2 = \Delta H_2 - \Delta H_1$$

$$\text{So } \Delta H^\circ_{\text{reaction}} = \{\text{sum of } \Delta H_f^\circ[\text{products}]\} - \{\text{sum of } \Delta H_f^\circ[\text{reactants}]\}$$

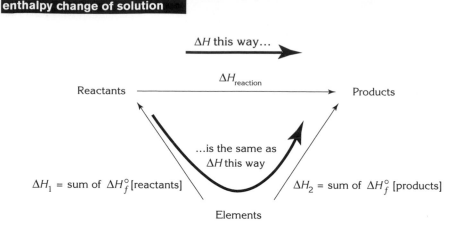

ΔH this way...

$\Delta H_{\text{reaction}}$

Reactants $\xrightarrow{\hspace{3cm}}$ Products

...is the same as
ΔH this way

ΔH_1 = sum of ΔH°_f[reactants]

ΔH_2 = sum of ΔH°_f [products]

Elements

Outline of a thermochemical cycle for calculating standard enthalpies of reaction from standard enthalpies of formation

Worked example

Calculate the enthalpy change for the reduction of iron(III) oxide by carbon monoxide.

$\Delta H^{\circ}_f [Fe_2O_3] = -824$ kJ/mol

$\Delta H^{\circ}_f [CO] = -110$ kJ/mol

$\Delta H^{\circ}_f [CO_2] = -393$ kJ/mol

Notes on the method

Write the balanced equation for the reaction.

All the enthalpy changes are given except ΔH° for the reaction.

Note that by definition ΔH°_f[element] = 0 kJ/mol

Pay careful attention to the signs. Put the value and sign for a quantity in brackets when multiplying, adding or subtracting enthalpy values.

Answer

$Fe_2O_3(s) + 3CO(g) \longrightarrow 2Fe(s) + 3CO_2(g)$

So $\Delta H^{\circ}_{\text{reaction}}$ = {sum of ΔH°_f[products]} – {sum of ΔH°_f[reactants]}

$\Delta H^{\circ}_{\text{reaction}}$ = {2 × ΔH°_f[Fe] + 3 × ΔH°_f[CO$_2$]} – {ΔH°_f[Fe$_2$O$_3$] + 3 × ΔH°_f[CO]}

= {0 + 3 × (–393 kJ/mol)} – {(–824 kJ/mol) + 3 × (–110 kJ/mol)}

= (–1179 kJ/mol) – (–1154 kJ/mol) = –25 kJ/mol

enthalpy change of solution: the *enthalpy change* when one mole of a substance dissolves in a stated amount of water under standard conditions.

$NaCl(s) + aq \longrightarrow Na^+(aq) + Cl^-(aq)$ $\qquad \Delta H^{\circ}_{\text{solution}} = +4$ kJ/mol

In this equation "+ aq" is short for adding water.

The enthalpy change of solution is the difference between the energy needed to separate the ions from the crystal lattice (*lattice energy*) and the energy given out as the ions are hydrated (sum of the *hydration* enthalpies).

$\Delta H^{\circ}_{\text{solution}} = \Delta H^{\circ}_{\text{hydration cation}} + \Delta H^{\circ}_{\text{hydration anion}} + \Delta H^{\circ}_{\text{lattice}}$

$$Na^+(g) \quad + \quad Cl^-(g)$$

$$\Delta H^\circ_{lattice} = +788 \text{ kJ/mol}$$

$$\Delta H^\circ_{hyd}[Na^+] + \Delta H^\circ_{hyd}[Cl^-]$$
$$= -784 \text{ kJ/mol}$$

$$Na^+(aq) + Cl^-(aq)$$

$$\underline{Na^+Cl^-(s)} \; |+aq \quad \Delta H^\circ_{solution} = +4 \text{ kJ/mol}$$

Energy (enthalpy) level diagram for sodium chloride dissolving in water. Note that the enthalpy change for going from the crystal to the gaseous ions is the lattice energy. This is the second type of Born–Haber cycle for analyzing enthalpy changes involving ionic compounds.

Note that the enthalpy change of solution is a small difference between two large enthalpy changes. Also note that the lattice energy and the hydration enthalpies tend to be affected in the same way by changes in the sizes of the ions and their charges. The smaller the ions and the larger the charges the greater the lattice energy but also the greater the hydration enthalpies. This means that it is not easy to make use of enthalpy cycles to account for trends in solubilities of ionic compounds down group I or group II in the periodic table.

Also note that an ionic salt may well dissolve in water even if $\Delta H^\circ_{solution}$ is slightly positive. Dissolving makes a significant difference in the numbers of ways that particles and energy can be arranged. The *entropy change* of the system, ΔS°_{system} may be positive and large enough to outweigh a negative entropy change in the surroundings.

enthalpy change of vaporization: the *enthalpy change* when one mole of a liquid turns to a *vapor* at its boiling point. Evaporation separates the particles in a liquid so values for specific enthalpy of vaporization give a measure of the strength of the bonding between particles in liquids. Substances with strong ionic or metallic bonding have much higher boiling points and enthalpies of vaporization than substances consisting of molecules with weak *intermolecular forces*.

Physicists generally use specific latent heat values in studying the energy changes when liquids evaporate. Specific latent heats of vaporization are measured in joules per kilogram rather than joules per mole.

entity: a term which can refer to any distinct particle such as an atom, molecule, ion, electron or radical. The term is used in precise definitions such as the definition of *amount of substance* in moles. It is important to be precise when specifying a particular entity. These, for example, are four distinct chemical entities: a hydrogen atom, H; a hydrogen molecule, H_2; a hydrogen ion, H^+; and a hydride ion, H^-.

entropy (S): a thermochemical quantity which makes it possible to predict the direction of changes (see also *standard molar entropy*). Change happens in the direction which leads to a total increase in entropy. When considering chemical reactions it is convenient to calculate the total entropy change in two parts: the entropy change of the system and the entropy change of the surroundings.

$$\Delta S^\circ_{total} = \Delta S^\circ_{system} + \Delta S^\circ_{surroundings}$$

The entropy of a system measures the number of ways, W, of arranging the molecules and sharing the energy between the molecules. Nothing seems to be happening to a closed flask of gas kept at a constant temperature. *Kinetic theory* tells us, however, that the molecules are in rapid motion, colliding with each other and with the walls of the container. The gas stays the same while constantly rearranging its molecules and redistributing the energy. The "number of ways," W, is large for a gas. Gases generally have higher entropies than liquids which have higher entropies than solids.

The relationship between S and W was derived by Ludwig Boltzmann (1844–1906), the Austrian physicist who was the first person to explain the laws of thermodynamics in terms of the behavior of atoms and molecules in motion.

$$S = k \ln W$$

where S is the entropy of the system, k the Boltzmann constant and $\ln W$ the natural logarithm of the number of ways of arranging the particles and energy in the system.

The entropy change of the surroundings during a chemical reaction is determined by the size of the *enthalpy change*, ΔH, and the temperature, T. The relationship is:

$$\Delta S_{surroundings} = -\frac{\Delta H}{T}$$

The minus sign is included because the entropy change is bigger the more energy is transferred to the surroundings. For an exothermic reaction, which transfers energy to the surroundings, ΔH is negative, so $-\Delta H$ is positive.

For many exothermic reactions at about room temperature the value of $-\Delta H/T$ is much larger than ΔS_{system} which means that ΔS_{total} is positive. This explains why exothermic reactions generally tend to go (are *feasible*), but there are exceptions.

Chemists often find it more convenient to work with *free energy changes*.

environmental chemistry: the study of chemical changes in the environment. Chemists use a range of analytical methods to estimate the amounts of elements in the environment. They apply the theory of *reaction kinetics,* the *equilibrium law*, and *thermochemistry* to explain the changes they observe.

Environmental chemists use models to summarize their findings and to make predictions. The models show how the various elements cycle through the environment (see *carbon cycle* and *nitrogen cycle*). The models show the main reservoirs for an element and the size of the flows of an element from one reservoir to another.

Knowledge of the scale of natural changes shows whether or not human activity is likely to disturb natural changes locally, regionally or globally. So environmental chemistry helps scientists to assess the seriousness of various types of pollution (see *acid rain, incineration, landfill, ozone, greenhouse effect, steady state systems* and *waste*).

environmental issues are of growing concern to chemists, the chemical industry and all who use chemicals as scientists have shown the harm which some chemicals can do to living things. Analytical chemists have contributed greatly to our knowledge of what happens to chemicals in the environment by developing very sensitive analytical techniques, such as chromatography and spectroscopy, to measure minute traces of substances which were once undetectable.

A challenge for chemists in recent years has been to redesign chemical processes to reduce both the energy required and the waste by-products formed. This is both more profitable and less damaging to the environment.

enzymes are *protein* molecules which are the *catalysts* for biochemical reactions that would otherwise be very slow. Saliva contains the enzyme amylase which aids digestion by speeding up the *hydrolysis* of starch to *sugar*. Catalase is an enzyme which protects living cells from a buildup of hydrogen peroxide by speeding up its decomposition to water and oxygen. Some laundry detergents contains proteases – enzymes which break down the proteins such as blood on dirty clothes.

Enzymes are highly specific. Each enzyme catalyzes a particular reaction. An enzyme has an *active site* which is just the right shape and size for the *substrate* molecules. Each enzyme works best at a particular temperature and pH and is less effective under other conditions. (See also *inhibitor*.)

epoxy resins are used as adhesives. The *resin* is a sticky liquid *polymer* with relatively short chains.

Mixing the resin with a hardener starts a chemical reaction which *cross-links* the chains to make a hard, strong, glass-like polymer. The higher the temperature the faster the reaction.

epoxyethane: an important chemical intermediate used to manufacture *surfactants*, *solvents* and *lubricants*. Epoxyethane is made by mixing *ethene* with air or oxygen and passing the mixture at about 300°C under pressure over a *heterogeneous catalyst*. The catalyst is finely divided silver, spread on the surface of an inert material such as alumina. There is a highly exothermic *side reaction* between ethene and oxygen forming carbon dioxide and steam. This means that very careful temperature control is essential to prevent the process from running out of control.

$$2\,CH_2{=}CH_2(g) \;+\; O_2(g) \longrightarrow 2CH_2{-}CH_2(g)$$

The three-membered ring in the molecule is highly strained, so that epoxyethane is reactive. The reactivity makes epoxyethane a hazardous chemical, so it is generally made and stored where it will be used.

Epoxyethane reacts vigorously with water, producing a range of products depending on the conditions.

$nCH_2{-}CH_2$

H$_2$O → HO $\!+\!$ CH$_2$CH$_2$O $\!+\!_n$ H

When $n{=}1$ the product is 1,2-ethanediol which is used to make polyesters and is the main ingredient of *antifreeze*. When n is >4 the product may be useful as a solvent or a non-ionic surfactant

ROH alcohol → HO $\!+\!$ CH$_2$CH$_2$O $\!+\!_n$ R

When $n{=}1$ or 2 and ROH is a primary alcohol the product is a solvent

Reactions of epoxyethane with water and alcohols

equilibrium constant: an expression that can be used to predict the amounts of reactants and products present when a *reversible reaction* reaches a state of *dynamic equilibrium*.

In general, for a reversible reaction at equilibrium:

$$aA + bB \rightleftharpoons cC + dD$$

$$K_c = \frac{[C]^c[D]^d}{[A]^a[B]^b}$$

This is the form for the equilibrium constant, K_c when the *concentrations* of the reactants and products are measured in moles per liter. [A], [B] and so on are the equilibrium concentrations.

The concentrations of the chemicals on the right-hand side of the equation appear on the top line of the expression. The concentrations of reactants on the left appear on the bottom line. Each concentration term is raised to the power of the number in front of its formula in the equation.

For gas reactions it is often more convenient to use partial pressures as the measure of concentrations. The equilibrium constant is then K_p.

Equilibrium constants are constant for a particular temperature. The variation of K with temperature accounts for the *temperature effect on equilibria*.

Worked example

Calculate the value of K_c for the reaction: $PCl_5(g) \rightleftharpoons PCl_3(g) + Cl_2(g)$ given that when 8.4 mol of $PCl_5(g)$ is mixed with 1.8 mol $PCl_3(g)$ and allowed to come to equilibrium in a 10 L container the amount of $PCl_5(g)$ at equilibrium is 7.2 mol.

Notes on the method
Write down the equation. Underneath write first the initial amounts, then write the amounts at equilibrium. Use the equation to calculate the amounts not given.

Calculate the equilibrium concentrations given the volume of the container. Substitute in the expression for K_c.

Answer

Equation	$PCl_5(g)$	$\rightleftharpoons$	$PCl_3(g)$	+	$Cl_2(g)$
Initial amounts/mol	8.4		1.8		0
Equilibrium amounts/mol	7.2 (= 8.4 − 1.2)		1.8 + 1.2 = 3.0		1.2
Equilibrium concentrations/ mol/L	7.2 ÷ 10 = 0.72		3.0 ÷ 10 = 0.3		1.2 ÷ 10 = 0.12

$$K_c = \frac{[PCl_3(g)][Cl_2(g)]}{[PCl_5(g)]} = \frac{0.3 \text{ mol/L} \times 0.12 \text{ mol/L}}{0.72 \text{ mol/L}} = 0.05 \text{ mol/L}$$

equivalence point: the point during any *titration* when the amount in moles of one reactant added from a buret is just enough to react exactly with all of the measured amount of chemical in the flask as shown by the *balanced equation*.

In a well planned titration the color change observed at the *end point* corresponds exactly with the equivalence point.

errors of measurement are not mistakes but unavoidable differences between measured values and the true values. Often the true value is not known, so analysts have to assess the uncertainty in their measurements (see *uncertainty of measurement*).

There are random errors which cause repeat measurements to vary and to scatter around a mean value. Averaging a number of readings helps to take care of random errors. The smaller the random errors the more precise the data (see *precision of data*).

There are systematic errors which affect all measurements in the same way, making them all lower or all higher than the true value. Systematic errors do **not** average out. Identifying and eliminating systematic errors is important for increasing the *accuracy of data*. Systematic errors can be corrected by *calibration* with another instrument that is known to be more reliable, but may be too expensive for general use.

There are also errors arising from the choice of measuring instrument. The accuracy of an instrument such as a buret, pipet or measuring cylinder is limited by its quality and by how accurately the chemist can take a reading from the scale. Instrumental errors can be reduced by choosing a suitable measuring device.

esters: sweet-smelling compounds found in perfumes and in fruit *flavors*. Ethyl butanoate has the odor of pineapples. Ripe fruits contain mixtures of esters. Esters are also used as *plasticizers* and *solvents*.

The general formula for an ester is RCOOR', where R and R' are *alkyl* or *aryl* groups.

Ethyl acetate is only slightly soluble in water.

methyl acetate (methyl ethanoate) ethyl acetate (ethyl ethanoate) ethyl benzoate

Names and structures of esters

An ester forms when a carboxylic acid reacts with an alcohol in the presence of a drop of concentrated sulfuric acid to act as a catalyst. The reaction is reversible. It takes an excess of the alcohol to convert most of the acid to ester.

Formation of ethyl acetate from acetic acid and ethanol

Another way to make an ester is to mix an alcohol with either an *acyl chloride* or an *acid anhydride*.

Hydrolysis splits an ester into an acid and an alcohol. Acids or bases can catalyze the hydrolysis. Base catalysis is generally more efficient because it is not reversible. The acid is produced as its negative ion which does not react with the alcohol.

ester salt

Base catalyzed hydrolysis of ethyl acetate

Lithium aluminum hydride ($LiAlH_4$) in dry ether reduces esters to alcohols more easily than it reduces carboxylic acids.

$$CH_3C \overset{O}{\underset{OCH_2CH_3}{\diagup}} \xrightarrow[\text{dry ether}]{\text{LiAlH}_4 \text{ in}} CH_3CH_2OH$$

Reduction of an ester to an alcohol by lithium aluminum hydride

Compounds with more than one ester link include *triglycerides* in fats and oils as well as *polyester* fibers. Alkaline hydrolysis of fats and oils produces soaps.

Several compounds used in medicine are esters including *aspirin*, *acetominophen* and the local *anesthetics* novocaine and benzocaine. The insecticides malathion and pyrethrin are also esters.

ethanol: the best known member of the family of *alcohols*. It is the alcohol in beer, wine and spirits.

The chemical industry makes ethanol by the hydration of ethene in the presence of an acid catalyst. An alternative source of the alcohol is *fermentation* and in some countries this is the source of ethanol for use as a *biofuel*.

Ethanol can be used as a fuel on its own in motor vehicles, but engines have to be modified. Gasohol is a blend of 90% *gasoline* with 10% ethanol, which has the advantage that it can be used in unmodified engines. Gasohol allows countries to reduce crude oil imports by mixing conventional gasoline with ethanol made by fermenting sugars or starch from crops.

Methylated spirit consists of ethanol (alcohol) mixed with about 5% methanol to make it undrinkable. Industrial methylated spirit is used as a *solvent* commercially and in laboratories. Rubbing alcohol is industrial methylated spirit with other additives including castor oil. The methylated spirits sold in hardware stores as a solvent and fuel is additionally colored blue by a dye.

ethene (CH_2=CH_2): the simplest *alkene*. It is a reactive gas. The petrochemical industry makes ethene on a very large scale by *thermal cracking* of natural gas or naphtha from the *fractional distillation* of oil.

Ethene is the starting material for the manufacture of a vast range of petrochemicals. The main products are *addition polymers* such as polythene, PVC and polystyrene.

ethers: organic compounds with the general formula R—O—R', where R and R' are *alkyl* or *aryl* groups. The most familiar ether is diethyl ether which has a sweet smell and is an *anesthetic*.

$$CH_3-O-CH_3 \qquad CH_3CH_2-O-CH_2CH_3 \qquad CH_3-O-CH_2CH_3$$

dimethyl ether	diethyl ether	methoxyethane
(methoxymethane)	(ethoxyethane)	

Names and structures of ethers

Ethers are more *volatile* than alcohols because they have no hydrogen bonds. The vapor of diethyl ether can be hazardous because it forms explosive mixtures in air.

Ethers are also much less soluble in water than alcohols. Diethyl ether is a good solvent for covalent compounds but not for salts so it is one of the liquids used for *solvent extraction*.

Ethers are relatively unreactive because they do not have O—H bonds. They do not react with acids or alkalis, oxidizing agents, sodium or phosphorus pentachloride. This makes ethers, such as diethyl ether, useful as non-aqueous solvents for reactions such as *lithium aluminum hydride* (LiAlH$_4$) reductions and the formation and use of *Grignard reagents*.

eutectic mixture: the mixture with the lowest freezing point of all the possible mixtures. Metals can form eutectic mixtures as can solutions of salts such as sodium nitrate in water.

For pairs of metals, such as tin and lead, the eutectic mixture is the *alloy* with the lowest melting point of all the possible alloys. An alloy with the eutectic composition freezes (or melts) at one temperature like a pure metal. The melting point of the eutectic, however, is lower than the melting points of the pure metals.

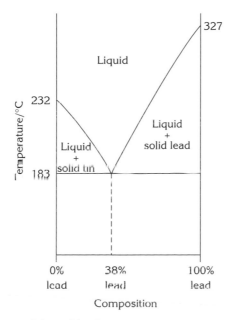

Phase diagram for mixtures of tin and lead

The eutectic mixture of tin and lead melts at 183°C. This low temperature makes the eutectic useful as solder.

eutrophication happens when the water in rivers or lakes is enriched by fertilizers from farmland or by nutrients from sewage and this makes it possible for algae to multiply rapidly. Thick layers of algae block out the light from plants growing below the surface so that they cannot produce oxygen as fast as usual. Then bacteria start to break down the mass of algae using up the remaining oxygen in the water. Other organisms such as fish die because they are starved of oxygen.

evaporation happens at the surface of a liquid as it turns to a gas. This is an example of a *change of state*. As a liquid such as water evaporates, the faster moving molecules near the surface escape from the pull of intermolecular forces and break away to become a vapor. Liquids tend to cool as they evaporate because the faster moving molecules break away so that the average kinetic energy of the remaining molecules falls.

Evaporation is an *endothermic* process. Energy must enter from the surroundings to keep a substance at a constant temperature as it evaporates. So liquids feel cold as they evaporate on the skin.

excited state: the state of an atom or molecule when one or more of its electrons is raised to a higher energy above the stable *ground state*. Heating, electricity or electromagnetic radiation can provide the energy to excite atoms or molecules. Energy is released back to the surroundings when the electrons fall back to the stable energy levels of the ground state. This happens during *flame tests*. Heat from a Bunsen flame excites the electrons in metal atoms from the sample. As the electrons drop back to the lower energy levels they give off radiation with a wavelength determined by the size of the energy jump (see *quantum theory*).

exothermic reactions give out energy to their surroundings. Freezing and condensing are exothermic *changes of state*.

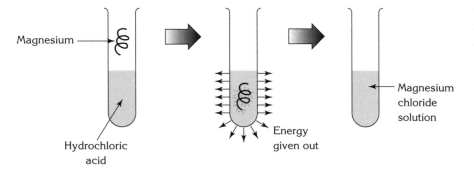

Diagrams to illustrate an exothermic reaction

For an exothermic reaction the enthalpy change, ΔH, is negative. During an exothermic reaction more energy is given out as new bonds form in the products than is needed to break bonds in the reactants.

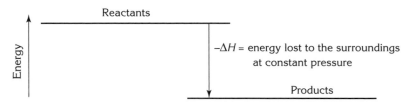

Energy level diagram *for an exothermic reaction*

Most *spontaneous* reactions are exothermic but there are exceptions so the *free energy change* (or total *entropy change*) must be examined to make reliable predictions about the direction and extent of change.

explosions are reactions which go with a bang. Gunpowder explodes if ignited in a confined space (see *fireworks*). The explosion is caused by the rapid buildup of pressure as the reaction produces a large volume of hot gas from a small volume of solid.

The "pop" during the test-tube test for hydrogen is a small explosion. This reaction explodes because it is a *radical chain reaction* with a step in which one radical reacts to produce two radicals which can then both react. The branching chain reaction produces more and more reactive radicals so the reaction goes faster and faster.

Explosives are compounds or mixtures of chemicals designed to produce explosions. Explosions are used in quarrying, mining, road building and to demolish old buildings. They are also used in warfare and as rocket propellants.

The Nobel prize was endowed with the wealth of the Swedish chemist Alfred Nobel (1833–96) who discovered that the dangerous explosive nitroglycerine could be used safely if absorbed in the mineral kieselguhr. He sold this product as "dynamite." He also developed gun cotton (cellulose nitrate) and gelignite (nitroglycerine in nitrocellulose).

High explosives only detonate if set off by a sudden shock which is usually supplied by a detonator. Mercury fulminate can be used as a detonator because it explodes instantly when ignited.

extrusion: a process for shaping a metal, such as *aluminum*, or a *plastic*, such as PVC, by forcing it through a shaped nozzle (die). Cooks use extrusion in a similar way when decorating a cake with colored frosting or when mincing meat. A plastics extruder works by melting pellets of the polymer and then using a rotating screw to push the molten material through a die.

face-centered cubic structure: one of the two *close packed* crystal structures of metals. Metals with the face-centered cubic structure are calcium, aluminum, cobalt, nickel and copper.

Faraday constant (F): the electric charge on one mole of electrons. The Faraday constant is used in calculations to determine the amount of change during *electrolysis*.

$$F = eL = 96\ 480\ \text{C/mol}$$

where e is the charge on one electron and L is the *Avogadro constant*. It is possible to measure F and e experimentally so this relationship can be used to arrive at a value for the Avogadro constant.

fats belong to the family of compounds call *lipids*. Fats are esters of glycerol (*triglycerides*) which are solid at around room temperature (below 20°C) because they contain a high proportion of saturated *fatty acids*. Solid triglycerides are generally found in animals. The triglycerides from plants are mostly liquids at around room temperature; these are *vegetable oils*.

fatty acids: carboxylic acids with long hydrocarbon chains which are combined with glycerol to make *triglycerides* in *fats* and *vegetable oils*. *Saturated* fatty acids do not have double bonds in the hydrocarbon chain. There is one or more double bond in an *unsaturated* fatty acid.

Fatty acid	Formula	Type
palmitic	$CH_3(CH_2)_{14}COOH$	saturated
stearic	$CH_3(CH_2)_{16}COOH$	saturated
oleic	$CH_3(CH_2)_7CH = CH(CH_2)_7COOH$	unsaturated
linoleic	$CH_3(CH_2)_4CH = CHCH_2CH = CH(CH_2)_7COOH$	unsaturated

Animal fats have a higher proportion of saturated fats in their triglycerides. In lard, for example, the main fatty acids are palmitic acid (28%), stearic acid (8%) and only 56% oleic acid. In olive oil the main fatty acids are oleic acid (80%) and linoleic acid (10%).

Polyunsaturated fats contain fatty acids with two or more double bonds in the chain.

Some fats or oils provide fatty acids which are essential to the human diet, including linoleic and linolenic acids.

f-block elements: the *lanthanide* and *actinide elements* in periods 6 and 7 of the periodic table. For these elements the last electron added to the atomic structure goes into one of the seven *f*-orbitals in the fourth or fifth shells. For convenience two rows of *f*-block elements are usually shown underneath the *periodic table* to cut down its overall width.

feasibility: a reaction is feasible if it tends to happen – even if the reaction is very, very slow. In this sense the reaction between methane (in natural gas) and oxygen is feasible at room temperature but, because of a high *activation energy*, the reaction is so slow that it does not happen. Put a match to the mixture of air and gas and the reaction is rapid.

The test of feasibility is *free energy change* for the reaction. If ΔG is negative, the reaction tends to happen.

For many chemical reactions, ΔG approximately equals ΔH, so chemists often use the sign of ΔH as a guide to feasibility. This can be misleading if the *entropy change* for the reaction system is large.

For redox reactions, *standard electrode potentials* offer an alternative way of deciding whether or not reactions tend to happen. $\Delta G \propto -E_{cell}$, so if the cell emf is positive the redox reaction in the cell will tend to happen.

For other reactions such as acid–base reactions, the easiest guide to the direction and extent of change is the equilibrium constant, K_c where $\Delta G \propto -\ln K_c$.

What this shows is that ΔG, E_{cell} and K_c values can all answer the same questions for a reaction:

- Will the reaction happen?

- How far will it go?

In practice chemists use the quantity which is most convenient to measure. Knowing the value of one of the three quantities it is possible to calculate the other two because they are all related.

It is always important to bear in mind that even if a reaction is feasible it may be very slow.

feedstock: the raw material for a process in the *chemical industry*.

Fehling's solution is used to distinguish *aldehydes* from *ketones* and to identify *reducing sugars*. Aldehydes and reducing sugars give a positive result. Ketones and other sugars do not. The deep blue solution turns greenish and then loses its blue color as an orange-red precipitate of copper(I) oxide appears.

Fehling's reagent does not keep so it is made when required by mixing two solutions. One solution is copper(II) sulfate in water. The other solution is a solution of tartrate ions in strong alkali. The tartrate salt is included to form a complex with copper(II) ions so that they do not precipitate as copper(II) hydroxide with the alkali.

fermentation converts sugars to alcohol (ethanol) and carbon dioxide. Fermentation is an example of *anerobic respiration*. Fermentation is catalyzed by a group of enzymes from yeast referred to as zymase. In brewing the aim is to produce alcohol in beer. In baking fermentation produces the gas bubbles which make the dough rise.

$$C_6H_{12}O_6(aq) \longrightarrow 2CO_2(g) + 2C_2H_5OH(aq)$$

glucose   carbon   ethanol
                  dioxide

ferrites: compounds such as the magnetic oxide of iron, Fe_3O_4, which can be used to make permanent magnets. They are *ferromagnetic*. Ferrites are *ceramics* so that they can be molded into complex shapes before being fired. They also have the great advantage that they do not conduct electricity.

ferromagnetism: the property of materials which can be used to make permanent magnets. The three ferromagnetic metals are iron, cobalt and nickel. Alnico is a magnetic alloy of these three metals.

fertilizers supply plants with the mineral salts they need for growth. Plants need three major nutrients: nitrogen, phosphorus and potassium which must be available in the soil in a soluble form so that they can be taken up by roots.

Manure and compost are traditional fertilizers used in "organic farming." They release nutrients slowly as they decompose through the action of bacteria and fungi in the soil.

Intensive agriculture relies on fertilizers made from minerals and the air. "Straight fertilizers" contain one of the three elements nitrogen (N), phosphorus (P) or potassium (K). "Compound fertilizers" contain two or more of these elements.

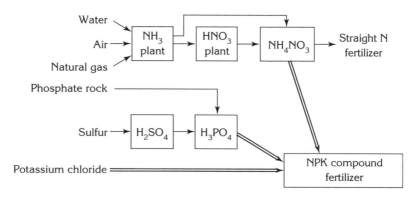

Flow diagram for manufacture of nitrogen and NPK compound fertilizers

The use of manufactured fertilizers greatly increases crop yields. However the use of large amounts of fertilizer increases the risk of nutrients leaching from the soil into the rivers, lakes and groundwater. This can lead to *eutrophication.*

fiber reactive dyes are strongly bound to the fabrics they dye because they react with the molecules in the fibers, forming covalent bonds. They are fast *dyes* which do not fade during washing.

Structure of a fiber reactive dye which forms covalent bonds with amino groups in the protein *molecules of wool*

fibers: long thin threads which make up materials such as asbestos, wool and polyester fabrics. Most of the fibers used for textiles consist of natural or synthetic polymers. Wool, silk and hair are natural fibers made of *protein.* Cotton, linen and kapok are plant fibers made of the *carbohydrate* cellulose.

Semisynthetic fibers are made by modifying cellulose. Heating cellulose from wood with sodium hydroxide and then carbon disulfide produces a solution called "viscose". Forcing this solution through very small holes in a spinneret produces fine filaments. The filaments turn back into cellulose fibers as they flow though a tank of sulfuric acid. Cloth made from these fibers is known as rayon.

Treating cellulose from wood or cotton with acetic anhydride produces cellulose acetate. Cellulose acetate fibers are made by spinning a solution of the polymer in acetone. Fibers form as the solvent evaporates.

Examples of synthetic polymers used to make fibers are *polyester*, *polyamides* and the *addition polymer*, polypropylene. Synthetic polymers are spun into fibers by forcing the hot molten polymer through a spinneret. Fibers form as the liquid cools and solidifies.

Fibers of inorganic materials are increasingly important for making *composite* materials. Examples are the fibers of glass and graphite.

fillers: materials mixed with *plastics* or *rubbers* to modify their properties. Fillers can make plastics stronger, tougher, harder or easier to mold. Fillers are also used to cut costs by bulking up plastics with something cheaper.

filtration separates an insoluble solid from a liquid. The liquid which passes through the filter is the filtrate. The solid retained on the paper is the residue.

Filtering through a funnel with a filter paper folded into a cone can be very slow. For efficient filtration it is important that the paper is folded carefully and fitted accurately into the funnel. The paper is then wetted with water, or other solvent, before starting to filter. The funnel should never be filled to the top of the paper.

One way of filtering faster is to fold the circle of paper more times to produce a fluted filter paper so that the whole surface is available for filtration.

An even quicker method is *vacuum filtration*.

The use of a small plug of mineral wool or cotton wool in a small funnel cuts losses when filtering off a drying agent from an organic liquid before final distillation.

fire extinguishers put out fires by excluding air or by cooling to slow down burning. The choice of fire extinguisher depends on the type of fire. Water is the cheapest way to put out burning wood, paper or rubbish. Water cools the fire as it evaporates and produces steam which helps to exclude air. Water is dangerous on electrical fires and useless on burning hydrocarbons or metals. Burning oils float on water and continues burning so water just spreads the fire. Burning metals react vigorously with water. Alternative fire extinguishers use carbon dioxide gas or nonflammable liquids such as halons and dry powders.

fireworks make *redox reactions* entertaining. The main active ingredient of any firework is gunpowder. Gunpowder is a mixture of two *fuels*, carbon and sulfur, with an *oxidant*, potassium nitrate. Gunpowder burns rapidly because the fuels are finely powdered and well mixed with the source of oxygen. Gunpowder explodes with a bang when confined because as the fuels burn they produce a large volume of the gases carbon dioxide and sulfur dioxide (see *explosions*). One cubic centimeter of gunpowder burns fast to make about 300 cm^3 of hot gas. The blue fuse which sets off fireworks is paper impregnated with potassium nitrate so that it will burn even when it is windy or damp.

The colors in fireworks come from tiny pellets of salts coated with gunpowder. The colors are the same as in flame tests: sodium for yellow, copper for blue-green and strontium for red.

first law of thermodynamics: see *thermodynamics (laws of)*.

first order reaction: a reaction is first order with respect to a reactant if the rate of reaction is proportional to the concentration of that reactant. The concentration term for this reactant is raised to the power one in the *rate equation*.

$$\text{Rate} = k\,[X]^1 = k\,[X]$$

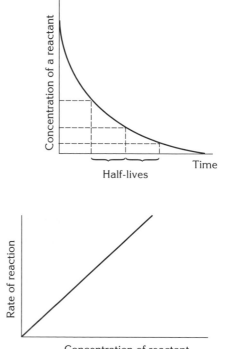

Variation of concentration of a reactant plotted against time for a first order reaction The gradient of this graph at any point measures the rate of reaction. The half-life *for a first order reaction is a constant so it is the same wherever it is read off the curve. It is independent of the initial concentration.*

Variation of reaction rate with concentration for a first order reaction. The graph is a straight line through the origin showing that the rate is proportional to the concentration of the reactant.

The rate of reaction of 2-bromo-2-methylpropane with hydroxide ions is first order with respect to the haloalkane but *zero order* with respect to hydroxide ions.

$$\text{Rate} = k[(CH_3)_3CBr]$$

fission means splitting. The word is used in two contexts in chemistry: bond breaking (*homolytic* and *heterolytic bond breaking*) and the splitting of atomic nuclei (nuclear fission).

flame tests help to detect some metal ions in salts. An element gives a colored flame if its *atomic spectrum* includes lines in the visible region of the spectrum.

Flame tests are particularly useful in qualitative analysis to distinguish metal ions which otherwise behave in a similar way. Both magnesium and calcium ions, for example, are precipitated by sodium hydroxide to give white precipitates which are insoluble in excess of the alkali. Calcium compounds give an orange-red flame but magnesium compounds do not color a flame.

Metal ion	Color
lithium	bright red
sodium	bright yellow
potassium	pale mauve
calcium	orange–red
strontium	scarlet
barium	yellowish–green
copper(II)	blue–green

flammable substances are hazardous because they easily catch fire in air.

Substances can be considered highly flammable if they:

- may spontaneously catch fire in air
- are liquids with a *flash point* below 21°C but above 0°C
- are solids which may catch fire and keep burning after brief contact with a flame
- are gases which burn in air if ignited at normal pressure
- react to form flammable gases (such as hydrogen) in contact with water or water vapor.

flash point: the lowest temperature at which a small flame can ignite the *vapor* from a volatile liquid or solid.

flavorings: mixtures of chemicals which give food, candy, toothpaste and other products their flavors. Many natural fruit flavors are subtle blends of *esters* and other chemicals. Chemists can imitate fruit flavors by mixing esters.

The *stereochemistry* of a molecule can affect its flavor.

float glass process: the industrial process for manufacturing large, uniform and smooth sheets of *glass*. Molten glass from a furnace at 1500°C pours onto a bath of molten tin where it stays liquid long enough for both surfaces to become flat and parallel. The glass cools and solidifies as it moves over the bath of tin. Once the glass is solid at about 600°C it can move onto rollers and gradually travel through an oven where it cools slowly to room temperature. This is *annealing*.

fluids: materials which flow. Liquids or gases are fluids because the atoms or molecules are not held in fixed positions but are free to move around.

fluoridation involves adding traces of fluoride compounds to water supplies to prevent tooth decay. The benefits were discovered in areas where water supplies naturally contain fluoride ions.

fluorine (F): a pale yellow gas made up of F_2 molecules. It is the most reactive of the halogens in *group VII* of the periodic table. Its electron configuration is [He] $3s^23p^5$.

Fluorine is the most *electronegative* of all elements so it forms ionic compounds with metals. Fluorine is the most powerful *oxidizing agent* and its oxidation state is –1 in all its compounds. It oxidizes other elements to their highest positive oxidation state. Sulfur, for example, forms SF_6.

Uses of fluorine include the manufacture of a wide range of compounds consisting of only carbon and fluorine (see *fluorocarbons*). The most familiar of these is the very slippery, non-stick *addition polymer* poly(tetrafluorethene), better known as PTFE.

fluorite structure: the cubic crystal structure of the ionic compound calcium fluoride, CaF_2. Each positive ion is surrounded by eight nearest neighbors at the corners of a cube and each negative ion is surrounded by four positive ions. So the *coordination number*s are 8 for the positive ion and 4 for the negative ion.

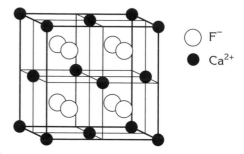

Structure of calcium fluoride showing 8:4 coordination. The structure consists of a face-centered cubic array of positive ions with negative ions in the tetrahedral holes

Other compounds with this structure are the fluorides of Sr, Ba, Cd, Pb and Hg.

fluorescent substances can absorb energy from ultraviolet (or other) radiation and immediately re-emit the energy as visible light. The inside surfaces of fluorescent lamp bulbs or tubes are coated with compounds such as magnesium tungstate and zinc silicate. These substances fluoresce when irradiated by the ultraviolet light from the mercury vapor in the lamp.

The optical brighteners in *detergents* are fluorescent. They absorb ultraviolet light from the Sun and re-emit the energy as blue light. This compensates for any yellowing of the fabric with age.

Fluorescence is an example of *luminescence*.

fluorocarbons: compounds of fluorine and carbon only. They are very *inert*, thermally stable and nonflammable. They are also nontoxic. They are electrical and thermal insulators. They are used during the manufacture and testing of electronic components and also as refrigerants, coolants and lubricants. One fluorocarbon is a good solvent for oxygen and, in emergencies, can act as "artificial blood."

foams consist of gases finely dispersed in liquids or solids. Foams are *colloids*. The mass of bubbles in a liquid foam holds the liquid and gas in place. This is useful in foam fire extinguishers which can blanket a fire with carbon dioxide held by the foam. It is also useful in shaving cream which retains the liquid soap on the skin.

Solid foams can be rigid but have a low density. Expanded polystyrene and foamed polyurethanes are examples. These materials are excellent thermal insulators because they trap a large volume of gases and are very poor thermal conductors.

Foams improve the texture of food including bread, meringues and cakes. In bread the gas bubbles form by *fermentation*, in cakes they are produced by heating baking powder and in meringues by whipping air into egg white before cooking.

food additives: natural or synthetic chemicals added to food, which can act as:

- colors to make food look more attractive
- preservatives to prevent the growth of microorganisms
- antioxidants to prevent oxygen from the air from making food unfit to eat – oxidation, for example, turns fats and oils rancid
- *emulsifiers* and stabilizers to control food texture
- sweeteners to improve the taste (see *sweetness*).

forensic science: the application of scientific techniques to gather evidence for a court of law. Forensic scientists make increasing use of sophisticated analytical techniques such as *mass spectrometry, chromatography* and various forms of *spectroscopy*.

formation constants, K_{stab}: equilibrium constants for the formation of *complex ions*. The greater the value of the formation constant the more stable the complex. The values for formation constants show that *chelate* complexes formed by bidentate and hexadentate *ligands* are more stable than complexes formed by monodentate ligands.

Complex	K_{stab}
$[Ag(NH_3)_2]^+$	1.6×10^7
$[Cu(NH_3)_4]^{2+}$	1×10^{12}
$[Ni\ en_3]^{2+}$ where en is the bidentate ligand ethylene diamine	2×10^{18}
CuY^{2-} where Y is the hexadentate ligand *EDTA*	6×10^{18}

Formation constants show the position of equilibrium when a new ligand replaces water molecules in the hydrated ions.

$$[Co(H_2O)_6]^{2+}(aq) + 6NH_3(aq) \rightleftharpoons [Co(NH_3)_6]^{2+}(aq) + 6H_2O(aq)$$

For simplicity this is often written as:

$$Co^{2+}(aq) + 6NH_3(aq) \rightleftharpoons [Co(NH_3)_6]^{2+}(aq)$$

The usual rules for writing an equilibrium constant and its units apply.

$$K_{stab} = \frac{[Co(NH_3)_6^{2+}]}{[Co^{2+}][NH_3]^6} = 1 \times 10^5$$

formula: used by chemists to represent the composition and structure of elements and compounds. (See *empirical formula, molecular formula, structural formula* and *stick structure.*)

formulation: the list of ingredients for a mixture of chemicals such as for a *medicine*. Medical drugs are not usually pure chemical compounds. The other additives in the formulation may have a range of purposes such as making the drug soluble, suspending it in water, thickening the mixture, preserving the medicine or improving its flavor.

fossil fuels: nonrenewable energy resources which we now burn to utilize the energy from the Sun which was stored up millions of years ago by *photosynthesis*. The fossil fuels are coal, oil and natural gas.

Fossil fuels contribute to the formation of *acid rain* when they burn. The fuels contain varying amounts of sulfur compounds. The proportion of sulfur compounds in crude oil, for example, varies from less than 1% up to 7%, depending on its origin. When crude oil is distilled, sulfur compounds tend to be concentrated in the heavy fractions such as fuel oils. Oil refining removes much of this *sulfur* and there is very little in gasoline. Sulfur compounds burn to produce sulfur dioxide. Nitrogen oxides form by direct combination of nitrogen with oxygen in the air as fuels burn at high temperatures in furnaces or engine cylinders.

fractional distillation: a method for separating mixtures of liquids with different boiling points. On a laboratory scale, the process takes place with a distillation apparatus fitted with a glass fractionating column between the flask and the still-head. Separation is improved if the column is packed with inert glass beads or rings to increase the surface area where rising vapor can mix with condensed liquid running back to the flask. The column is hotter at the bottom and cooler at the top. The thermometer reads the boiling temperature of the compound passing over into the condenser.

Suppose the flask contains a mixture of two liquids. The boiling liquid in the flask produces a vapor which is richer in the more volatile of the liquids (the one with the lower boiling point).

Most of the vapor condenses in the column and runs back. As it does so it meets more of the rizing vapor. Some of the vapor condenses. Some of the liquid evaporates. In this way the mixture repeatedly evaporates and condenses as it rises up the column. It is like carrying out a series of simple distillations. This can be represented as a series of steps in the boiling-point composition diagram.

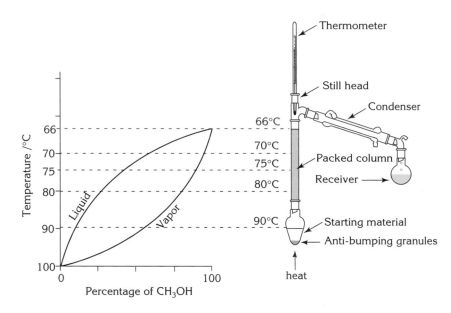

Boiling-point composition diagram for mixtures of methanol and water. Methanol boils at a lower temperature so it the more volatile substance. Boiling any mixture of the liquids produces a vapor which is richer in methanol.

fractional distillation of oil: a large-scale, *continuous process* for separating crude oil into fractions. A furnace heats the oil which then flows into a fractionating tower containing about 40 "trays" pierced with small holes. Condensed vapor flows over the trays and runs down into the tray below. Rizing vapor mixes with liquid on a tray as it bubbles up through the holes.

The column is hotter at the bottom and cooler at the top. Rizing vapor condenses when it reaches the tray with liquid at a temperature below its boiling point. Condensing vapor releases energy which heats the liquid in the tray which in turn evaporates the more volatile compounds in the mixture.

Fractional distillation of crude oil

With a series of trays the outcome is that the *hydrocarbons* with small molecules rise to the top of the column while larger molecules stay at the bottom. Fractions are drawn off from the column at various levels.

Some components of crude oil have boiling points too high for them to vaporize at atmospheric pressure. Lowering the pressure in a *vacuum distillation* column reduces the boiling points of the hydrocarbons and makes it possible to separate them.

fragmentation patterns: see *mass spectrometry.*

free energy change, ΔG: the thermochemical quantity used by chemists to decide whether a reaction tends to happen and how far it will go. ΔG is the test for the *feasibility* of a reaction. If ΔG is negative the reaction is feasible.

The advantage of ΔG values for chemists is that tables of standard free energies of formation can be used to calculate the standard free energy change for any reaction. The calculations follow exactly the same steps as the calculations to calculate standard *enthalpy changes for reactions* from standard *enthalpy changes of formation.*

The idea of *entropy* underlies the quantity "free energy." Any change tends to happen if the total entropy change is positive. Working with entropy values is generally less convenient because of the need to consider the entropy changes in the surroundings as well as in the reaction mixture. This was the reason Willard Gibbs suggested the concept "free energy," which he defined by: $\Delta G = -T\Delta S_{total}$

$$\Delta S_{total} = \Delta S_{system} + \Delta S_{surroundings}$$

For a change at constant temperature and pressure:

$$\Delta S_{surroundings} = \frac{-\Delta H}{T}$$

and so: $\Delta S_{total} = \Delta S_{system} + \dfrac{-\Delta H}{T}$

Hence $-T\Delta S_{total} = -T\Delta S_{system} + \Delta H$

From Gibb's definition this becomes: $\Delta G = \Delta H - T\Delta S_{system}$

Often the $T\Delta S_{system}$ term is relatively small compared to the enthalpy change so that:

$\Delta G \approx \Delta H.$

This is the reason that chemists often use ΔH values to decide whether or not a reaction will tend to go. The approximation becomes less justified at higher temperatures when T is bigger and so $T\Delta S_{\text{system}}$ is bigger.

freezing point: the temperature at which a liquid turns to a solid. A pure liquid has a sharp freezing point which is the same as its melting point. A *phase* diagram for the liquid shows how the freezing point varies with pressure.

Dissolving a solute in a liquid lowers its freezing point. In dilute solutions the lowering of freezing point is directly proportional to the *molality* of the solution.

> freezing depression = $k_f \times$ molality

The constant k_f is the freezing point constant of the solvent.

Adding antifreeze to the water in an engine lowers the freezing point and so prevents the coolant from freezing in winter. When there is a threat of ice, salt is scattered on roads because a mixture of salt and water freezes at temperatures well below 0°C.

frequency of electromagnetic radiation: the number of complete waves passing any point per second. The SI unit of frequency, ν, is the hertz (Hz). Frequencies of *electromagnetic radiation* vary from about 10^5 Hz for radiowaves up to 10^{20} Hz for gamma rays.

All electromagnetic radiation travels at the same speed, c, in a vacuum. The *frequency*, *wavelength*, λ, and speed are simply related by: $c = \nu\lambda$.

Quantum theory shows that the higher the frequency the higher the energy of the quanta (photons) of radiation.

Friedel–Crafts reaction: a method for forming C—C bonds to build up the carbon skeleton by adding a side chain to an *arene* such as benzene. In a Friedel–Crafts reaction a *haloalkane* or an *acyl chloride* undergoes an *electrophilic substitution* reaction with an arene. The reaction takes place in the presence of a catalyst such as aluminum chloride (a *Lewis acid*).

One example is the substitution of a group with three carbon atoms in place of a hydrogen atom in the benzene ring, using 2-chloropropane.

Friedel–Crafts alkylation with a chloroalkane. Benzene reacting with 2-chloropropane.

The reaction with an acyl chloride produces a *ketone*.

Friedel–Crafts acylation with an acyl chloride. Benzene reacting with acetyl chloride.

This important type of reaction was discovered and developed jointly by the French organic chemist Charles Friedel (1832–1899) and the American, James Crafts (1839–1917). The reaction is used both on a laboratory and on an industrial scale. A Friedel–Crafts reaction to make ethyl benzene is the first step in the production of the *addition polymer* polystyrene.

The purpose of the of the aluminum chloride is to produce the electrophile which attacks the benzene ring.

Formation of the electrophile

Electrophilic attack and substitution

The mechanism of Friedel–Crafts alkylation

froth flotation: a process for separating valuable mineral from a rock made up of several minerals. The rock is finely crushed to separate the minerals and then stirred with a solution of *surfactants* as a stream of air bubbles rises up through the mixture. The surfactants are selected to make sure that the particles of metal *ore* are caught up by the bubbles and rise to the surface where they can be skimmed off. The unwanted minerals (or gangue) sink and flow away as a stream of waste.

fuel cell: an electrochemical cell which is continuously supplied with fuel and oxidizing agent. A fuel cell produces electric power from a fuel, directly, without having to burn it and then uses the energy to drive a turbine and spin a generator. In a power station about 70% of the energy released when fuels burn is wasted. A fuel cell running on hydrogen or methanol and oxygen provides a much more efficient source of electricity.

Schematic diagram of a hydrogen–oxygen fuel cell. The cell continues to supply electric power so long as it is supplied with the hydrogen fuel and with oxygen.

Hydrogen–oxygen fuel cells with strong alkali as the electrolyte are used in the space shuttle. In proton-exchange membrane cells the electrolyte is a polymer which can conduct hydrogen ions in one direction from one side to the other. In this type of cell the electrodes are made of conducting carbon cloth impregnated with platinum, which acts as the catalyst.

fuels burn in air or oxygen to release energy. Power stations often use the *combustion* of *fossil fuels* to raise steam and generate electricity. Most of the fuels for transportation are made from fractions produced by the *fractional distillation of crude oil*. There is growing interest in the use of *biofuels*.

Gasoline is a blend of *hydrocarbons* based on the gasoline fraction (hydrocarbons with 5–10 carbon atoms). Jet fuel is produced from the kerosene fraction (hydrocarbons with 10–16 carbon atoms). Fuel for diesel engines is made from diesel oil (hydrocarbons with 13–25 carbon atoms).

Fuels have to be refined to remove components which would harm engines or cause air pollution when they burn.

The energy density of a fuel is the amount of energy available from a kilogram of fuel. This is especially important for fuels used for transportation. The higher the energy density of fuels the smaller the mass of the fuel in the tanks of motor vehicles or airplanes.

The energy density of a fuel can be calculated given its standard enthalpy change of combustion. The energy density of hexane (a hydrocarbon in gasoline) is 48 400 kJ/kg.

Ethanol is a biofuel used on its own, or in gasohol. ΔH_c°[ethanol] = −1367 kJ/mol.

The molar mass of ethanol, C_2H_5OH = 46 g/mol.

So burning 46 g (= 4.6×10^{-2} kg) ethanol gives out 1367 kJ energy.

The energy density of ethanol = 1367 kJ ÷ 4.6×10^{-2} kg = 29 700 kJ/kg.

full equation: see *chemical equations*.

fullerenes: molecular *allotropes* of carbon. The best known example is the "bucky ball," *buckminsterfullerene*, C_{60}. Other fullerenes have formulas C_{28}, C_{32}, C_{50}, and C_{70}.

functional group: the atoms and bonding which give a series of organic compounds its characteristic properties and is responsible for most of the reactions. The hydrocarbon chain which makes up the rest of any organic molecule is generally inert to most common reagents such as acids and alkalis.

Examples of functional groups include the:

- two carbon atoms joined by a double bond in *alkenes*
- —Cl in chloroalkanes
- —OH in *alcohols*
- —NH_2 in *amines*
- —COOH in *carboxylic acids*.

fundamental particles: the particles that make up an *atom*. In chemistry the fundamental particles are protons, neutrons and electrons. Physicists with high-energy particle accelerators have shown that protons and neutrons are not fundamental but in turn made up of quarks.

	Relative mass	Charge
proton	1	+1
neutron	1	0
electron	$\dfrac{1}{1870}$ (negligible)	−1

fusion means either melting or joining. (See *enthalpy change of melting* and *nuclear fusion*.)

galvanic cell: an *electrochemical cell* in which a *spontaneous* chemical reaction is used to generate an electric current.

gamma radiation (γ): high-energy *electromagnetic radiation* given off during *radioactive decay*. The new nucleus formed when a radioactive atom emits an *alpha* or *beta particle* is often in an excited state; it gives off gamma rays as it loses energy. Emission of gamma rays does not cause a change in the structure of the nucleus.

The wavelengths of gamma rays are shorter than those of *X-rays*. Gamma radiation is *ionizing radiation* which is more penetrating than alpha or beta radiation and is stopped only by several centimeters of lead.

gas: see *kinetic theory of gases*.

gas constant: the constant R in the *ideal gas* equation $PV = nRT$. The value of the constant depends on the units used for pressure and volume. If all quantities are in SI units with the pressure in N/m^2 (pascals) and volume in cubic meters (m^3), then $R = 8.314$ J/K·mol.

gas laws: the laws which describe the behavior of *ideal gases* and are summarized by the ideal gas equation. The gas laws include *Boyle's law*, *Charles' law* and *Avogadro's law*.

gas–liquid chromatography (GLC): a sensitive analytical technique for analyzing mixtures of liquids. In a modern gas–liquid apparatus, the stationary phase is a thin film of liquid *adsorbed* on the inside surface of a coiled capillary tube about 30 meters long inside an oven. A sample is injected into the hot column from a syringe. The mobile phase, which is a gas, carries the vapors through the column. The components in the mixture separate as

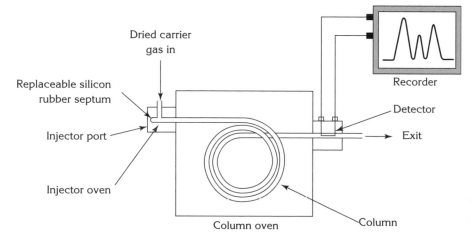

The main features of gas–liquid chromatography

they pass through the column. They are detected as they emerge and the signal from the detector is fed to a chart recorder.

The chart from the recorder shows how long it took each component of the mixture to pass through the column. This is the retention time. The areas under the peaks on the printout give a measure of the proportions of the components in a mixture.

A GLC machine can be calibrated by injecting measured amounts of known compounds and recording their retention times.

Applications of GLC include:

- tracking down the source of oil pollution from the pattern of peaks, which acts like a fingerprint for any batch of oil
- measuring the level of alcohol in urine or blood samples from drivers
- detecting and measuring pesticides in river water.

gasoline consists mainly of a mixture of *hydrocarbons* and is the fuel for motor engines. Gasoline has to be carefully blended if modern engines are to start reliably and run smoothly. The proportion of *volatile* hydrocarbons added to gasoline is higher in winter to help cold starting but lower in summer to prevent vapor forming before the fuel gets to the carburetor.

The starting point for making gasoline is the gasoline fraction from the *fractional distillation* of oil. Hydrocarbons in this fraction are mainly *alkanes* with 5 to 10 carbon atoms. Fractional distillation does not provide enough gasoline and produces more than enough of the heavier fractions, so oil refineries run *catalytic crackers* to make more hydrocarbons of the right size.

For smooth running gasoline has to burn smoothly without *knocking*. The octane number of a fuel measures its performance. The higher the compression of fuel and air in the engine cylinders, the higher the octane number has to be to stop knocking.

The octane number scale was devised by Thomas Midgley (1889–1944) and he discovered antiknock additives based on lead, which were used for many years. Leaded fuel is now being phased out. There are two main objections to adding lead compounds to gasoline. The first is that lead compounds are *toxic* and harmful, especially to young children, when released into the air from car exhausts. The second is that lead compounds "poison" the catalysts in *catalytic converters*.

The oil companies now produce high octane fuel by increasing the proportions of both branched alkanes and arenes plus blending in some oxygen compounds. The four main approaches are:

- *cracking* which not only makes more small molecules but also forms hydrocarbons with branched chains
- *isomerization* which turns straight chain alkanes into branched chain compounds by passing them over a platinum catalyst
- *reforming* which turns cyclic alkanes into *arenes* such as benzene and methyl benzene
- adding *alcohols* or *ethers* such as MTBE (methyl tertiary butyl ether).

Structure of the ether MTBE which has an octane number of 120. Adding MTBE and other methods raise the octane number of gasoline from about 70 to 95 as required for unleaded gasoline.

$$CH_3 - \underset{\underset{CH_3}{|}}{\overset{\overset{CH_3}{|}}{C}} - O - CH_3$$

gas tests are used to identify gases produced during *qualitative analysis*.

Gas	Test	Observations
Hydrogen	Burning splint	Burns with a "pop"
Oxygen	Glowing splint	Splint bursts into flame (relights)
Carbon dioxide	Limewater (aqueous calcium hydroxide)	Turns milky white
Hydrogen chloride (hydrogen bromide and iodide react in the same way)	Smell Blue litmus Ammonia vapor (from a drop of concentrated ammonia solution on a glass rod)	Pungent Turns red Thick white smoke
Chlorine	Color Smell Effect on moist blue litmus paper Moist starch-iodide paper	Greenish-yellow Pungent – bleach-like Turns the paper red and then bleaches it Turns blue-black
Sulfur dioxide	Smell Blue litmus Acid dichromate	Pungent Turns red Turns green
Hydrogen sulfide	Smell Burning splint Lead(II) acetate	"Rotten eggs" Gas burns – yellow deposit of sulfur Turns brown-black
Ammonia	Smell Red litmus	Pungent Turns blue
Nitrogen dioxide	Color Blue litmus	Orange-brown Turns red
Water vapor	Appearance Anhydrous cobalt(II) chloride paper	"Steams" in the air Turns from blue to pink

gas volume calculations use the gas laws and equations to calculate the volumes of gases involved in reactions. Gas volume calculations are straightforward when the reactants and products are all gases. Equal volumes of gases contain equal amounts in moles when measured under the same conditions – this is *Avogadro's law*. As a result the ratio of the gas volumes in a reaction must be the same as the ratio of the numbers of moles in the equation (see *Gay-Lussac's law of combining volumes*).

Worked example

What volume of oxygen reacts with 60 cm³ methane and what volume of carbon dioxide forms if all gas volumes are measured under the same conditions?

Notes on the method

Write the balanced equation.

Note that below 100°C the water formed condenses to an insignificant volume of liquid.

Answer

The equation for the reaction is:

$$CH_4(g) + 2O_2(g) \longrightarrow CO_2(g) + 2H_2O(l)$$

$\quad$ 1 mol $\qquad$ 2 mol $\qquad\qquad$ 1 mol

So 60 cm³ methane reacts with 120 cm³ oxygen to form 60 cm³ carbon dioxide.

The other approach to gas volume calculations is based on the fact that the volume of a gas, under given conditions, depends only on the amount of gas in moles. It does not matter which gas is involved so long as the gas is behaving (at least approximately) like an *ideal gas*.

$\quad$ volume of gas (cm³) – amount of gas (mol) × molar volume (cm³/mol)

The molar volume for gases at stp (273 K and 101.3 kPa) is 22 400 cm³.

Worked example

What volume of hydrogen is produced under standard conditions when 1.95 g zinc reacts with excess acid?

Notes on the method

Start by writing the equation for the reaction. Convert the quantity of zinc to an amount in moles.

Answer

The equation for the reaction is:

$$Zn(s) + 2HCl(aq) \longrightarrow ZnCl_2(aq) + H_2(g)$$

The amount of zinc – 1.95 g ÷ 65 g/mol = 0.03 mol

$\quad$ 1 mol zinc produces 1 mol hydrogen.

Volume of hydrogen = 0.03 mol × 22 400 cm³/mol = 672 cm³

Gay-Lussac's law of combining volumes states that when gases are involved in reactions, the ratios of the volumes of gases are simple whole numbers so long as all measurements are taken at the same temperature and pressure. For example 50 cm³ hydrogen react with 50 cm³ chlorine to form 100 cm³ of hydrogen chloride. The ratios are 1:1:2. The law is named after the French chemist Joseph Gay-Lussac (1778–1850).

Avogadro's law accounts for Gay-Lussac's observations. If equal volumes of gases contain equal numbers of molecules, under the same conditions, it follows that the ratios of the volumes are also the ratios of the number of molecules as the equation shows.

$$H_2(g) + Cl_2(g) \longrightarrow 2HCl(g)$$

gel: a *colloid* in which a liquid is finely dispersed in a solid. Table jelly is a typical gel with water loosely held in a network of large gelatine molecules. Warming or even shaking can break up the solid network so that the gel liquefies.

geometrical isomerism: molecules with the same molecular and structural formulas but different shapes (geometries). *Alkenes* and other compounds with C = C *double bonds* may have geometrical isomers because there is no rotation around the double bond. The isomers are labelled *cis* and *trans*. In the *cis* isomer, similar functional groups are on the same side of the double bond. In the *trans* isomer similar functional groups are on opposite sides of the double bond.

cis 2-butene   trans 2-butene

Geometric isomers of 2-butene. They are distinct compounds with different melting points, boiling points and densities.

Some inorganic *complexes* also have geometrical isomers.

cis isomer   trans isomer

Cis and trans isomers of the planar complex $[Ni(NH_3)_2Cl_2]$. The cis form is blue–violet. The trans form is green.

The anti-cancer drug, *cisplatin* is the *cis* isomer of a flat (planar) complex.

germanium (Ge): an element with some metallic and some nonmetallic features. It is a metalloid; shiny like a metal but hard and brittle like a nonmetal. It is the element below silicon but above tin in *group IV* of the periodic table with the *electron configuration* $[Ar]3d^{10}4s^24p^2$. Germanium is a *semi-conductor*.

Gibbs free energy: see *free energy change*. The concept is named after the American physical chemist, Josiah Gibbs (1839–1903). Hence the symbol ΔG.

glasses: *ceramic* materials which are rigid like solids but which are not crystalline. Glass is made by melting one or more oxides in a furnace. The liquid glass is cooled until thick enough to mold and then shaped and cooled further until it sets solid.

It is possible to make glass from pure silicon dioxide (SiO_2) but it melts at 1700°C. Molten silicon dioxide is a thick, sticky liquid (*viscous*). As the liquid cools back to its melting point, the atoms cannot move freely enough to return to the ordered arrangement of crystalline silica. They become "frozen" into a disordered state.

Most glass is made from silica mixed with other oxides which melt at a lower temperature than pure silica. Windows, bottles and drinking glasses are made of soda lime glass. Lead glass is used for decorative cut glassware. *Borosilicate glass* is used to make ovenware and laboratory glassware.

166

glass electrode: an electrode used in combination with a *reference electrode* to make a pH probe for a pH meter. A glass electrode has a thin walled bulb made of special glass which responds to changes in pH. A salt bridge connects the glass electrode to a reference electrode to make a complete electrochemical cell.

global warming: the gradual warming of the Earth's atmosphere which many scientists believe is leading to climate change. The theory is that global warming is the result of an increased concentration of gases such as carbon dioxide in the air which is enhancing the *greenhouse effect*.

Graham's law of effusion says that the rate of *effusion* of a gas is *inversely proportional* to the square root of its *molar mass* at constant temperature and pressure.

$$\frac{\text{rate of effusion of gas A}}{\text{rate of effusion of gas B}} = \sqrt{\left(\frac{\text{molar mass of B}}{\text{molar mass of A}}\right)}$$

This means that a gas with low molar mass such as hydrogen ($M_r = 2$) effuses faster than a gas with a greater molar mass such as oxygen ($M_r = 32$). According to Graham's law, hydrogen molecules effuse four times faster than oxygen molecules. A balloon filled with hydrogen deflates much more rapidly than a balloon filled with oxygen.

graphite: one of the *allotropes* of carbon, consisting of a giant structure of atoms. Carbon has by far the highest melting point (over 3800 K) of any element. Graphite has an almost ideal combination of physical properties for use as a high temperature *ceramic* except that it has to be used in a non-oxidizing environment. Graphite bonded with clay is used to make crucibles for melting metals. Large blocks of graphite are used as *refractories* to line furnaces.

gravimetric analysis: a method of quantitative analysis for finding the composition and formulas of compounds based on accurate weighing of reactants and products. A familiar example, in school, is to find the formula of magnesium oxide by weighing a piece of magnesium in a crucible and then weighing the magnesium oxide produced after strong heating in air. Chemists have developed sophisticated and precise methods of gravimetric analysis which include *combustion analysis* to find the formulas of organic compounds.

green chemistry involves the design of chemical processes and products that reduce or eliminate the use and generation of hazardous substances. So the aim of green chemistry is to decrease pollution. This includes the development of methods of clean *synthesis* which are more selective and so produce fewer by-products that end up as waste. Research into new *catalysts* has an important role to play in this aspect of green chemistry.

Another aspect to green chemistry is the move to nontoxic solvents which are less harmful if released into the environment. This helps to cut the use of harmful *volatile organic compounds*.

greenhouse effect: the effect of some gases in the air which keeps the surface of the Earth about 30°C warmer that it would be if there were no *atmosphere*. Without the greenhouse effect there would be no life on Earth.

When radiation from the Sun reaches the Earth's atmosphere about 30% is reflected into space, 20% is absorbed by gases in the air and about half reaches the surface of the Earth.

The warm surface radiates energy back into space but at longer, infrared wavelengths. Some of the infrared radiation is absorbed and warms up the atmosphere. This is the greenhouse effect.

Grignard reagents

The gases in the air which absorb infrared radiation are called greenhouse gases. Nitrogen and oxygen make up most of the air but they are not greenhouse gases. The main natural greenhouse gases are carbon dioxide, methane, dinitrogen oxide and water vapor. *CFCs* are also greenhouse gases.

The concentrations of greenhouse gases in the air are rising because of human activity such as burning *fossil fuels* and agriculture. This is enhancing the greenhouse effect. There is growing evidence that this is responsible for *global warming*.

Grignard reagents: reactive organic compounds containing magnesium atoms which are used to form $C-C$ bonds in organic synthesis. They are examples of *organometallic compounds*. The reagents were discovered by the French organic chemist Victor Grignard (1871–1935).

Grignard showed that *haloalkanes* react with magnesium in dry ether (ethoxyethane).

$$RBr + Mg \longrightarrow RMgBr,$$ where R represents an alkyl group.

The $C-Mg$ bond in a Grignard reagent is polarized with the metal atom at the positive end of the dipole and the carbon atom at the negative end. As a result, the carbon atom attached to magnesium is nucleophilic and $C-C$ bonds form when Grignard reagents take part in *nucleophilic substitution* or *nucleophilic addition* reactions. This is what makes the reagents so useful in synthesis.

Summary of the reactions of Grignard reagents

ground state: the stable state of an atom or molecule with the electrons in the lowest available energy levels.

group: a vertical column of elements in the *periodic table*. Elements in the same group have similar chemical properties because they have the same outer *electron configuration*.

There are trends in properties down a group as the number of full inner electron shells increases and the atoms get larger. Generally the metallic characteristics of elements increase down a group. In *groups I and II* the metals get more reactive down the group.

Nonmetal characteristics decrease down a group. Down *group VII* the halogens get less reactive. *Group IV* shows a trend from nonmetals at the top to metals at the bottom with germanium, a *metalloid*, in the middle.

IUPAC now recommends that the groups should be numbered from 1 to 18. Groups I and II become groups 1 and 2. Groups 3 to 12 are the vertical families of *d-block elements*, the groups traditionally numbered III to VIII then become groups 13 to 18.

group I elements belong to the family of *alkali metals* with similar chemical properties because they all have one electron in an outer *s-orbital*. These elements are more similar to each other than the elements in any other group. Even so, because of the increasing number of full, inner shells, there are trends in properties down the group from lithium to cesium. The element in period 7, francium, is very rare and all its isotopes are radioactive.

lithium, Li	[He]2s^1	rubidium, Rb	[Kr]5s^1
sodium, Na	[Ne]3s^1	cesium, Cs	[Xe]6s^1
potassium, K	[Ar]4s^1		

The metals are powerful *reducing agents*. They react by losing the outer s electron to form M^+ ions. The first ionization energies decrease down the group as the increasing number of full shells means that the outer electrons get further away from the same effective nuclear charge (see *shielding*).

Atomic and ionic radii increase down the group. For each element the 1+ ion is smaller than the atom because of the loss of the outer shell of electrons. The tendency to react and form ions increases down the group.

The small size of the lithium ion means that it has a relatively high *polarizing power* and is heavily *hydrated* in solution. As a result lithium is in some ways not typical of the group as a whole. It resembles magnesium in some ways. This is an example of the *diagonal relationship*.

The metals are soft and easily cut with a knife. They are shiny when freshly cut but quickly dull in air as they react with moisture and oxygen. The metals are stored under oil.

All the metals burn brightly in oxygen on heating to form ionic oxides. Lithium forms a simple oxide, Li_2O. Sodium forms mainly the *peroxide*, Na_2O_2. Potassium forms a superoxide, KO_2. The oxides are basic; they react with acids to form salts.

$$Li_2O(s) + 2HCl(aq) \longrightarrow 2LiCl(aq) + H_2O(l)$$

All the metals react with water to form hydroxides and hydrogen. The rate and violence of the reaction increases down the group. Lithium reacts steadily with cold water. Cesium reacts explosively.

All the metals react vigorously with chlorine to form colorless, ionic chlorides, M^+Cl^-, which are soluble in water. The crystal structures depend on the size of the metal ion (see *cesium chloride structure* and *sodium chloride structure*).

The hydroxides are:

* similar in that they all have the formula MOH, and are soluble in water forming alkaline solutions (they are *strong bases*)

* different in that their solubility increases down the group.

The carbonates are similar in that they all have the formula M_2CO_3, and, with the exception of lithium, do not decompose on heating (see *thermal stability of carbonates*). The carbonates of sodium and potassium are soluble, forming alkaline solutions because the carbonate ion is a *base*.

169

The nitrates are:

- similar in that they all have the formula MNO_3, are colorless crystalline solids, are very soluble in water and decompose on heating,

- different in that they become more difficult to decompose down the group. Lithium nitrate, like magnesium nitrate, decomposes on heating to the oxide, nitrogen dioxide and oxygen. The nitrates of sodium and potassium need strong heating to decompose and they form the nitrite:

$$2KNO_3(s) \longrightarrow 2KNO_2(s) + O_2(g)$$

group II elements belong to the family of *alkaline earth metals* with similar chemical properties because they all have two electrons in an outer *s-orbital*.

beryllium, Be [He]$2s^2$	*strontium*, Sr [Kr]$5s^2$
magnesium, Mg [Ne]$3s^2$	*barium*, Ba [Xe]$6s^2$
calcium, Ca [Ar]$4s^2$	

The first member of the group, beryllium, is not a typical member of group II. Because of the small size of the Be^{2+} ion, its chemistry is in many ways more like the chemistry of aluminum than of magnesium. This is an example of the *diagonal relationship*.

The following similarities and differences refer to the elements Mg, Ca, Sr and Ba. The symbol M is here used to represent any one of these elements.

The metals are *reducing agents* which react by losing their two s electrons to form M^{2+} ions. The first and second *ionization energies* decrease down the group as the increasing number of full shells means that the outer electrons are further away from the same effective nuclear charge (see *shielding*).

Atomic radii and *ionic radii* increase down the group. For each element the 2+ ion is smaller than the atom because of the loss of the outer shell of electrons. The tendency to react and form ions increases down the group.

The metals are harder and denser than group I metals and have higher melting points. In air the surface of the metals is covered with a layer of oxide

All the metals burn brightly in oxygen on heating to form white, ionic oxide, $M^{2+}O^{2-}$. The oxides are basic. They react with acids forming salts.

$$BaO(s) + 2HNO_3(aq) \longrightarrow Ba(NO_3)_2(aq) + H_2O(l)$$

All the metals react with water to form hydroxides and hydrogen or acids to form salts and hydrogen. Magnesium reacts only slowly with hot water. Barium reacts quite fast even with cold water. Barium is so reactive with air and moisture that it is generally stored under oil like the alkali metals.

All the metals react vigorously with chlorine to form colorless, ionic chlorides, MCl_2. Unlike the group I chlorides, the chlorides of this group are usually *hydrated*. They are soluble in water.

The hydroxides are:

- similar in that they all have the formula $M(OH)_2$ and are to some degree soluble in water, forming alkaline solutions

- different in that their solubility increases down the group.

The carbonates are:

- similar in that they all have the formula MCO_3, are insoluble in water and decompose on heating: e.g. $CaCO_3(s) \longrightarrow CaO(s) + CO_2(g)$
- different in that they become more difficult to decompose down the group (see the *thermal stability of carbonates*).

The nitrates are:

- similar in that they all have the formula $M(NO_3)_2$, are colorless crystalline solids, are very soluble in water and decompose to the oxide on heating:

$$2Mg(NO_3)_2 \longrightarrow 2MgO(s) + 4NO_2(g) + O_2(g)$$

- different in that they become more difficult to decompose down the group.

The sulfates are:

- similar in that they are all colorless solids with the formula MSO_4
- different in that they become less soluble down the group.

group IV: a group of elements which shows a trend from *nonmetals* at the top to *metals* at the bottom. All these elements have four electrons in their outer shell. The characteristic oxidation states of the group are +4 and +2. The +2 state becomes more important down the group and is the more stable state of lead. This is an example of the *inert pair effect*.

carbon, C	[He]$2s^22p^2$	tin, Sn	[Kr]$4d^{10}5s^23p^2$
silicon, Si	[Ne]$3s^23p^2$	lead, Pb	[Xe]$4d^{14}6s^23p^2$
germanium, Ge	[Ar]$3d^{10}4s^23p^2$		

The two common *allotropes* of carbon (diamond and graphite) consist of covalent giant structures. Silicon and germanium have diamond-like giant structures and are *semiconductors*.

The room temperature allotrope of tin has a metallic structure. The low temperature allotrope, grey tin has the nonmetallic diamond structure. Lead has a metallic structure.

The two metals, tin and lead, have lower melting points than the nonmetals with giant structures.

The group IV compounds in the +4 state generally behave more like the compounds of nonmetals:

- the +4 oxides (CO_2, SiO_2, SnO_2 and PbO_2) are acidic or, in the case of tin and lead, amphoteric with a bias towards *acidic oxide* behavior
- the +4 chlorides are molecular liquids which (with the exception of CCl_4) are rapidly hydrolyzed by water.

The compounds of tin and lead in the +2 state are more typical of metallic compounds:

- the +2 oxides (SnO and PbO) are *amphoteric* but with a bias towards *basic oxide* properties
- the +2 chlorides are white solids: $PbCl_2$ is an ionic solid.

The +4 state is more stable for tin, so that tin(II) compounds are reducing agents. Tin(II) chloride reduces iron(III) to iron(II) for example.

The +2 state is the more stable state for lead. So PbO_2 is a strong oxidizing agent. Lead(IV) oxide will oxidize chloride ions to chlorine gas.

group VII elements belong to the family of *halogens* with similar chemical properties because they all have seven electrons in the outer shell – one less than the next noble gas in group 8. The element *astatine* in group VII is very rare and highly radioactive. The halogens are all toxic.

| *fluorine*, F | [He]$2s^22p^5$ | *bromine*, Br | [Ar]$3d^{10}4s^24p^5$ |
| *chlorine*, Cl | [Ne]$3s^23p^5$ | *iodine*, I | [Kr]$4d^{10}5s^25p^5$ |

The symbol X is often used to represent a halogen.

The halogens all consist of diatomic molecules, X_2, linked by a single *covalent bond* and all are volatile. *Intermolecular forces* increase down the group as the numbers of electrons in the atoms increase, so melting points and boiling points rise down the group. Iodine atoms are the most polarizable. Fluorine and chlorine are gases at room temperature, bromine is a liquid and iodine a solid.

Halogen atoms are highly electronegative. They form ionic compounds or compounds with *polar covalent bonding*. *Electronegativity* decreases down the group. Bonding in aluminum fluoride is ionic, but anhydrous aluminum chloride is a covalent solid.

The halogens are powerful *oxidizing agents*. Fluorine is the strongest oxidizing agent and iodine the weakest in the group. (See *displacement reactions*.)

Hot iron burns in chlorine forming iron(III) chloride. The metal reacts much less vigorously with iodine forming iron(II) iodide. (See also *halide ions* and *hydrogen halides*.)

group VIII: a family of colorless, unreactive elements called the noble gases. The gases exist as single atoms. The *intermolecular forces* between the atoms are very weak, so at room temperature they behave very much like *ideal gases*. They have very low melting and boiling points. The strength of the forces between the atoms increases as the number of electron shells increases. The larger atoms are more *polarizable*, so melting points and boiling points rise down the group.

First *ionization energies* fall down the group as the outer electrons get further away from the nucleus. The ionization energies of the elements near the bottom of the group are low enough for them to be able to take part in chemical reactions and form compounds.

Helium, *neon* and *argon* do not form compounds with other elements. *Krypton* combines with fluorine. *Xenon* forms compounds with fluorine and oxygen.

Haber process: the process for *ammonia manufacture* developed by the German physical chemist Fritz Haber (1868–1934).

half-cell: an electrode dipping into a solution of ions. Two half-cells connected by a salt bridge make up an *electrochemical cell*.

half-equation: an *ionic equation* used to describe either the gain or the loss of electrons during a redox process. Half-equations help to show what is happening during a *redox reaction*. Two half-equations combine to give an overall balanced equation for a redox reaction.

Iron(III) ions can oxidize iodide ions to iodine. This can be shown as two half-equations:

electron gain (reduction): $Fe^{3+}(aq) + e^- \longrightarrow Fe^{2+}(aq)$

electron loss (oxidation): $2I^-(aq) \longrightarrow I_2(s) + 2e^-$

The number of electrons gained must equal the number lost. So the first half-equation must be doubled to arrive at the overall balanced equation:

$2Fe^{3+}(aq) + 2e^- \longrightarrow 2Fe^{2+}(aq)$

$\underline{2I^-(aq) \longrightarrow I_2(s) + 2e^-}$

$2Fe^{3+}(aq) + 2I^-(aq) \longrightarrow 2Fe^{2+}(aq) + I_2(s)$

half-life (radioactive): the time for half the atoms in a sample of a radioactive isotope to decay away. It is also the time for the count rate for alpha or beta particles from a sample to fall by half. Half-lives for radioactive isotopes can be as short as a fraction of a second or as long as millions of years. The half-lives of radioactive isotopes are unaffected by changes in temperature or pressure or the presence of catalysts. The half-life remains the same whether the atoms are in the elemental state or in one of its compounds.

half-life (chemical): chemists measure the half-lives of chemical reactions. The half-life of a reaction is the time for the concentration of one of the reactants to fall by half. Half-lives help to identify first order reactions. At a constant temperature, the half-life of a *first order reaction* is the same wherever it is measured on a concentration/time graph. So the half-life is independent of the initial concentration.

halide ions: the ions of the halogen elements. They include the fluoride, F^-, chloride, Cl^-, bromide, Br^- and iodide, I^-, ions. (See also *silver halides*.)

Warming sodium chloride with concentrated sulfuric acid produces clouds of hydrogen chloride gas. This *acid–base* reaction can be used to make hydrogen chloride.

$NaCl(s) + H_2SO_4(l) \longrightarrow HCl(g) + NaHSO_4(s)$

Both sulfuric acid and hydrogen chloride are strong acids. The reaction goes from left to right because the hydrogen chloride is a gas and escapes from the reaction mixture, so the reverse reaction cannot happen.

This type of reaction cannot be used to make hydrogen bromide or hydrogen iodide, because bromide and iodide ions are strong enough reducing agents to reduce sulfur from the +6 state to lower *oxidation states*.

The reactions of halide ions with sulfuric acid show that there is a trend in the strength of the halide ions as *reducing agents*:

- chloride ions do not reduce sulfuric acid at all, so the only gaseous product is hydrogen chloride gas
- bromide ions turn to orange bromine molecules as they reduce H_2SO_4 to SO_2 (mixed with some hydrogen bromide gas)
- iodide ions are the strongest reducing agents; they turn into iodine molecules as they reduce H_2SO_4 to S and H_2S; scarcely any hydrogen iodide forms.

So the trend as reducing agents is: $I^- > Br^- > Cl^-$. In *group VII*, iodine is the weakest oxidizing agents so it has the least tendency to form negative ions. Conversely iodide ions are the ones that most readily give up electrons and turn back into iodine molecules. (See also *displacement reactions*.)

hallucinogens: drugs which powerfully affect the brain and cause people to experience hallucinations. Some hallucinogens occur naturally in plants such as the cactus, peyote, which produces mescaline. Other hallucinogens are synthetic chemicals such as LSD and MDMA (the *amphetamine* called ecstasy). There is much that is not understood about the effects of drugs on the brain but it seems that chemicals such as LSD and MDMA affect the nerve cells which release a chemical transmitter called serotonin. Part of the scientific case against the use of MDMA is that there is evidence from studies on animals that it can damage and destroy the nerve fibers of cells that produce serotonin in the brain.

haloalkanes: compounds formed by replacing one or more of the hydrogen atoms in *alkanes* with halogen atoms.

Names and structures of some haloalkanes. A primary haloalkane has the halogen atom at the end of the chain. A secondary compound has the halogen atom somewhere along the chain but not at the ends. A tertiary haloalkane has the halogen atom at a branch in the chain.

Chloromethane, bromomethane and chloroethane are gases at room temperature. Most other haloalkanes are colorless liquids which do not mix with water.

Halogen atoms are more electronegative than carbon atoms so a carbon—halogen bond is polar. The characteristic reactions of haloalkanes are *nucleophilic substitution* reactions.

$CH_3CH_2CH_2CH_2Br$

heat with NaOH(aq) or NaOH(aq) → $CH_3CH_2CH_2CH_2OH$ alcohol

heat with KCN in ethanol → $CH_3CH_2CH_2CH_2CN$ nitrile

heat with NH_3 in ethanol → $CH_3CH_2CH_2CH_2NH_2$ primary amine (secondary and tertiary amines also form)

Nucleophilic substitution reactions of 1-bromobutane

The *hydrolysis* of haloalkanes makes it possible to distinguish chloro-, bromo- and iodo-compounds. Heating the compound with alkali releases halide ions. Acidifying with nitric acid and then adding silver nitrate produces a precipitate of the silver halide.

The rates of reaction of haloalkanes are in the order: RI > RBr > RCl where R represents an alkyl group.

The C—I bond is the longest and the weakest (as measured by the mean *bond energy*). The C—Cl bond is the shortest and the strongest. Bond polarity does not appear to be a factor in determining the rates because chlorine is the most electronegative of the elements so the C—Cl bond is the most polar.

An *elimination reaction* happens on warming a haloalkane with a solution of potassium hydroxide in ethanol. The reaction goes more readily in compounds with a halogen atom at a branch in the carbon chain.

Haloalkanes are important intermediates in the manufacture of other chemicals. They are also used as:

• solvents (e.g. dichloromethane)
• refrigerants (e.g. hydrofluorocarbons, which are replacing *CFCs*)
• pesticides (e.g. bromomethane)
• fire extinguishers (e.g. CBr_2ClF).

There are growing restrictions on the uses of many haloalkanes because of concern about their hazards to health, their persistence in the environment and their effect on the *ozone layer*.

halogenation: see *chlorination*, *bromination* and *iodination*.

halogens: the family name of the *group VII* elements. The halogens are *fluorine*, *chlorine*, *bromine*, *iodine* and *astatine*. The name halogen means "salt-former" and is based on the fact that the elements combine with most metals to form salts (*halides*).

hardness: a property of materials which shows how easy they are to dent or scratch (see *Mohs' scale*). The term is also used in the expression *hard water*.

hard water does not easily lather with *soap* but instead forms a greasy scum. Water is hard if it contains calcium or magnesium ions. Scum is a *precipitate* formed when soap mixes with water containing these ions.

$$Ca^{2+}(aq) + 2C_{17}H_{35}CO_2^-(aq) \longrightarrow Ca(C_{17}H_{35}CO_2)_2(s)$$

stearate ions in soap   insoluble precipitate (scum)

Water is temporarily hard if it contains the hydrogencarbonates of calcium or magnesium. This type of hardness is removed by boiling. Boiling reverses the reaction which produced the hard water as rain containing carbon dioxide trickled through limestone, chalk or dolomite.

$$Ca(HCO_3)_2(aq) \underset{\text{formation of hard water}}{\overset{\text{boiling}}{\rightleftharpoons}} CaCO_3(s) + H_2O(l) + CO_2(g)$$

The solid calcium carbonate precipitates as scale which coats heating elements and gradually blocks the pipes in heating systems.

Permanent hardness is not removed by boiling. The mineral gypsum, $CaSO_4$, is slightly soluble in water and makes water permanently hard.

Ion exchange resins and other methods of *water treatment* soften water by removing calcium and magnesium ions.

hazard warning signs identify chemicals which can harm people. The signs have been agreed upon internationally and they have very specific meanings. Words beside or below the signs indicate the degree of hazard. Very toxic and *toxic substances*, for example, carry the same sign but are distinguished by the words alongside.

 Toxic   Corrosive

Oxidizing   Flammable

Hazard warning signs for chemicals

heat exchangers are widely used in the chemical industry to transfer energy from a hot liquid or gas to a cooler liquid or gas.

A laboratory water condenser is a simple example of a heat exchanger. The water flowing in the jacket around the inner tube cools the hot vapor from a *distillation* column. The water flows in the opposite direction to the vapor.

An industrial heat exchanger is normally made of steel rather than glass and has many tubes carrying the hotter *fluid* through the surrounding container of cooler fluid.

Heat exchangers are important to the economics of industry. In *sulfuric acid manufacture* by the contact process the reaction of sulfur dioxide with oxygen to make sulfur trioxide is highly exothermic. The gases must be cooled after each pass through the *catalyst* to maximize the yield. Instead of letting the energy go to waste, the hot gases are cooled in heat exchangers in which water turns to steam. The steam is then used to generate electricity. Typically a sulfuric acid plant has no fuel bills because it can generate all the electricity it needs. This helps to make the process commercially viable.

heavy metals: metals such as cadmium, mercury and lead which have relatively high relative atomic masses. The term does not have a precise chemical meaning but crops up in descriptions of pollution caused by *toxic* heavy metal ions.

heavy water: the common name for deuterium oxide, D_2O. This is water in which both the atoms linked to oxygen are the *deuterium* isotope of hydrogen.

helium (He): the first member of the group of noble gases, *group VIII*. Helium is the second most common element in the universe after hydrogen. The energy of stars such as the Sun comes from *nuclear fusion* which turns hydrogen nuclei into helium nuclei. Helium was detected on the Sun by *spectroscopy* during a total eclipse in 1868 over 20 years before it was isolated and identified on Earth.

On Earth one of the main sources of helium is *natural gas*. Some natural gas wells produce up to 7% helium. The gas boils at 4.2 K and is the coldest liquid available to study the properties of materials at low temperatures.

Helium is used to provide an inert atmosphere for welding and for growing crystals of pure *semiconductors* such as germanium and silicon.

Divers breathe a mixture of helium with oxygen. Helium is less soluble in blood than nitrogen and so this mixture reduces the risk of divers suffering from the "bends" when they come to the surface.

Helium is less dense than air under the same conditions of temperature and pressure. It is a much safer gas than hydrogen for filling airships and weather balloons.

hemoglobin: the red globular *protein* in blood which carries oxygen from the lungs to the cells in body tissues. A hemoglobin molecule consists of four *polypeptide* chains, each with a nitrogen atom forming a dative bond to an Fe^{2+} ion in a heme group. So there are four heme groups in each hemoglobin molecule.

A heme group

Nitrogen atoms in the porphyrin ring form four more dative bonds to the Fe^{2+} ion, leaving one remaining site on the metal ion which can accept a pair of electrons from an oxygen molecule (in oxyhemoglobin which is bright red) or a water molecule (in deoxyhemoglobin which is dull red).

The reactions between heme groups and oxygen or water are reversible, allowing hemoglobin to pick up and release oxygen. The reaction with *carbon monoxide* is irreversible, which explains why the gas is dangerously toxic.

Henderson–Hasselbalch equation: an equation which helps chemists to explain the behavior of *buffer solutions* and indicators. The equation is derived by rearranging the expression for K_a of a weak acid and then putting both sides of the equation into logarithmic form. The advantage of taking logs is to produce an equation which can be easily used to calculate *pH* values, which are themselves logarithmic.

For a weak acid HA:

$$HA(aq) + H_2O(l) \rightleftharpoons H_3O^+(aq) + A^-(aq)$$

$$K_a = \frac{[H_3O^+(aq)][A^-(aq)]}{[HA(aq)]}$$

This rearranges to give:

$$[H_3O^+(aq)] = \frac{K_a[HA(aq)]}{[A^-(aq)]}$$

Taking logs and substituting pH for $-lg\ [H_3O^+(aq)]$ and pK_a for $-lg\ K_a$, gives:

$$pH = pK_a + lg\ \frac{[A^-(aq)]}{[HA(aq)]} \quad \text{because} \quad -lg\ \frac{[HA(aq)]}{[A^-(aq)]} = +lg\ \frac{[A^-(aq)]}{[HA(aq)]}$$

In a mixture of a weak acid and its salt, the weak acid is only slightly ionized while the salt is fully ionized, so it is often accurate enough to assume that all the anions come from the salt present and all the unionized molecules from the acid.

Hence: $pH = pK_a + lg\ \frac{[salt]}{[acid]}$, which is the Henderson–Hasselbalch equation.

So in a buffer solution with [acid] = [salt], $pH = pK_a + lg\ 1 = pK_a$, since $lg\ 1 = 0$.

Diluting a buffer solution does not change [salt]/[acid] so the pH does not change.
The equation also shows why it takes a big change in the ratio [salt]:[acid] in a buffer solution to make a significant change in pH. Since $log\ 1 = 0$, $log\ 10 = 1$ and $log\ 100 = 2$, the ratio has to change by a factor of ten to make the pH change by one unit.

The equation also explains why *acid–base indicators* change color roughly in the range $pH = pK_{in} \pm 1$. When $pH = pK_a$, [HIn] = [In$^-$] and the two different colors of the indicator are present in equal amounts the indicator is midway through its color change.

Add a few drops of acid and the pH falls. The characteristic acid color of the indicator is distinct when $[HIn] \approx 10 \times [In^-]$. At this point $pH = pK_a + log\ 0.1 = pK_a - 1$, since $log\ 0.1 = -1$

Add a few drops of alkali and the pH rises. The characteristic alkaline color of the indicator is distinct when $[In^-] \approx 10 \times [HIn]$. At this point $pH = pK_a + log\ 10 = pK_a + 1$, since $log\ 10 = +1$.

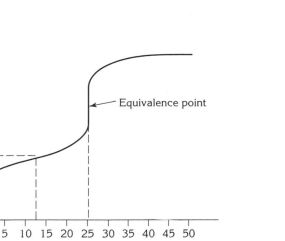

Curve showing the change of pH when a 25 cm³ 0.1 mol/L acetic acid, a weak acid, is titrated with a strong base. Halfway to the end point half the acid has been converted to its salt, so [acid] = [salt] and pH = pK$_a$ = 4.8

Henry's law describes the effect of changing the pressure on the solubility of a gas in a liquid. The law applies only to gases which are slightly soluble. It does not apply to the solubility in water of gases such as ammonia or hydrogen chloride which react as they dissolve.

The law states that the mass of gas that dissolves in a certain volume of a liquid at constant temperature varies with the *partial pressure* of the gas above the liquid.

The law applies to the solubility of gases such as nitrogen, oxygen and *helium* in water and can be used to explain the dangers faced by divers breathing gas mixtures under pressure.

Henry's law is a special case of the *equilibrium law* and can be described in the following form since the partial pressure of a gas is a measure of its concentration:

$$\frac{\text{concentration of gas in solution at equilibrium}}{\text{concentration of gas above the liquid}} = K \text{ (a constant)}$$

herbicides: agrochemicals which kill weeds and other unwanted plants. Some herbicides, such as glyphosate, are broad spectrum which means that they kill almost every plant they contact. Others are narrow spectrum meaning that they are more selective: atrazine is effective for controlling weeds in corn fields while allidochlor is selective for grasses.

Hess's law makes it possible to calculate enthalpy changes which cannot be measured. The law states that the enthalpy change for a reaction is the same whether it takes place in one step or a series of steps. So long as the reactants and the products are the same the overall enthalpy change will be the same whether the reactants are converted to products directly or through two or more intermediate reactions.

Hess's law is used to calculate:

- *enthalpy changes of formation* from enthalpy changes of combustion
- *enthalpy changes of reaction* from enthalpy changes of formation
- lattice energies in *Born–Haber cycles*.

heterocyclic compounds: compounds with rings made of carbon atoms with atoms of at least one other element such as nitrogen, sulfur or oxygen. The bases in *DNA* are heterocyclic compounds. Adenine is an example. Many *alkaloids* are also heterocyclic compounds.

The structure of the heterocyclic compound nicotine. Nicotine is present in the leaves of tobacco plants. Pure nicotine, a colorless liquid, is highly toxic. Nicotine from tobacco plants is used as an insecticide but the compound is better known as the addictive drug in tobacco smoke.

nicotine

heterogeneous catalyst: a *catalyst* which is in a different *phase* from the reactants. Generally a heterogeneous catalyst is a solid while the reactants are gases or in solution in a solvent.

The advantage of a heterogeneous catalyst is that it can be held in a reaction vessel as the reactants flow in and the products flow out. There is no difficulty in separating the products from the catalyst.

Heterogeneous catalysts are used in large scale continuous processes such as *catalytic cracking*. In the Haber process for *ammonia manufacture* the nitrogen and hydrogen gases flow through a reactor containing small lumps of iron.

Platinum metal alloyed with other metals such as rhodium is an important catalyst. It is used to oxidize ammonia during *nitric acid manufacture*. It is also used in *catalytic converters*. In a catalytic converter the expensive catalyst is finely divided and supported on the surface of a ceramic to increase the surface area in contact with the exhaust gases.

Possible mechanism for the hydrogenation of an alkene in the presence of a nickel catalyst

Similarly, the expensive silver catalyst used to manufacture *epoxyethane* is finely spread over the surface of alumina.

Impurities in the reactants can poison catalysts so that they become less effective. Carbon monoxide poisons the iron catalyst used in the Haber process. Lead compounds in the exhaust from a car engine poison catalytic converters so lead-free gasoline must be used.

Heterogeneous catalysts work by adsorbing reactants at *active sites* on the surface of the solid. Nickel acts as a catalyst for the addition of hydrogen to $C = C$ in unsaturated compounds by adsorbing hydrogen molecules which probably split up into single atoms held on the surface of the metal crystals.

If a metal is to be a good catalyst for a *hydrogenation* reaction it must not adsorb the hydrogen so strongly that the hydrogen atoms become unreactive. This happens with tungsten. Equally, if adsorption is too weak there will not be enough adsorbed atoms for the reaction to go at a useful rate, as is the case with silver. The strength of adsorption must have a suitable intermediate value. Suitable metals for hydrogenation are nickel, platinum and palladium.

heterogeneous equilibrium: an equilibrium system in which the substances involved are not all in the same *phase*. An example is the equilibrium state involving two solids and a gas, formed on heating calcium carbonate in a closed container.

$$CaCO_3(s) \rightleftharpoons CaO(s) + CO_2(g)$$

The concentrations of solids do not appear in the expression for the equilibrium constant. Pure solids have a constant "concentration."

$$K_c = [CO_2(g)] \text{ or } K_p = p_{CO_2}$$

The same applies to the equilibrium between a solid and its saturated solution in water (see *solubility product constant*).

heterolytic bond breaking (dissociation): the type of bond breaking which produces ionic *intermediates* in reactions. In heterolytic fission, a *covalent bond* breaks so that one of the atoms joined by the bond takes both of the shared pair of electrons while the other is left with none. This type of bond breaking happens during *electrophilic* and *nucleophilic substitution* and *addition reactions* and is favored when reactions take place in *polar solvents* such as water.

$$H\!:\!\overset{\times\times}{\underset{\times\times}{Br}}\!\!\times \longrightarrow H^+ \quad \overset{\times\times}{\underset{\times\times}{:Br:}}\overset{-}{}$$

$$H\frown Br \longrightarrow H^+ + Br^-$$

Heterolytic bond breaking

hexadentate ligand: a molecule or ion which forms six *coordinate bonds* with the central metal ion in a complex. The common example is *EDTA*. (See also *chelates*.)

hexagonal close packed structure (HCP): see *close packed structures*.

high-performance liquid chromatography (HPLC): a sophisticated version of *liquid chromatography*. The technique is able to separate components in a mixture which are very similar to each other. The mobile phase is a solvent of very high purity. The stationary phase consists of very small particles of a solid such a silica packed into a long steel tube. The use of fine particles increases the surface area helping to make the separation efficient. A pump provides the very high pressure needed to force the solvent through the tightly packed column. One application of HPLC is to study what happens to drugs as they are metabolized in the body.

181

Hofmann degradation

Hofmann degradation: a reaction which turns an *amide* into a primary *amine* while removing a carbon atom from the molecule. The reaction shortens the carbon chain, hence the term "degradation." The reaction was discovered by August von Hofmann (1818–92). He was a pioneering German organic chemist who was head of the Royal College of Chemistry in London for 20 years from 1845.

Amides react in this way when treated with bromine and concentrated sodium hydroxide solution.

Equation for the Hofmann degradation of an amide to a primary amine with one less carbon atom

homogeneous catalyst: a *catalyst* which is in the same phase as the reactants. Typically the reactants and the catalyst are dissolved in the same solution.

Transition metal ions can be effective homogeneous catalysts because they can gain and lose electrons as they change from one *oxidation state* to another. The oxidation of iodide ions by persulfate ions, for example, is very slow.

$$S_2O_8^{2-}(aq) + 2I^-(aq) \longrightarrow 2SO_4^{2-}(aq) + I_2(aq)$$

The reaction is catalyzed by iron(III) ions in the solution. A possible mechanism is that iron(III) is reduced to iron(II) as it oxidizes iodide ions to iodine. The $S_2O_8^{2-}(aq)$ ions oxidize the iron(II) back to iron(III) ready to oxidize some more of the iodide ions and so on.

Acid catalysis and *base catalysis* are other examples of homogeneous catalysis.

homogeneous equilibrium: an equilibrium in which all the substances involved are in the same *phase*.

homolytic bond breaking: the type of bond breaking which produces *radicals* with unpaired electrons. In homolytic fission, a *covalent bond* breaks so that the atoms joined by the bond separate, each taking one of the shared pair of electrons.

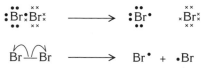

Homolytic bond breaking

hormones: chemical messengers carried throughout the body in the bloodstream. Hormones produce a response from particular target cells. Hormones are produced by glands and pass directly into the blood. Examples of hormones are *insulin* from the islet cells in the pancreas, adrenaline from the adrenal glands near the kidneys and sex hormones produced by the testes in men and the ovaries or other organs in women.

Hund's rule: one of the rules which help to predict electron configurations of atoms (see *aufbau principle*). The rule states that the most stable arrangement of electrons in an atom is the one with as many unpaired electrons as possible all with parallel *spins*. For example, in a nitrogen atom, the three *p*-electrons have parallel spins and each occupy a separate *p* energy level.

Electron configuration of a nitrogen atom   N   1s ⇅   2s ⇅   2p ↑ ↑ ↑

hydration takes places when water molecules bond to ions or add to molecules. Water molecules are *polar* and so they are attracted to both positive and negative ions.

It is hard to see why the charged ions in a crystal of sodium chloride separate and go into solution in water with only a small energy change. Where does the energy come from to overcome the attraction between the ions? The explanation is that the ions are so strongly hydrated by the polar water molecules that the sum of the *enthalpy changes of hydration* nearly balances the *lattice energy*.

Some metal ions are hydrated in crystals as well as solution. In many hydrated salts the metal is present as a complex ion. This is true of copper(II) ions in blue copper sulfate. Cobalt(II) ions also bond to water molecules when blue (anhydrous) cobalt(II) chloride turns pink with water.

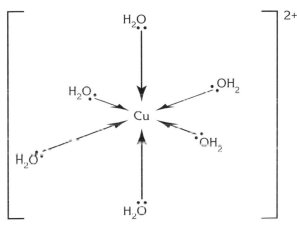

Hydrated copper(II) ion. Here coordinate bonds hold the water molecules to the metal ions

Hydrogen ions are hydrated in solution. They do not float around freely, but bond to water molecules forming *hydronium ions*.

The term hydration is also used to describe the addition of water to molecules such as *alkenes*. Ethene, for example, is hydrated when mixed with steam and passed over a phosphoric acid catalyst.

$$CH_2 = CH_2(g) + H_2O(g) \xrightarrow[\text{heat under pressure}]{\text{phosphoric acid catalyst}} CH_3CH_2OH(l)$$

hydrazine (N_2H_4): a colorless fuming liquid which is manufactured by oxidizing ammonia with sodium hypochlorite in the presence of gelatin. Hydrazine, like ammonia, is a base and a reducing agent. It has a *lone pair of electrons* on each nitrogen atom.

Structure and shape of a hydrazine molecule

Hydrazine with one hydrogen replaced by a methyl group can be used as a rocket fuel in space; it catches fire on contact with liquid N_2O_4.

hydrides: compounds of elements with hydrogen. The hydrides of the group I and II metals are ionic and contain the hydride ion, H^-.

$$2Na(s) + H_2(g) \longrightarrow 2Na^+H^-(s)$$

Ionic hydrides are ionic crystals. They are rapidly hydrolyzed by water because the hydride ion is a strong base.

$$H^-(s) + H_2O(l) \longrightarrow H_2(g) + OH^-(aq)$$

The hydrides of nonmetals are covalently bonded molecular liquids and gases. They include the *hydrogen halides*, *water*, hydrogen sulfide and *ammonia*. There are very many hydrides of carbon including the *alkanes*, *alkenes* and *arenes*.

The properties of the hydrides of the three highly *electronegative* elements nitrogen, oxygen and fluorine are affected by *hydrogen bonding*.

Hydrogen also forms *interstitial hydrides* with some *d-block elements*.

hydrocarbons: compounds which consist of just carbon and hydrogen. Important classes of hydrocarbons are the *alkanes*, *alkenes*, *alkynes* and *arenes*.

hydrochloric acid: a solution of hydrogen chloride gas in water. Hydrogen chloride is a strong acid so the solution in water is fully ionized into aqueous hydrogen ions (*hydronium ions*) and chloride ions. The concentration of commercial concentrated hydrochloric acid is about 12 mol/L. Dilute hydrochloric acid (about 2 mol/L) is commonly used as a laboratory reagent. Hydrochloric acid has the advantage of being cheaper and safer to dilute than sulfuric acid. It is a non-oxidizing acid unlike dilute nitric acid.

hydrogen: the commonest element in the universe. Stars, like the Sun, consist mainly of hydrogen and get their energy from *nuclear fusion* which turns hydrogen into *helium*.

Hydrogen atoms are the smallest of all atoms consisting, normally, of one proton and one electron in the 1s *orbital*. There are two other *isotopes*: *deuterium* and *tritium*.

Life on Earth depends on hydrogen. Most *organic compounds* contain hydrogen and life requires water – hydrogen oxide.

Hydrogen is an important industrial chemical. It is produced from steam and natural gas for *ammonia manufacture*. It is also one of the products of the *electrolysis of brine*. Hydrogen is used to *hydrogenate* vegetable oils for margarine and similar spreads.

Hydrogen forms a wide range of *hydrides* with other elements.

hydrogen bonding: a type of attraction between molecules which is much stronger than other types of *intermolecular force*, but much weaker than covalent bonding.

Hydrogen bonding affects molecules in which hydrogen is covalently bonded to one of the three highly *electronegative* elements nitrogen, oxygen and fluorine.

The hydrogen atoms lie between two highly electronegative atoms. They are hydrogen bonded to one of them and covalently bonded to the other. The *covalent bond* is highly polar. The small hydrogen atom ($\delta+$) can get close to the other electronegative atom ($\delta-$) to which it is strongly attracted.

The three atoms associated with a hydrogen bond are always in a straight line.

Hydrogen bonding in hydrogen fluoride

Hydrogen bonding accounts for:

- the relatively high boiling points of ammonia, water and hydrogen fluoride which are out of line for the trends in the properties of the other hydrides in groups V, VI and VII

- the open structure of *ice*

- the tertiary structure of *proteins*

- the pairing of bases in a *DNA* double helix

- the pairing up (*dimerization*) of carboxylic acids in a non-aqueous solvent.

Boiling points for the hydrides of the elements in groups IV, V, VI and VII

hydrogen electrode: a *half-cell* of great theoretical importance but of limited practical significance. By definition the *standard electrode potential* of the hydrogen electrode is zero, $E^\circ(\frac{1}{2}H_2|H^+) = 0.00$ V.

A standard hydrogen electrode sets up an equilibrium between hydrogen ions in solution (1 mol/L) and hydrogen gas (at 1 bar pressure) all at 298 K on the surface of a platinum electrode coated with platinum black.

By convention, when a standard hydrogen electrode is the left-hand electrode in an electrochemical cell, the cell e.m.f is the electrode potential of the right-hand electrode.

Hydrogen electrodes are inconvenient and so most measurements are taken with alternative *reference electrodes* acting as secondary standards.

$$Pt[H_2(g)] \mid 2H^+(aq) \parallel Zn^{2+}(aq) \mid Zn(s)$$

Conventional diagram to represent the cell which defines the standard electrode potential of the $Zn^{2+}(aq)|Zn(s)$ electrode

hydrogen emission spectrum: the spectrum of radiation from excited hydrogen gas. There is a bluish glow when a high voltage is applied across the electrodes at each ends of a glass tube containing hydrogen at low pressure. A spectroscope produces a line spectrum which can be recorded photographically.

Each line in the *atomic spectrum* corresponds to electrons dropping from a higher energy level to a lower level. The series of lines are named after the people who first discovered or studied them.

The pattern of energy levels in a hydrogen atom with just one electron is simpler than in other atoms with more than one electron. In a hydrogen atom all the orbitals in the same shell have the same energy. So the energy levels correspond to the main shells.

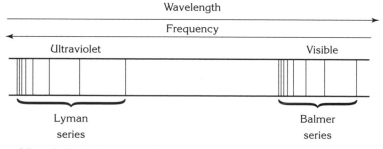

The series of lines in the hydrogen emission spectrum

The lines of the Balmer series are in the visible region. The lines of the Lyman series are in the ultraviolet. The bigger the energy jumps the higher the *frequency* of *electromagnetic radiation* emitted.

The energy gaps between energy levels get smaller as electrons recede from the nucleus. As a result, in each series, the differences between the energy jumps get smaller and smaller until they converge. They converge at the high frequency (big jump) end. The biggest jump in the Lyman series is for an electron dropping back from the very edge of an atom to the lowest energy level. The size of this jump is the ionization energy for a hydrogen atom. So the ionization energy of hydrogen can be calculated from the frequency of the convergence limit of the Lyman series since (according to *quantum theory*) $\Delta E = h\nu$.

hydrogen halides: compounds of hydrogen with the *halogens*. They are all colorless, molecular compounds with the formula HX where X stands for F, Cl, Br or I.

The bonds between hydrogen and the halogens are *polar*. The H — F bond is so polar that the properties of the compound are affected by *hydrogen bonding*.

Hydrogen chloride, hydrogen bromide and hydrogen iodide are similar in that they are:

- colorless gases at room temperature which fume in moist air
- very soluble in water, forming acid solutions (hydrochloric, hydrobromic and hydriodic acids)
- *strong acids* so they ionize completely in water.

Hydrogen chloride, hydrogen bromide and hydrogen iodide show some trends down *group 7* in that they:

- become less thermally stable – heating does not decompose hydrogen chloride but a hot wire will decompose hydrogen iodide into hydrogen and iodine

- become easier to oxidize to the halogen – hydrogen iodide is a strong *reducing agent*.

Hydrogen fluoride is rather different from the other hydrogen halides:

- it is a liquid at room temperature (boiling point 19.9°C) because of hydrogen bonding

- it is a *weak acid*

- its solution in water (hydrofluoric acid) reacts with glass and can be used for etching.

hydrogen ions (H⁺): hydrogen atoms which have lost an electron. Since a hydrogen atom consists of one proton and one electron this means that a hydrogen ion is just a *proton*. In water, hydrogen ions do not float around freely; they become attached to water molecules forming hydronium ions, H_3O^+. A lone pair on the oxygen atom forms a dative bond with the hydrogen ion.

Bonding in a hydronium ion

Hydrogen ions (protons) transfer from an acid to a base during an *acid–base reaction*. This gives rise to the *Brønsted–Lowry* definition of an acid as a proton donor and a base as a proton acceptor.

hydrogen peroxide (H_2O_2), when pure is a pale blue liquid. Like water it is a *hydride* of oxygen and it is a *hydrogen bonded*, molecular liquid.

Bonding and shape of a molecule of hydrogen peroxide

Hydrogen peroxide is usually supplied in solution in water.

Only in peroxides does oxygen have the oxidation state −1, in between 0 and −2. So hydrogen peroxide can act as both an *oxidizing agent* and as a *reducing agent*.

Hydrogen peroxide decomposes slowly on standing in the dark, but much more rapidly in the presence of a catalyst such as powdered manganese dioxide or the *enzyme* catalase. This is a *disproportionation* reaction with oxygen starting in the −1 state and ending up in the −2 and 0 states. Hydrogen peroxide, as it were, oxidizes and reduces itself.

$$2H_2O_2(aq) \longrightarrow 2H_2O(l) + O_2(g)$$

A
B
C
D
E
F
G
H
I
J
K
L
M
N
O
P
Q
R
S
T
U
V
W
X
Y
Z

$$0 - O_2$$
$$-1 - H_2O_2 \qquad O_2^{2-}$$
$$-2 - H_2O \qquad O^{2-}$$

Oxidation states of oxygen

Hydrogen peroxide oxidizes cobalt(II) to cobalt(III) in the presence of ammonia, and chromium(III) to chromium(VI) under alkaline conditions. As an oxidizing agent hydrogen peroxide is also used to bleach paper and textiles. It is used to manufacture the bleaches used in laundry detergents. In water treatment it destroys microorganisms. The great advantage of hydrogen peroxide as a bleach is that it turns into water when it reacts, so it does not contaminate or pollute.

$$H_2O_2(aq) + 2H^+(aq) + 2e^- \longrightarrow 2H_2O(l)$$

Hydrogen peroxide is not a powerful reducing agent but it will decolorize an acid solution of permanganate ions. A *potassium permanganate* titration is one way of determining the concentration of a solution of hydrogen peroxide.

$$H_2O_2(aq) \longrightarrow O_2(g) + 2H^+(aq) + 2e^-$$

hydrogenation: a reaction which adds hydrogen to a compound. Hydrogenation converts liquid vegetable oils to the solid fats used as ingredients of *margarine*. In these reactions hydrogen adds to double bonds in the hydrocarbon chains of unsaturated *fatty acids*. Hydrogen adds to $C = C$ double bonds at room temperature in the presence of a platinum or palladium catalyst or on heating in the presence of a nickel catalyst. (See also *heterogeneous catalyst*.)

hydrolysis: a reaction in which a compound is split apart in a reaction involving water. Hydrolysis reactions are often catalyzed by acids or alkalis.

Hydrolysis of the ester ethyl acetate catalyzed by alkali

Hydrolysis is used to make *soap* from fats and oils. This is an example of *ester* hydrolysis.

Inorganic examples of hydrolysis include *hydrolysis of nonmetal halides* and the *hydrolysis of salts*.

hydrolysis of nonmetal halides splits the compounds into an *oxoacid* (or hydrated oxide of the nonmetal) and hydrogen halide. Nonmetal chlorides are typically molecular liquids at room temperature. Examples are CCl_4, $SiCl_4$ and PCl_3. Most of these compounds do not mix with water but react with it. For example:

$$SiCl_4(l) + 2H_2O(l) \longrightarrow SiO_2(s) + 4HCl(g)$$

The exception is CCl_4 which does not react with water. This is an example of *kinetic inertness*. The four large chlorine atoms prevent water molecules from using their *lone pairs of electrons* to attack the small central carbon atom.

The hydrolysis of phosphorus bromide is one of the laboratory methods for making hydrogen bromide gas:

$$PBr_3(l) + 3H_2O \longrightarrow H_3PO_3(s) + 3HBr(g)$$

In a similar way, hydrogen iodide can be made by adding water to hydrolyze the PI_3 in a mixture of red phosphorus and iodine.

hydrolysis of salts changes the pH of a solution of a salt by altering the concentrations of H_3O^+ and OH^- ions.

Solutions of the salts of *aluminum(III)*, iron(III) and chromium(III) are acidic because of hydrolysis. The hydrated aluminum(III) ion, for example, is an acid. Al^{3+}, because of its *polarizing power*, draws electrons towards itself, so that the water molecules can more easily give away protons than free water molecules.

$$[Al(H_2O)_6]^{3+}(aq) + H_2O(l) \longrightarrow [Al(H_2O)_5OH]^{2+}(aq) + H_3O^+(aq)$$

The hydrated ions in M^{2+} salts are much less acidic because of the smaller polarizing power of these metal ions.

Solutions of the salts of weak bases are acidic. This includes the salts of ammonia and amines. The NH_4^+ ion in an ammonium salt is an acid because ammonia (the weak base) does not have a strong hold on the protons it accepts.

$$NH_4^+(aq) + H_2O(l) \rightleftharpoons NH_3(aq) + H_3O^+(aq)$$

Solutions of the salts of weak acids are alkaline. This includes the salts of:

- carbonic acid (carbonates)
- sulfurous acid (sulfites)
- acetic acid (acetates)
- hydrogen sulfide (sulfides).

The negative ions from weak acids are strong bases; they take hydrogen ions from water molecules and turn back into the acid.

$$CH_3CO_2^-(aq) + H_2O(l) \rightleftharpoons CH_3COOH(aq) + OH^-(aq)$$

hydronium ions: aqueous hydrogen ions or hydrated *protons* formed when an *acid* dissolves in water. A lone pair of electrons on a water molecule forms a *coordinate* bond with a hydrogen ion from an acid.

The formula of the hydronium ion is H_3O^+. It is often convenient to write $H^+(aq)$ instead, but remember that the hydrogen ion is *hydrated*.

hydrosphere: the watery part of the Earth's surface. This includes the oceans, rivers, lakes, groundwater and glaciers. Over 80% of the hydrosphere is in the oceans and shallow seas. Almost all the rest of the water is trapped in sediments.

The oceans are linked to the rest of the hydrosphere by the water cycle: water evaporates from the sea into the atmosphere, falls as rain or snow onto land and then returns to the oceans via rivers and lakes.

hygroscopic substances absorb water from the air. The chemicals used as drying agents in *desiccators* are hygroscopic; they include anhydrous calcium chloride and calcium oxide. It is difficult to weigh hygroscopic substances accurately which makes them unsuitable as *primary standards* for titrations.

ibuprofen: a drug used in *medicines* formulated to treat pain, inflammation and fever.

The structure of ibuprofen. Note the asymmetric carbon atom marked with an asterisk. Ibuprofen is a chiral compound.

Ibuprofen was developed to reduce inflammation in the joints of people suffering from rheumatoid arthritis. Testing showed that the drug was also an effective pain reliever and a rival to aspirin. The research and development program took 30 years.

ice: the solid state of water which is remarkable because, at the freezing point, it is less dense than liquid water and so floats. Ice owes its open cage-like structure to *hydrogen bonding*. Each water molecule is hydrogen bonded to four others. The hydrogen bonds are 0.177 nm long which is much longer than the length of a covalent O — H bond (0.094 nm). The distance between the oxygen atoms is equal to the sum of the lengths of a hydrogen bond and a normal O — H bond. Note that in every instance the three atoms O — H ········ O are in a straight line.

Structure of ice

The view from some angles shows the oxygen atoms in ice arranged in puckered hexagons – a pattern which is reflected in the hexagonal shapes of snowflakes.

When ice melts the open structure partly collapses as some of the hydrogen bonds break, so at 0°C the liquid is denser than the solid.

ideal gases: gases which obey the ideal gas equation $pV = nRT$, where p is the pressure in N/m^2 (pascals), V is the volume in m^3, T is the temperature on the *Kelvin* scale in K, n the *amount* in moles and R the *gas constant*. It is important to convert all the quantities to *SI units* before substituting values in the equation.

The ideal gas equation incorporates all the *gas laws*. For example, with a fixed amount of gas n is constant, so at constant temperature $pV = $ constant, which is *Boyle's law*.

In practice *real gases* deviate from ideal gas behavior. Under laboratory conditions the gases which are close to behaving like ideal gases are the ones which are well above their boiling points such as the *noble gases*, nitrogen, oxygen and hydrogen. For these gases at room temperature and pressure the assumptions of *kinetic theory* are nearly true. The attractive forces between these nonpolar molecules are small enough to be ignored and the volume of the molecules is insignificant compared to the total volume of gas.

The ideal gas equation can be used to determine the *molar mass* of gases and of other substances which evaporate easily.

ideal mixture: a mixture of liquids which obeys *Raoult's law*. Only very similar compounds form ideal mixtures, such as the two *alkanes* hexane and heptane or the two *alcohols*, 1-propanol and 2-propanol.

A mixture of two liquids A and B is only ideal if the *intermolecular* forces in pure A, in pure B and in the mixture of A and B are all the same strength. A mixture of liquids deviates from Raoult's law if the intermolecular forces in the mixture are stronger or weaker than in the pure liquids.

ignition temperature: the lowest temperature at which a substance will spontaneously catch fire in air. There is no need for a flame to ignite a gas or *vapor* at or above its ignition point. Examples of ignition temperatures in air are: hexane, 487°C, and ethanol, 558°C.

immiscible liquids: liquids which do not mix. As a rule *nonpolar* liquids such hexane and other hydrocarbons do not mix with polar liquids such as water.

Oily hydrocarbon liquids are usually less dense than water so they form a separate, colorless layer on top of water. Liquid organic halogen compounds such as dichloromethane are often denser than water so that they give rise to a separate liquid layer underneath water.

Liquids which are immiscible are useful in *solvent extraction*.

immobilized enzymes: *enzymes* bound to a solid so that they can be easily recovered from a reaction mixture by filtering or centrifuging. Alternatively the solid holding the enzyme can be contained in a column and the raw material (*substrate*) passed through the column turning into products in the presence of the trapped enzyme *catalyst*. Three methods of linking an enzyme to a solid support are:

- ionic bonding between charged amino side-chains on the protein molecules of the enzyme and charged groups on the support

- trapping the enzyme in a tangle of polymer molecules

- using covalent bonds to attach the enzyme to the insoluble solid.

Immobilized enzymes are used to convert starch from corn into glucose and then to convert much of the glucose to fructose which is sweeter than glucose or sucrose (see *sweetness*).

Using immobilized enzymes has meant a big decrease in the cost of high fructose syrup used to make colas and other soft drinks.

incineration: one of the methods available for disposing of *waste*. Burning plastic waste or domestic refuse in a modern incinerator releases energy to generate electricity.

Just dumping trash in *landfill* is expensive and a waste of resources. Incineration greatly reduces the volume of waste to be dumped in landfill sites. It also replaces *fossil fuels*, thus cutting the overall emission of greenhouse gases.

There has been strong opposition to burning waste. Old incinerators were inefficient, operated at too low a temperature and often were not designed to generate electricity. Incomplete combustion produces *carbon monoxide* which is toxic. Another worry is that if waste contains chlorine compounds such as PVC an incinerator will give off corrosive or toxic chemicals such as hydrogen chloride or *dioxins*. High temperature combustion at 800–1250°C, however, helps to achieve complete combustion and limit the emission of toxic chemicals. Modern incinerators have elaborate gas cleaning systems to control the emission of harmful gases and dust particles.

indicators are used to detect the endpoint during titrations. *Acid–base indicators* change color over a range of pH values. Starch is used as an indicator in *iodine-thiosulfate* redox titrations. Other indicators are available for redox titrations and for *complex-forming titrations*.

induced dipoles: see *intermolecular forces*.

inductive effect: a term to describe the extent to which electrons are pulled away from, or pushed towards, an atom by the atoms or group to which they are bonded. The word crops up sometimes in accounts of reaction mechanisms.

More *electronegative* atoms pull electrons away from a carbon atom, so the carbon carries a slight positive charge. This means that that carbon atom is open to attack by nucleophiles which happens when carbon is bonded to oxygen, chlorine or bromine.

Other groups, especially *alkyl groups*, have a slight tendency to push electrons towards the carbon atom to which they are bonded. One of the effects of this kind of inductive effect is that a secondary *carbocation* is more stable than a primary carbocation. This helps to account for the products formed during *electrophilic addition* of compounds such as HBr to unsymmetrical alkenes (see *Markovnikov's rule*).

Primary alkyl *amines* are stronger bases than ammonia because of the inductive effect. The push of electrons from the alkyl group to the nitrogen atom helps to stabilize the ion formed when the amine accepts a proton.

methylamine   methylammonium ion

inert chemicals: a chemical is inert if it has no tendency to react under given circumstances. The lighter noble gases, helium and neon live up to the original name for group VIII. They are inert towards all other reagents.

Nitrogen is a relatively unreactive gas which can be used to create an "inert atmosphere" free of oxygen which is much more reactive. Nitrogen is not inert in all circumstances. It reacts, for example, with hydrogen in the *Haber process* and with oxygen at high temperatures to form *nitrogen oxides*.

Sometimes there is no tendency for a reaction to happen because the reactants are *stable*. This is so if the *free energy change* for the reaction is positive.

Sometimes there is no reaction even though thermochemistry suggests that it should happen. The free energy change is negative so the change is *feasible*. A high activation energy means that the rate of reaction is very, very slow. Methane, for example, does not burn in oxygen at room temperature even though it would be energetically favorable for it to do so. This is an example of a substance being kinetically inert.

inert gases: the older name for the group of *noble gases* which was used until the discovery of compounds formed by xenon.

inert pair effect: describes the observation that, for groups III, IV and V of the periodic table, the oxidation state which is 2 below the highest state becomes increasingly stable down the group. In group 4, for example, the main oxidation state for carbon and silicon is +4. The +2 oxidation state becomes important in the chemistry of tin and lead. In the chemistry of lead the +2 state is even more stable relative to the +4 state than it is in the chemistry of tin. What this means is that two of the four electrons in the outer shell become less available for bonding down the group, hence the term "inert pair effect." This effect is best regarded as a reminder of a trend rather than an explanation.

infrared (IR) spectroscopy: an analytical technique used to identify functional groups in organic molecules. Most compounds absorb IR radiation. The wavelengths they absorb correspond to the natural frequencies at which vibrating bonds in the molecules bend and stretch. It is *polar covalent bonds* such as $O-H$, $C-O$ and $C=O$ which absorb strongly as they vibrate.

Bonds vibrate in particular ways and absorb radiation at specific wavelengths. This means that it is possible to look at an IR spectrum and identify particular functional groups.

Molecules with several atoms can vibrate in many ways because the vibrations of one bond affect others close to it. Complex pattern of vibrations can be used as a "fingerprint" to be matched against the recorded IR spectrum in a database. Comparing the IR spectrum of a product of synthesis with the spectrum of the pure compound can be used to check that the product is pure. (See page 194.)

inhibitor: a chemical which stops or slows down a specific chemical reaction. Inhibitors are sometimes called negative *catalysts*. Liquid *monomers* such as styrene are often supplied with a dissolved inhibitor to prevent polymerization in the bottle.

Enzyme inhibitors are very important in the control of metabolism. Competitive inhibition reduces the rate of a reaction catalyzed by an enzyme in the presence of a molecule with a similar shape to the normal *substrate*. The inhibitor binds to the *active site* of the enzyme and stays there unchanged, stopping normal enzyme activity.

IR spectra of ethanol, ethanal and acetic acid. The units along the bottom are wavenumbers (in cm⁻¹). IR wavenumbers range from 400 cm⁻¹ to 4500 cm⁻¹. Spectroscopists find the numbers more convenient than wavelengths. The wavenumber is the number of waves in 1 cm. Transmittance on the vertical axis measures the percentage of radiation which passes through the sample. The troughs appear at the wavenumbers at which the compound absorbs strongly.

initial rate method: a method for finding the *order* of a chemical reaction and the form of its *rate equation*. This method is used for reactions with several terms in the rate equation which makes other methods hard to use.

The method is based on finding the rate immediately after the start of a reaction. This is the one point when all the concentrations are known.

The procedure is to make up a series of mixtures in which all the initial concentrations are the same except one. A suitable method is used to measure the change of concentration with time for each mixture (see *rates of reaction*). The results are used to plot concentration/time graphs. The initial rate for each mixture is then found by drawing tangents to the curve at the start and calculating their gradients.

Worked example

The initial rate method was used to study the reaction A + 2B $\longrightarrow$ C + D. The initial rate was calculated from five graphs plotted to show how the concentration of B varied with time for different initial concentrations of reactants.

Experiment	Initial concentration of A, mol/L	Initial concentration of B, mol/L	Initial rate of reaction, mol/L·s
1	0.10	0.05	2×10^{-4}
2	0.20	0.05	8×10^{-4}
3	0.30	0.05	18×10^{-4}
4	0.20	0.10	8×10^{-4}
5	0.20	0.20	8×10^{-4}

What is:

- the rate equation for the reaction?
- the value of the rate constant?

Notes on the method

Recall that the rate equation cannot be determined from the balanced equation for the reaction.

First study the experiments in which the concentration of A varies but the concentration of B stays the same. How does doubling or tripling the concentration of A affect the rate?

Next study the experiments in which the concentration of B varies but the concentration of A stays the same. How does doubling the concentration of B affect the rate?

Substitute values for any one experiment in the rate equation to find the value of the rate constant, k. Take care with the units.

Answer

From experiments 1, 2 and 3: doubling $[A]_{initial}$ increases the rate by a factor of 4 (2^2). Tripling $[A]_{initial}$ increases the rate by a factor of 9 (3^2).
So rate $\propto [A]^2$.

From experiments 2, 4 and 5: doubling $[B]_{initial}$ does not change the rate.
So rate $\propto [B]^0 = 1$.

The reaction is second order with respect to A and zero order with respect to B. The rate equation is:

$$\text{rate} = k\,[A]^2$$

Rearranging this equation, and substituting values from experiment 2:

$$k = \frac{\text{rate}}{[A]^2} = \frac{8 \times 10^{-4} \text{ mol/L·s}}{(0.2 \text{ mol/L})^2}$$

$$k = 0.02 \text{ L/mol·s}$$

A B C D E F G H I J K L M N O P Q R S T U V W X Y Z

inorganic chemistry: the study of the chemical elements and their compounds. In an advanced level course the study of the elements concentrates on the reactions of selected elements and of their simpler compounds such as oxides, chlorides and hydrides. Inorganic chemistry includes the chemistry of the element carbon and a few of its compounds such as carbon monoxide, carbon dioxide, carbon tetrachloride and the carbonates. The chemistry of most other carbon compounds belongs to *organic chemistry*.

Most of the reactions of inorganic chemicals belong to one of these four types:

• *acid–base*

• *redox*

• *ionic precipitation*

• *complex forming* (or *ligand substitution*).

insulin: the *hormone*, produced by special cells in the pancreas, which helps to control the level of glucose in the blood. Digestion of food after a meal raises the level of glucose in blood. In response the pancreas secretes insulin which circulates throughout the body and speeds up the rate at which cells take up glucose.

Insulin is a *protein* consisting of two folded chains of amino acids which are *cross-linked* by disulfide links.

interhalogen compound: a compound formed when one halogen combines with another. For example, passing a stream of chlorine gas over solid iodine produces a red–brown liquid, iodine monochloride, ICl. With excess chlorine this turns to a yellow solid, iodine trichloride, ICl_3, which is unstable and easily decomposes back to ICl.

intermediate bonding: bonding which is neither purely ionic nor purely covalent. In most compounds the bonding between the atoms of different elements is to some extent intermediate between the two extremes.

The *electronegativity* values for two elements are a guide to the extent to which the bonds between them will be covalent, ionic or intermediate between the two. Electronegativity values are particularly useful for discussing *polar covalent bonds*.

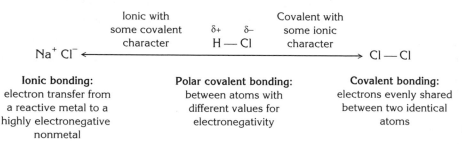

A spectrum from purely ionic to purely covalent bonding

Where the bonding is largely ionic, the *polarization of ions* helps to explain why there may be a degree of electron sharing.

intermediates in reactions: atoms, molecules, ions or *radicals* which do not appear in the balanced equation but which are formed during one step of a reaction, then used up in the next step. Chemists can use spectroscopy to detect intermediates which only exist for a short time during a reaction.

Examples of reaction intermediates are the radicals formed during a *radical chain reaction* or the carbocations formed during *nucleophilic substitution* by the S_N1 mechanism and during *electrophilic addition* to alkenes.

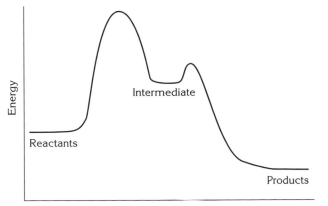

Energy changes in the course of a reaction. Note the energy dip as the intermediate forms. The dip is not deep enough for the intermediate to exist as a stable product. Contrast this with the energy profile for a reaction with just a transition state between reactants and products.

intermolecular forces: weak attractive forces between molecules. Without intermolecular forces there could be no molecular liquids or solids. Also, *real gases* would behave more like *ideal gases.*

Weak intermolecular forces arise from electrostatic attractions between *dipoles*, including attractions between:

- molecules with *permanent dipoles* such as hydrogen chloride

- a permanent dipole in one molecule and a dipole induced in a neighboring molecule, such as the attraction between hydrogen bromide and an ethene molecule

- temporary dipoles created fleetingly in nonpolar atoms or molecules (London forces).

Examples of substances with nonpolar atoms and molecules include the noble gases, iodine and alkanes. When nonpolar atoms or molecules meet there are fleeting repulsions and attractions between the nuclei of the atoms and the surrounding clouds of electrons. Temporary displacements of the electrons lead to temporary dipoles. It is the attractions between these transient dipoles which give rise to the tendency for the molecules to cohere.

The greater the number of electrons the greater the polarizability of the molecule and the greater the possibility for temporary, induced dipoles. This explains why the boiling points rise down *group VII* (the halogens) and *group VIII* (the noble gases). For the same reason the boiling points in *alkanes* increase with the increasing number of carbon atoms in these hydrocarbon molecules.

In polymers the intermolecular forces become larger as molecules get longer so that there is a larger area of contact over which intermolecular forces can operate. The strength of *thermoplastic polymers* depends on the total strength of intermolecular forces between tangled, long chain molecules.

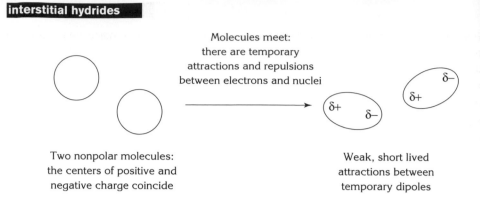

Molecules meet:
there are temporary
attractions and repulsions
between electrons and nuclei

$\delta-$

$\delta+$ $\delta-$

$\delta+$

Two nonpolar molecules:
the centers of positive and
negative charge coincide

Weak, short lived
attractions between
temporary dipoles

An illustration of the origins and effect of temporary induced dipoles

Weak intermolecular forces due to attractions between temporary and permanent dipoles are often called *van der Waals forces*. Forces of this kind are roughly a hundred times weaker than covalent bonds.

The properties of some molecular compounds are affected by stronger intermolecular forces called *hydrogen bonding*. Hydrogen bonds are about ten times stronger than van der Waals forces.

interstitial hydrides: compounds formed by *d-block elements* with hydrogen. The hydrogen atoms fit into the spaces between the atoms in the metal crystal. The holes are the interstices of the crystal, hence the name.

The amounts of hydrogen in the lattice vary with the temperature and pressure. The compounds do not have a definite composition. The hydride of titanium has the variable formula TiH_x, where x is less than 2.

Palladium reacts particularly well with hydrogen and can absorb nearly a thousand times its own volume of the gas at room temperature.

Interstitial hydrides release hydrogen gas on heating so they have been investigated as a possible safe way of storing hydrogen.

iodination: a reaction in which an iodine atom replaces a hydrogen atom in an organic molecule. One example is the iodination of acetone – a reaction which crops up in the study of reaction rates. The reaction is *acid catalyzed*.

$$CH_3COCH_3(aq) + I_2(aq) \longrightarrow CH_2ICOCH_3(aq) + H^+(aq) + I^-(aq)$$

iodine (I): a lustrous grey-black solid at room temperature, formula I_2, which *sublimes* when gently warmed to give an purple vapor. Iodine is a *halogen*, the fourth element in *group VII*, with the *electron configuration* $[Kr]4d^{10}5s^25p^5$.

Iodine consists of diatomic molecules with pairs of atoms held together by single *covalent bonds*. The molecules are nonpolar so the *intermolecular forces* are relatively weak, but stronger than in bromine because the atoms are larger and have more electrons. This makes iodine molecules more *polarizable* than bromine molecules. (See *crystal structures of nonmetals*.)

Iodine, like the other halogens, is an *oxidizing agent* but a less powerful oxidizing agent than bromine.

Iodine reacts with metals to form iodides. Because of the polarizability of the large iodide ion, the iodides formed with small cations or highly charged cations are essentially covalent (see *polarization of ions*). Examples are lithium iodide, magnesium iodide and aluminum iodide.

Iron(III) iodide and copper(II) iodide do not exist because iodide ions reduce the metals ions to their lower oxidation state. The only iodides of these metals are iron(II) iodide and copper(I) iodide.

Iodine oxidizes hydrogen on heating forming hydrogen iodide. Unlike the reactions of chlorine and bromine, this is a reversible reaction which has been studied in detail to establish the *equilibrium law*.

$$H_2(g) + I_2(g) \rightleftharpoons 2HI(g)$$

Hydrogen iodide is a fuming, acidic gas. Like the other *hydrogen halides* it is very soluble in water and a strong acid.

Iodine is only very slightly soluble in water. It is much more soluble in a solution of potassium iodide because of the formation of the *triiodide ion*. Iodine solution in aqueous potassium iodide is a brownish-yellow color. Iodine dissolves freely in nonpolar solvents such as hexane forming a solution with the same color as iodine vapor.

As a weaker oxidizing agent, iodine converts thiosulfate ions to tetrathionate ions. This is a quantitative reaction used in *iodine-thiosulfate titrations*.

Iodine and its compounds are used to make pharmaceuticals, photographic chemicals and dyes. Iodine is needed in the diet so that the thyroid gland in the neck can make the *hormone* thyroxine which regulates growth and metabolism. In many regions sodium iodide is added to table salt to supplement the iodine in the diet and drinking water and so prevent goiter.

iodine–thiosulfate titrations: a useful analytical method for measuring *amounts* of *oxidizing agents*. The method is based on the fact that oxidizing agents convert iodide ions to iodine quantitatively. Among the oxidizing agents which do this are iron(III) ions, copper(II) ions, chlorine molecules and hypochlorite ions in *bleach*, dichromate ions with acid, iodate ions with acid and permanganate ions with acid.

$$2I^-(aq) \longrightarrow I_2(aq) + 2e^-$$

The iodine stays in solution in excess potassium iodide turning a yellow–brown (see *triiodide ion*).

The iodine produced is then titrated with a standard solution of sodium thiosulfate which reduces iodine molecules back to iodide ions. This too happens quantitatively exactly as in the equation.

$$I_2(aq) + 2S_2O_3^{2-}(aq) \longrightarrow 2I^-(aq) + S_4O_6^{2-}(aq)$$

The greater the amount of oxidizing agent added, the more the iodine formed and so the more thiosulfate needed to react with it. When thiosulfate is added from a buret the color of the iodine gets paler. Near the end point the solution is a very pale yellow. Adding a little soluble starch solution as an indicator near the end point gives a sharp color change from blue–black to colorless.

Worked example

Calculate the concentration of a solution of sodium thiosulfate standardized by this method. A 0.0642 g sample of potassium iodate, KIO_3, was dissolved in water in a conical flask. Excess of potassium iodide was dissolved in the solution which was then acidified with dilute sulfuric acid. The iodine formed was titrated with the solution of sodium thiosulfate from a buret. The volume of sodium thiosulfate solution needed to decolorize the blue iodine–starch color at the end point was 24.50 cm³.

Notes on the method

Write the equations and determine the amount in moles of $S_2O_3^{2-}$ equivalent to 1 mol IO_3^-.

There is then no need to consider the amounts of iodine in the calculations.

Look up the *molar mass* of potassium iodate: $M_r(KIO_3) = 214.0$ g/mol.

Use your calculator only at the last step of the calculation to avoid repeated rounding errors.

Answer

The equations for producing iodine:

$$IO_3^-(aq) + 5I^-(aq) + 6H^+(aq) \longrightarrow 3I_2(aq) + 3H_2O$$

The equation for the reaction during the titration:

$$I_2(aq) + 2S_2O_3^{2-}(aq) \longrightarrow 2I^-(aq) + S_4O_6^{2-}(aq)$$

So 1 mol IO_3^- produces 3 mol I_2 which then reacts with 6 mol $S_2O_3^{2-}$.

$$\text{The amount of } KIO_3 \text{ at the start } = \frac{0.0642 \text{ g}}{214.0 \text{ g/mol}}$$

6 mol $S_2O_3^{2-}$ react with the iodine formed by 1 mol IO_3^-.

So the amount of thiosulfate in 24.5 cm³ (= 0.0245 L) solution

$$= 6 \times \frac{0.0642 \text{ g}}{214.0 \text{ g/mol}}$$

So the concentration of the sodium thiosulfate solution

$$= 6 \times \frac{0.0642 \text{ g}}{214.0 \text{ g/mol}} \times \frac{1}{0.0245 \text{ L}}$$

$$= 0.0735 \text{ mol/L}$$

ion exchange is a process for swapping ions in a solution and is used to soften and deionize water. An ion exchange resin consists of small beads of a *polymer* which has been modified so that along the chains there are ionic groups. In a cation exchange resin the polymer has negatively charged groups which attract positive ions.

A water softener contains a cation exchanger in which all the negatively charged sites are attracting sodium ions. As hard water flows over the resin beads the calcium ions in the *hard water* change places with the sodium ions. The softened water can then be used for washing without forming insoluble precipitates with *soaps* and other *detergents*.

Ion exchange is a reversible process. Once all the sodium ions in a bed of resin have been used up it can be recharged by running a concentrated solution of sodium chloride through it to replace calcium ions with sodium ions, ready to soften more water. The salt added to a dishwasher is needed to make sure that the water softener in the machine is regenerated in this way.

Ca^{2+} ions in hard water

Ion-exchange resin bead holding Na$^+$ ions

Na$^+$ ions in softened water

Using ion exchange to soften water

It takes two ion exchange resins to deionize water: a cation resin holding hydrogen ions and an anion resin with hydroxide ions. As water runs through the two resins the cation exchangers holds onto positive metal ions and replaces them with hydrogen ions. The anion exchanger holds on to the negative ions and replaces them with hydroxide ions.

Then H$^+$ (aq) + OH$^-$(aq) $\longrightarrow$ H$_2$O(l), so no ions are left in solution.

ionic bonding: one of three types of strong chemical bonding. Ionic bonding occurs in compounds of metals with nonmetals. Ionic compounds form particularly when the reactive *s-block* metals combine with the halogens of group VII or oxygen in group VI. The *crystal structures of ionic compounds* are three-dimensional lattices of ions.

Ionic bonding is the result of *electrostatic forces* of attraction between positive metal ions and negative nonmetal ions. The larger the charges on the ions the bigger the attractive force. The smaller the ions the closer the charges get to each other and the stronger the force of attraction.

Lewis electron dot formulas provide a balance sheet for keeping track of the electrons when ionic compounds form. These diagrams do not describe the mechanisms of the reactions.

Na$^•$ | $\overset{\times\times}{\underset{\times\times}{•Cl}}^{\times}$ $\longrightarrow$ Na$^+$ + $\overset{\times\times}{\underset{\times\times}{:Cl}}^{\times}{}^-$

sodium atom chlorine atom   sodium ion chloride ion
(2.8.1)   (2,8.7)   (2.8)   (2.8.8)

Ca$\overset{•}{:}$ + $\overset{\times\times}{\underset{\times\times}{\times F}}^{\times}$ $\overset{\times\times}{\underset{\times\times}{\times F}}^{\times}$ $\longrightarrow$ Ca^{2+} + $\overset{\times\times}{\underset{\times\times}{:F}}^{\times}{}^-$ $\overset{\times\times}{\underset{\times\times}{:F}}^{\times}{}^-$

calcium atom two fluorine atoms   calcium ion two fluoride ions
(2.8.8.2)   (2.7)   (2.8.8)   (2.8)

Lewis formulas for the formation of sodium chloride and calcium fluoride

Energy is needed to remove electrons from metal ions (see *ionization energy*). The energy changes on adding electrons to nonmetal atoms are quite small (see *electron affinity*). Ionic crystals are stable because of the large release of energy as the ions form a crystal lattice (see *lattice energy*). The *Born–Haber cycle* helps to analyze the stability of ionic crystals. This energy cycle also helps to explain why it is generally true that the atoms of s- and p-block elements gain or lose electrons to attain the electron configuration of the nearest *noble gas*.

ionic character of bonds: a phrase used to describe the extent to which bonds between the atoms of two different elements are polar. The greater the difference in electronegativity between the elements the more ionic the bonding. (See also *polar bonds*, *intermediate bonding* and *polarization of ions*.)

ionic equations describe chemical changes by showing only the reacting ions in solution. These equations leave out the *spectator ions* which remain in solution unchanged. For example, an ionic equation shows that the use of a barium salt to test for sulfate ions is essentially the same reaction whether the reagent is barium chloride or barium nitrate and whatever the source of the sulfate ions.

$$Ba^{2+}(aq) + SO_4^{2-}(aq) \longrightarrow BaSO_4(s)$$

Note that ionic equations are balanced both for atoms and charges.

An ionic equation also shows that the neutralization of any fully ionized *strong acid* by a fully ionized *strong base* is essentially the same process:

$$H^+(aq) + OH^-(aq) \longrightarrow H_2O(l)$$

ionic precipitation: a reaction which produces a solid precipitate on mixing two solutions. This type of reaction can be used to make an insoluble salt from two soluble salts. Silver bromide forms as a precipitate on mixing solutions of silver nitrate and potassium bromide. This is best represented by an *ionic equation* leaving out the *spectator ions*.

$$Ag^+(aq) + Br^-(aq) \longrightarrow AgBr(s)$$

This is the reaction used to make silver bromide for *photography*. Silver metal is dissolved in nitric acid to make silver nitrate, $Ag^+NO_3^-$. Then the solution of silver nitrate is mixed with potassium bromide (K^+Br^-) in a solution of gelatin. This produces a very fine precipitate of silver bromide (AgBr) suitable for making a photographic film.

Ionic precipitation reactions play a big part in *anion tests* and *cation tests*.

ionic radius: the radius of an ion in a crystal. Ionic radii are determined by X-ray diffraction methods. The radius of the positive ion of an element is smaller than its atomic radius. The radius of the negative ion of an element is larger than its atomic radius.

Na	Na⁺	F	F⁻
r_{atom} = 191 pm	r_{ion} = 102 pm	r_{atom} = 71 pm	r_{ion} = 133 pm
Mg	Mg²⁺	O	O²⁻
r_{atom} = 160 pm	r_{ion} = 72 pm	r_{atom} = 73 pm	r_{ion} = 140 pm

Comparison of the radii of atoms and ions

Trends in ionic radius help to account for patterns of bonding and properties in the periodic table.

$$N^{3-} \qquad O^{2-} \qquad F^- \qquad Na^+ \qquad Mg^{2+}$$

Ionic radii for selected elements. All the ions shown have the same electron configuration. They are isoelectronic.

Note these patterns:

- down any group the ionic radii increase as the number of inner full shells increases

- across a period, the radii of ions with the same electron configuration decrease as the nuclear charge increases.

Ions in solution are *hydrated*. The radius of the ion plus the hydrating water molecules is larger than the ion in a crystal. A smaller ion is likely to be more heavily hydrated than a larger ion with the same charge.

Ion	Radius in a crystal, pm	Radius when hydrated, pm
Li+	74	1000
Na+	102	790

ionization energy: a measure of the energy needed to remove an electron from a gaseous atom or ion. Ionization energies give evidence for the arrangement of electrons in atoms in shells and subshells. Ionization energies can help to explain which ions an element can form. Values for ionization energies are used in energy cycles, such as the *Born–Haber cycle*, to investigate bonding in compounds.

The first ionization energy for an element is the energy needed to remove one mole of electrons from one mole of gaseous atoms. Successive ionization energies for the same element measure the energy needed to remove a second, third, fourth electron and so on.

Ionization energies can be measured using a *mass spectrometer* or by studying the emission spectra of atoms (see *hydrogen emission spectrum*). In these ways it is possible to measure energy changes involving ions which do not normally appear in chemical reactions.

$$Na(g) \longrightarrow Na^+(g) + e^- \qquad \text{first ionization energy} = 496 \text{ kJ/mol}$$

$$Na^+(g) \longrightarrow Na^{2+}(g) + e^- \qquad \text{second ionization energy} = 4563/\text{kJ mol}$$

$$Na^{2+}(g) \longrightarrow Na^{3+}(g) + e^- \qquad \text{third ionization energy} = 6913 \text{ kJ/mol}$$

There are 11 electrons in a sodium atom so there are 11 successive ionization energies for this element. The electron configuration of the element is $1s^2 2s^2 2p^6 3s^1$. There is 1 electron in the outer shell which is furthest from the nucleus and shielded by ten inner electrons. There are eight electrons in the second shell and these are closer to the nucleus and only have two inner *shielding* electrons. The two inner electrons feel the full attraction of the nuclear charge and are closest to the nucleus. They are hardest to remove.

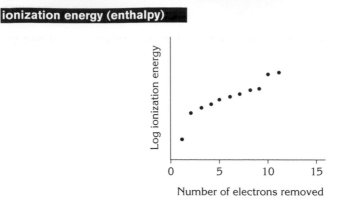

Plot of log (ionization energy) against number of electrons removed for sodium

So the successive ionization energies for an element rise and there are big jumps in value each time electrons start to be removed from the next shell in towards the nucleus.

In the periodic table, ionization energies tend to rise from left to right across a period because from one element to the next the added electron goes into the same main shell as the charge on the nucleus increases by one. The rise in values is not smooth but shows a 2–3–3 pattern, corresponding to the way that the s- and p-orbitals fill up.

The 2s orbital is full at Be, there is then a slight dip as the next electron goes into a p orbital with a slightly higher energy. The three p orbitals each have one electron in a nitrogen atom. The dip at oxygen happens as the next electron has to pair up with another electron. Nevertheless, the main factor is the charge on the nucleus which increases steadily across the period. This pattern repeats from Na to Ar.

Ionization energies decrease down a group. Down a group the number of full shells increases. The increased shielding effect balances out the increasing charge on the nucleus. The outer electrons get further and further away from the same effective nuclear charge, so they are easier to remove.

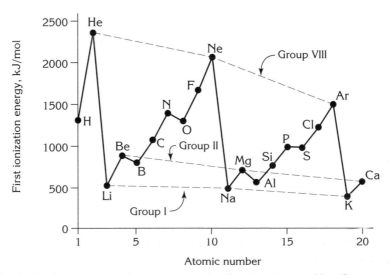

Plot of first ionization energy against proton number for the elements H to Ca.
Note that the trend is for the values of the first ionization energies to decrease down any group.

ionizing radiation includes all the types of radiation with enough energy per *photon* to ionize atoms. Ionizing radiation includes *alpha* and *beta particles*, *gamma rays* and *X-rays*. Living cells are very sensitive to ionizing radiation. High doses of ionizing radiation can kill living cells and even small doses can cause mutation. Exposure to ionizing radiation can, in time, give rise to cancer.

ions: see *anions*, *cations* and *ionic bonding*.

iron (Fe): a *d-block* metal which shows the characteristic behavior of the *transition metals*. Its *electron configuration* is: $[Ar]3d^64s^2$. The pure metal is soft and easily worked. It is strongly *ferromagnetic*. In most of its uses iron is mixed with other elements to make *steel*. Iron is so widely used that *corrosion of the metal* is a serious economic problem.

Iron can exist in more than one *oxidation state*. Iron dissolves in dilute acids to form iron(II) salts. The aqueous iron(II) ion is pale green. Oxidizing agents including oxygen in the air, chlorine or potassium permanganate, all oxidize iron(II) to iron(III). The aqueous iron(III) ion is yellow.

$$Fe^{2+}(aq) \longrightarrow Fe^{3+}(aq) + e^-$$

Adding aqueous sodium hydroxide to solutions of iron salts helps to distinguish iron(II) from iron(III) compounds (see *cation tests*). A greenish precipitate of iron(II) hydroxide turns brownish-red when exposed to the air, as it is oxidized to iron(III) hydroxide.

Iron ions form a variety of complexes in both the +2 and the +3 states. The hexaaquo complex of the iron(III) ion, $Fe(H_2O)_6^{3+}$, is acidic in solution for the same reason that the *aluminum(III) ion* is acidic.

A very sensitive test for iron(III) is to add a dilute solution of thiocyanate ions, SCN^-. The solution turns to a deep blood-red color due to the formation of iron complexes such as $[Fe(H_2O)_5(SCN)]^{2+}$. The SCN^- *ligand* takes the place of a water molecule in the aquo complex.

An example of an iron(III) complex with a *bidentate ligand* is $[Fe(C_2O_4)_3]^{3-}$, in which the ligand is the oxalate ion.

Iron and its compounds act as *homogeneous* and *heterogeneous catalysts*.

iron extraction produces the metal from its oxide ores in large blast furnaces.

Coke burning in air heats the furnace.

$$C(s) + O_2(g) \longrightarrow CO_2(g)$$

Coke also produces the *reducing agent* by reacting with carbon dioxide further up the furnace to make carbon monoxide.

$$C(s) + CO_2(g) \longrightarrow 2CO(g)$$

The carbon monoxide reduces the oxide ore to iron.

$$Fe_2O_3(s) + 3CO(g) \longrightarrow 2Fe(l) + 3CO_2(g)$$

Limestone ($CaCO_3$) decomposes to calcium oxide (CaO), which combines with silicon dioxide and other impurities to make a liquid *slag*. For example:

$$CaO(s) + SiO_2(s) \longrightarrow CaSiO_3(l)$$

The molten metal and slag run to the bottom of the furnace, where the slag floats on the metal so that it can be tapped off separately. Most of the iron is then turned into *steel*.

irritant chemicals do not destroy tissues, so they are less than corrosive. However skin or eyes may become inflamed or affected by sores if exposed to the chemical. This may happen quite quickly or after extended or repeated contact with the chemical.

isoelectric point: the pH at which the overall charge on an *amino acid* or *protein* is zero. At this pH amino acids or proteins do not move towards one or other of the electrodes during *electrophoresis*.

At low pH an amino acid has a positive charge because there is an extra hydrogen ion on the basic amino group. At high pH an amino acid is negatively charged because the carboxylic acid group has given away a hydrogen ion. At the pH of the isoelectric point the amino acid is present in solution as a *zwitterion* and the total charge is zero.

| in acid solution | zwitterion | in alkaline solution |

The charges on alanine at different pH values

Isoelectric points vary from one amino acid to another because of differences in their side chains.

A large protein molecule is at its least soluble in water at its isoelectric point when it is uncharged.

isoelectronic atoms and ions have the same number and arrangement of electrons. The following particles are all isoelectronic: N^{3-}, O^{2-}, F^-, Ne, Na^+, Mg^{2+} and Al^{3+}. They all have the electron configuration $1s^2 2s^2 2p^6$.

Methane and the ammonium ion are isoelectronic. Both have ten electrons, two in the first shell of the central atom and eight forming four covalent bonds with hydrogen atoms. They both have the same shape (see *shapes of molecules*).

isomerization: a process used in oil refining to convert straight chain *alkanes*, such as pentane, into branched *isomers* such as 2-methylbutane. The value of the process is that branched alkanes increase the octane number of *gasoline*. Isomerization happens when the hot hydrocarbon vapor passes over a *catalyst*.

isomers: compounds with the same molecular formula but different structures. Structural isomerism may arise because:

• the hydrocarbon chain is branched in different ways

butane

2-methylpropane

Isomers of C_4H_{10}

- the functional group is in a different position

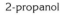

Isomers of C_3H_8O   1-propanol   2-propanol

- the functional groups are different.

Isomers of C_2H_6O   ethanol   dimethyl ether

(See also *geometrical isomerism* and *optical isomerism*.)

isomorphous compounds have the same type of chemical formula and crystallize with the same shape and structure. *Alums* are isomorphous. They form octahedral crystals.

isotactic polymer: an *addition polymer* with a regular structure. In isotactic poly(propene), for example, all the methyl side chains are on the same side of the carbon chain. The molecules coil into a regular helical shape and pack together to form a highly crystalline polymer which is very strong. This is the form of the polymer which is useful for durable fibers, tough moldings in motor vehicles and containers which can hold boiling water.

Structure of isotactic poly(propene) with all the methyl groups on the same side of the carbon chain

isotopes: atoms of the same element which have the same number of protons in the nucleus but a different number of neutrons. In other words the atoms of isotopes have the same *atomic number* but different *mass numbers*. The isotopes have the same chemical properties because they have the same number and arrangement of electrons (*electron configuration*).

The isotopes of an element can be separated and detected with a *mass spectrometer*. Peaks in a mass spectrum are often multiple because of the existence of isotopes.

Mass spectrum of chlorine

isotopic abundance: the proportions of the different *isotopes* of an element. Chlorine, for example, has two isotopes: chlorine-35 and chlorine-37. Naturally occurring chlorine contains 75% of chlorine-35 and 25% of chlorine-37.

$$\text{So the mean relative atomic mass} \; = \; \frac{(3 \times 35) + 37}{4} \; = \; 35.5$$

Most relative atomic mass values are not whole numbers because most elements have several naturally occurring isotopes.

isotopic labelling uses *isotopes* as markers to trace the movement of chemicals or to investigate what happens to particular atoms during chemical changes (see also *tracers*). Atoms of the normally abundant isotope of an element in a molecule are replaced by a different isotope.

Isotopes have the same chemical properties so it is possible to use them to follow what happens during a change without altering the normal course of the process. Radioactive isotopes can be tracked by detecting their radiation. The fate of nonradioactive isotopes can be followed by analyzing samples with a *mass spectrometer*.

Isotopic labelling has been used to investigate the mechanism of ester hydrolysis. Using water labelled with oxygen-18 (instead of the normal isotope oxygen-16) it was shown that it is the C — O bond in the ester which breaks and not the O — C_2H_5 bond. Mass spectrometry showed that the heavier oxygen atoms end up in the acid and not in the alcohol.

Use of labelling to investigate bond breaking during ester hydrolysis

isotropic solids have properties which are the same in all directions. Crystals of sodium chloride are isotropic because the ions are spherical and the crystal structure is cubic. Most substances are more or less *anisotropic*.

IUPAC (International Union of Pure and Applied Chemistry): the recognized authority for the names of chemical compounds, for chemical symbols and terms and the values of data such as relative atomic masses. IUPAC names are systematic names based on a set of rules which make it possible to determine the chemical structure of a compound from its name. Chemists increasingly use approved IUPAC names for simpler compounds but stick to traditional names when the systematic name is complex. The systematic name 2-hydroxy propane-1,2,3-tricarboxylic acid describes the structure but is cumbersome compared to the traditional name citric acid based on its occurrence in citrus fruits. (See *names of carbon compounds*, *names of complex ions* and *names of inorganic compounds*.)

joule (symbol J): the *SI unit* of energy. The quantities of energy transferred during chemical reactions are relatively large, so chemists generally measure energy changes in kilojoules (kJ). 1 kJ = 1000 J.

K_a: the symbol for an *acid dissociation constant* (or ionization constant).

K_c: the symbol for an *equilibrium constant*, with the concentrations measured in moles per liter.

The expression for K_c makes it possible to explain the effect of changing the concentration of one of the chemicals in an equilibrium mixture. There is an equilibrium, for example, in a solution of bromine in water:

$$Br_2(aq) + H_2O(l) \rightleftharpoons HOBr(aq) + Br^-(aq) + H^+(aq)$$
orange   colorless

At equilibrium: $K_c = \dfrac{[HOBr(aq)][Br^-(aq)][H^+(aq)]}{[Br_2(aq)]}$ where these are equilibrium concentrations

In dilute solution $[H_2O(l)]$ is constant.

Adding alkali removes $H^+(aq)$ from the right-hand side:

$$H^+(aq) + OH^-(aq) \longrightarrow H_2O(l).$$

This briefly upsets the equilibrium. For an instant after adding alkali:

$$\frac{[HOBr(aq)][Br^-(aq)][H^+(aq)]}{[Br_2(aq)]} < K_c$$

The system restores equilibrium as bromine reacts with water to produce more of the products while lowering the concentration of $Br_2(aq)$. There is very soon a new equilibrium. Once again:

$$\frac{[HOBr(aq)][Br^-(aq)][H^+(aq)]}{[Br_2(aq)]} = K_c$$

but now with new values for the various concentrations.

Speaking roughly, chemists say that adding alkali makes the "position of equilibrium shift to the right." The effect is visible because the orange color of the solution fades. This is as *Le Chatelier's principle* predicts. The advantage of using K_c is that it makes quantitative predictions possible.

K_p: the symbol for an equilibrium constant for an equilibrium involving gases with the concentrations measured by *partial pressures*.

Equilibrium		K_p	Units of K_p (using the SI unit of pressure, Pa)
$H_2(g) + I_2(g) \rightleftharpoons 2HI(g)$		$K_p = \dfrac{(p_{NH_3})^2}{(p_{H_2})(p_{I_2})}$	no units

(cont'd)

$$N_2(g) + 3H_2(g) \rightleftharpoons 2NH_3(g)$$ $$K_p = \frac{(P_{NH_3})^2}{(P_{N_2})(P_{H_2})^3}$$ Pa^{-2}

$$N_2O_4(g) \rightleftharpoons 2NO_2(g)$$ $$K_p = \frac{(P_{NO_2})^2}{(P_{N_2O_4})}$$ Pa

Worked example

When a mixture of 10.0 mol sulfur dioxide and 5.0 mol oxygen comes to equilibrium at 450°C and 200 kPa pressure, the amount of sulfur trioxide formed is 9.0 mol. Calculate K_p for the reaction.

Notes on the method

First write the equation for the reaction and determine the amount in moles of each gas at equilibrium. Note that 2 mol $SO_2(g)$ reacts with 1 mol $O_2(g)$ forming 2 mol $SO_3(g)$. 9.0 mol $SO_2(g)$ must react to form 9.0 mol $SO_3(g)$.

Hence calculate the mole fraction of each gas. This leads to the partial pressures for the gases at equilibrium. Substitute these values in the expression for K_p and then decide on the units.

Answer

The equation:	$2SO_2(g)$	$+$	$O_2(g)$	$\rightleftharpoons$	$2SO_3(g)$
Initial amounts	10.0 mol		5.0 mol		0.0 mol
Equilibrium amount	1.0 mol		0.5 mol		9.0 mol
Equilibrium mole fractions	$\frac{1.0}{10.5}$		$\frac{0.5}{10.5}$		$\frac{9.0}{10.5}$
Equilibrium partial pressures	$\frac{1.0}{10.5} \times 200$ kPa		$\frac{0.5}{10.5} \times 200$ kPa		$\frac{0.5}{10.5} \times 200$ kPa
	= 19.0 kPa		= 9.52 kPa		= 171.4 kPa

Hence $K_p = \dfrac{(P_{SO_3})^2}{(P_{SO_2})^2\,(P_{O_2})} = \dfrac{(171.4 \text{ kPa})^2}{(19.0 \text{ kPa})^2(9.52 \text{ kPa})} = 8.55$ k/Pa

K_w: the symbol for the *autoionization constant of water*.

Kekulé structure: a ring structure for benzene proposed by the German chemist Friedrich Kekulé in 1865. According to Kekulé the idea of a ring structure came to him while he was daydreaming in front of a fire.

Representations of the Kekulé structure for benzene

The problem with this structure is that it suggests that there should be two isomers of 1,2-dichlorobenzene.

Isomers of 1,2-dichlorobenzene suggested by the Kekulé structure. Isomers with this formula have never been separated. There is only one form of the compound.

In practice it has never been possible to separate isomers of disubstituted benzenes. To get around this problem Kekulé suggested that benzene molecules rapidly alternate between the two possible structures. This is not the modern explanation. Today, chemists use the idea of *delocalization of electrons* to account for the benzene structure.

kelvin (symbol K): the *SI unit* of temperature. Temperatures above *absolute zero* are measured in kelvins. Temperatures on the Celsius scale use the same size units but zero is set at the freezing point of water (273.15 K). Temperature differences measured on either scale are the same and given as kelvins (K).

kerosene: a mixture of hydrocarbons produced by the *fractional distillation of oil*. Part of the kerosene fraction is refined for use as kerosene, while some is cracked to make *gasoline*. Most of the kerosene fraction, however, is purified for use as the fuel for jet engines in aircraft.

ketones: *carbonyl compounds* in which the carbonyl group is attached to two *alkyl groups*. The carbonyl group is the *functional group* which gives ketones (and *aldehydes*) their characteristic properties. Ketones are named after the alkane with the same carbon skeleton by changing the ending "e" to "one." Where necessary a number in the name shows the position of the carbonyl group.

Oxidation of secondary alcohols with acidified potassium dichromate produces ketones which, unlike aldehydes, are not easily oxidized further.

Oxidation of 2-propanol to acetone

Fehling's solution and *Tollen's reagent* cannot oxidize ketones. There is no change when testing ketones with these reagents.

Chemists can identify ketones with the help of instrumental techniques such as *mass spectrometry* and *infrared spectroscopy*.

Sodium borohydride reduces ketones to secondary alcohols, as does *lithium aluminum hydride*.

2-propanol

Reduction of acetone to 2-propanol

Ketones, like aldehydes, undergo addition. These are *nucleophilic addition reactions.*

Addition reactions of acetone

kilogram (symbol kg): the *SI unit* of mass. 1 kg = 1000 g.

kinetic control: a term used when the product of a reaction is the one which forms faster rather than an alternative product which is more *stable* but forms more slowly.

The production of methanal from methanol and oxygen is an example of kinetic control.

$$2CH_3OH(g) + O_2(g) \longrightarrow 2HCHO(g) + 2H_2O(l)$$

This reaction is fast in the presence of a suitable catalyst at a temperature at which the alternative reaction cannot go. The alternative is burning to carbon dioxide and water. Combustion is more exothermic but has a higher *activation energy.*

$$2CH_3OH(g) + O_2(g) \longrightarrow 2CO_2(g) + 4H_2O(l)$$

Progress of reaction from reactants to products

Reaction profiles for alternative reactions illustrating kinetic and thermodynamic control

In the absence of a catalyst, and at a higher temperature, methanol burns to produce the products which are thermodynamically more stable. In these circumstances the system is under *thermodynamic control.*

kinetic inertness: a term used when a reaction does not occur, even though the reaction appears to be *feasible.* There is no change because the rate of reaction is too slow to be noticeable. There is a barrier preventing change – usually a high *activation energy.* The compound or mixture is inert.

Examples of kinetic inertness are:

- a mixture of methane (natural gas) and oxygen at room temperature
- a mixture of CCl_4 and water at room temperature (see *hydrolysis of nonmetal chlorides*)
- a solution of *hydrogen peroxide* in the absence of a catalyst
- *aluminum* metal in dilute hydrochloric acid.

Nothing happens in these mixtures even though the possible reactions tend to occur. In each example, thermochemistry shows that the reaction is feasible (as indicated by the *equilibrium constant*, or the *standard electrode potential*, or the *free energy change*) yet nothing happens.

Chemists refer to:

- systems with no tendency to react as "stable"
- systems which should react but do not do so for a rate (kinetic) reason as "inert."

kinetics of reactions: see *reaction kinetics.*

kinetic theory of gases: an explanation of why gases obey the *gas laws*. The theory is based on a model of an *ideal gas*. The assumptions are that:

- a gas consists of molecules in rapid, random motion
- the volume of the molecules can be neglected in comparison with the total volume of gas
- the molecules do not attract each other (no *intermolecular forces*)
- the collisions with the walls of any container are perfectly elastic.

By applying Newton's laws of motion and then assuming that the average kinetic energy of molecules is proportional to temperature, it is possible to derive the ideal gas equation from this model.

The model helps to explain why *real gases* deviate from ideal gas behavior.

The model of gas molecules in rapid random motion underlies the theories of *reaction kinetics* and chemical equilibrium.

knocking: a noise from a *gasoline* engine heard when the mixture of fuel and air ignites too early while still being compressed by the piston. This is pre-ignition.

Compression heats up the mixture of fuel and air. Pre-ignition means that the fuel starts burning before being ignited by a spark from the spark plug. Internal combustion engines run powerfully if the fuel starts to burn when the piston is at the right point in the cylinder so that the expanding gases force the piston down smoothly. Knocking is a sign that a fuel with a higher octane number is needed. Prolonged knocking damages an engine.

krypton (Kr): a *noble gas* which makes up less than 0.1% of *air*. The gas is separated from the other gases in the air by *fractional distillation* and used to fill fluorescent lights and flashbulbs.

labile complexes: a term used to describe *complex ions* in which it is easy to replace the *ligands*. *Ligand substitution* is difficult or impossible with inert complexes.

Chromium(III) and cobalt(III) complexes are *kinetically inert*. Most other complexes are labile.

landfill: where local authorities, public services and industries dispose of most of their *waste*. A modern landfill site is engineered to contain waste and the products of decomposition so that *leaching* does not carry hazardous chemicals into the soil and contaminate water sources. Water percolating through rotting landfill can dissolve harmful chemicals such as the ions of toxic metals which can be recovered from the liquids by *ion exchange* if waste disposal is managed correctly.

Microorganisms break down the organic materials which make up more than half of household waste. At first there is oxygen present and microorganisms convert complex carbohydrates to simple sugars. *Fermentation* begins as the conditions become anaerobic, turning carbohydrates, proteins and fats into organic acids, carbon dioxide and hydrogen. In time the microorganisms start to produce the gases methane and carbon dioxide. This landfill gas can be a hazard because methane is highly flammable but it can also be collected from pipes buried in the waste and burnt to generate electricity.

Some landfills are for general domestic trash but others are planned to contain more hazardous materials. After dumping, heavy vehicles compact the waste and the layers of waste are then covered with soil.

lanthanide elements: the series of 14 elements between lanthanum and hafnium in the sixth period of the periodic table. Across this series of elements the electrons fill the $4f$ energy levels. These elements were once called the "rare earths" but, apart from promethium, they are not rare. Cerium is five times as abundant as lead. The elements all have very similar chemical properties. They can be separated by *ion exchange* or *chromatography*. The main source of these elements is the ore monazite.

lattice: a term used to describe a network of atoms or ions in a crystal.

lattice energy: the *energy change* when one mole of an ionic compound turns to free gaseous ions.

$$NaCl(s) \longrightarrow Na^+(g) + Cl^-(g) \qquad \Delta H^\circ_{lattice} = +787 \text{ kJ/mol}$$

Note the difference between the lattice energy and the standard *enthalpy change of formation* of sodium chloride which refers to the formation of a compound from its elements, $\Delta H^\circ_f[NaCl(s)] = -411$ kJ/mol.

Lattice energy is one of the thermochemical quantities which cannot be measured directly. Lattice energies are calculated indirectly with the help of the *Born–Haber cycle*.

Lattice energies give a measure of the strength of the ionic bonding in a crystal. Ionic bonding is the result of the *electrostatic forces* between the oppositely charged ions. So, for a given crystal structure, the lattice energy increases if:

- the charge on the ions increases
- the ions get smaller (and so closer together).

Lattice energies appear in thermochemical cycles to explore the factors which determine the *enthalpy change of solution* of ionic salts.

layer lattice: a crystal lattice in which there is strong bonding within layers of atoms but relatively weak bonding between the layers. It is often easy to cleave materials with a layer lattice along planes parallel to the layers. Examples of substances with layer lattices are graphite, mica and molybdenum sulfide. Graphite and molybdenum sulfide are used as *lubricants* because of the ease with which layers of atoms slide over each other.

Le Chatelier's principle: a qualitative guide to the effect of changes in concentration, pressure or temperature on a system at equilibrium. The principle was suggested as a general rule by the French physical chemist Henri Le Chatelier (1850–1936).

The principle states that when the conditions of a system at equilibrium change, the position of equilibrium shifts in the direction that tends to counteract the change.

The principle can be used to discuss the conditions chosen for *ammonia manufacture*. The equilibrium system involved is:

$$N_2(g) + 3H_2(g) \rightleftharpoons 2NH_3(g) \qquad \Delta H = -92.4 \text{ kJ/mol}$$

There are four moles of gases on the left side of the equation but only two moles on the right. Le Chatelier's principle predicts that raising the pressure will make the equilibrium shift from left to right. This reduces the number of molecules and so tends to reduce the pressure. So increasing the pressure increases the proportion of ammonia at equilibrium.

The reaction is *exothermic* from left to right and so *endothermic* from right to left. Chatelier's principle predicts that raising the temperature will make the position of equilibrium shift in the direction which takes in energy (tending to cool the mixture). So raising the temperature lowers the proportion of ammonia at equilibrium.

Le Chatelier's principle can be misleading, especially when predicting the effects of changes in concentration or pressure. It is generally better to use the *equilibrium law* and the values of K_c or K_p to predict the effect of changes in concentration or pressure.

leaching: a process for separating a material from an insoluble solid using a suitable solvent. Leaching is one of the methods used to extract metals from low grade ores. Copper is obtained in this way from waste tips at old mines. These tips contain ores which have been oxidized by weathering. The finely divided ore is treated with sulfuric acid in a leach tank stirred by a flow of air.

$$Cu_2(OH)_2CO_3(s) + 4H^+(aq) \longrightarrow 2Cu^{2+}(aq) + 3H_2O(l) + CO_2(g)$$

Leaching also describes the extraction of nitrates and other fertilizers from soil by rainwater. Water draining through the soil can carry away soluble nitrate ions into lakes and streams, causing *eutrophication*. Precautions are necessary in *landfill* to prevent hazardous chemicals leaching from the decomposing wastes into the soil and then into underground water sources.

lead (Pb): a greyish metal often used on roofs of buildings because it resists corrosion and is malleable so that it can be bent to fit the shape and create a waterproof flashing where roof tiles meet walls. Lead is a dense metal and is an effective shield to stop X-rays and other radiations. Lead melts at a relatively low temperature for a metal and, when alloyed with tin, produces solder with an even lower melting point (see *eutectic mixture*).

Lead comes below tin in *group IV* of the periodic table. It forms compounds in two oxidation states: +2 and +4. The +2 state is more stable so that compounds in the +4 state are oxidizing agents.

Lead forms two soluble lead(II) compounds: the nitrate and the acetate. Lead ions are colorless in solution. Adding alkali to a solution of lead(II) ions gives a white precipitate of $Pb(OH)_2$ which dissolves in excess, showing this hydroxide is *amphoteric*.

The pigment chrome yellow consists of lead chromate which precipitates on mixing a solution of lead(II) ions with a solution of chromate(VI) ions. As well as being used by artists and printers, chrome yellow is the pigment in the yellow stripes painted on roads.

Lead is still used to produce the *gasoline* additive lead tetraethyl, $Pb(C_2H_5)_4$. Burning leaded gas has greatly increased the concentrations of lead in the environment. Many countries have decided to phase out leaded gas because lead compounds are toxic and can cause brain damage, especially in children.

lead–acid cell: an electrochemical cell which is rechargeable because the chemical changes at the electrodes are reversible. The working of the cells involves lead in three *oxidation states* as $Pb(0)$, $Pb(II)$ and $Pb(IV)$.

In a fully charged lead–acid cell the negative electrode consists of lead metal. The positive electrode consists of lead coated with lead(IV) oxide. The electrodes dip into a solution of sulfuric acid. These are the electrode processes as current flows:

- at the negative electrode:
$$Pb(s) + SO_4^{2-}(aq) \longrightarrow PbSO_4(s) + 2e^-$$
- at the positive electrode:
$$PbO_2(s) + SO_4^{2-}(aq) + 4H^+(aq) + 2e^- \longrightarrow PbSO_4(s) + 2H_2O(l)$$

The equations help to explain why the cell is rechargeable. The changes which happen as a current flows from the cell both produce an insoluble solid, lead sulfate. This traps the lead(II) ions beside the electrode instead of dissolving in the electrolyte. When the current is reversed to charge up the cell, the two reactions can be reversed, turning lead(II) back to lead metal at one electrode and back to PbO_2 at the other.

A 12-volt car battery usually consists of six 2 V lead–acid cells connected in series. Banks of lead–acid cells can be used to provide emergency standby power for lighting wherever a power outage could be serious. Lead–acid cells have the great advantage that they can be recharged many times and last for a long time.

leaving group: an atom or group of atoms which breaks away from a molecule during a *substitution reaction* or an *elimination reaction*.

Some groups are better at leaving than others. A water molecule is a better leaving group than a hydroxide ion. This explains why the substitution reactions of *alcohols* go better in acid conditions.

Nucleophilic substitution *of a hydroxide ion for a bromide ion from 1-bromobutane. The bromide ion is the leaving group.*

Nucleophilic substitution *of a bromide ion for a hydroxide ion in 2-propanol. The reaction goes faster under acid conditions. The — OH group in the alcohol is protonated, so the leaving group is a water molecule.*

Lewis acid/base theory: a theory which gives a very broad definition of acids and bases in terms of electron pairs. A Lewis acid is a molecule or ion which can form a bond by accepting a pair of electrons. A Lewis base is a molecule or ion which can form a bond by donating a pair of electrons.

The formation of a *hydronium ion* is a Lewis acid–base reaction between the proton (a Lewis acid) and a water molecule (a Lewis base). So, in this theory, it is the proton rather than the proton donor, which is an acid.

Lewis   Lewis
acid   base

Formation of a hydronium ion as the Lewis base, water, forms a coordinate bond *with a proton*

This much broader definition of acids and bases describes the formation of a complex ion as a reaction between a metal ion (a Lewis acid) and ligands (Lewis bases).

$$Ni^{2+} \quad + \quad 6\ :NH_3 \quad \longrightarrow \quad [Ni(NH_3)_6]^{2+}$$

Lewis   Lewis   Complex ion
acid   base

Complex formation as a Lewis acid–base reaction

Aluminum chloride ($AlCl_3$) is a Lewis acid. This accounts for its use as a catalyst in a Friedel–Crafts reaction.

Chemists normally use the *Brønsted–Lowry theory*. When they use the terms acid and base they mean "proton donor" and "proton acceptor." To signal that they are using the wider Lewis definition they refer to "Lewis acids" and "Lewis bases."

Lewis electron dot formulas

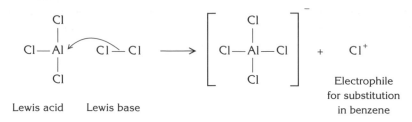

Aluminum chloride acting as a Lewis acid

Lewis electron dot formulas show the way in which the outer electrons of atoms are shared or transferred when *covalent bonds* and *ionic bonds* form. Dots, crosses and small circles help to keep count of the number of electrons contributed by each atom to bonding.

Lewis formulas for molecules show both bonding and lone pairs of electrons; they help to predict the *shapes of molecules*.

Despite the use of dots and crosses for the electrons coming from different atoms, all the electrons are the same. Once bonds form it is impossible to say which electron came from which atom.

ligand substitution reactions involve swapping one *ligand* for another in a *complex ion*. These ligand exchange reactions are often reversible.

The ligands NH_3 and H_2O are similar in size and are uncharged. With these ligands exchange reactions take place without change in the *coordination number* of the metal ion.

$$[Cu(H_2O)_6]^{2+}(aq) \; + \; 4NH_3(aq) \; \rightleftharpoons \; [Cu(NH_3)_4(H_2O)_2]^{2+}(aq) \; + \; 4H_2O(l)$$

Reversible ligand exchange reaction of copper(II) ions. Note that in this example substitution is incomplete. The aquo complex is pale blue but the ammine complex is a very deep blue.

The chloride ion is larger than uncharged ligands such as water, so that fewer chloride ions can fit around a central metal ion. Ligand exchange involves a change in coordination number.

$$[Cu(H_2O)_6]^{2+}(aq) \; + \; 4Cl^-(aq) \; \rightleftharpoons \; [CuCl_4]^{2-}(aq) \; + \; 6H_2O(l)$$

Ligand exchange with change of coordination number. The aquo complex is pale blue but the chloro complex is yellow.

ligands: the molecules or ions bound to the central metal ion in a complex ion. Examples of molecules which can act as ligands are water and ammonia. Examples of ions which act as ligands are hydroxide ions, chloride ions and cyanide ions. Ligands are Lewis bases. They have one or more *lone pairs of electrons* which can form *coordinate bonds* to the metal ion (see *monodentate ligands*, *bidentate ligands*, *hexadentate ligands* and *EDTA*).

limewater: a test reagent for detecting carbon dioxide. The reagent is a saturated solution of calcium hydroxide in water. The reagent is a colorless solution. Bubbling carbon dioxide through the solution produces a white precipitate of calcium carbonate (see *gas tests*).

$$Ca(OH)_2(aq) + CO_2(g) \longrightarrow CaCO_3(s) + H_2O(l)$$

limiting reactant: the chemical in a reaction mixture which is present in an amount which limits the theoretical yield (see *yield calculations*). Often in a chemical synthesis some of the reactants are added in excess to make sure that the most valuable chemical is converted as far as possible to the required product. The limiting reactant is the one which is not in excess and so is used up (if the reaction goes to completion).

lipids: a broad class of biological compounds which are soluble in organic solvents such as ethanol, but insoluble in water. Lipids are very varied and include *fatty acids*, *fats*, *vegetable oils*, phospholipids and *steroids*. Lipids release more energy per gram when oxidized than carbohydrates. This makes lipids important as a concentrated energy store in living organisms.

lipid/water solubility: a characteristic of *pharmaceuticals* which affects what happens to drugs in the body. Ideally drugs are taken by mouth in the form of tablets or capsules. This is only possible if the drug can survive in the stomach, which is highly acid. The drug then has to cross from the gut into the bloodstream and reach its target without being broken down in the liver or excreted by the kidneys.

Cell membranes consist largely of *lipids*. This means that a drug has to be sufficiently fat-loving (lipophilic) if it is to cross from the gut to the bloodstream. However, fat in the body will extract any drug which is too lipophilic from the bloodstream so that it ceases to be available as an active chemical.

A drug that is too water loving (hydrophilic) may be unable to cross the fatty barriers of cell membranes. It may also be quickly removed by the kidneys and excreted in the urine.

Functional groups that make a molecule hydrophilic are: OH, — NH_2, — COOH or — COO^-. Functional groups which make a molecule lipophilic include alkyl and aryl (hydro carbon) side chains.

liquids flow to take the shape of their container. The atoms, molecules or ions in a liquid are free to move about but with nothing like the freedom of the molecules of a *gas*. The particles of a liquid are generally slightly more widely spaced than in a solid but the density of packing is only slightly less. Liquids lack the order of crystalline solids. Liquids, like solids, are hard to compress. Unlike gases, they have a definite volume.

liquid chromatography was the first type of *chromatography* developed. The Russian botanist Michel Tswett developed this type of chromatography to separate plant pigments. In 1903 he separated leaf colors (chlorophylls, carotenes and xanthophylls) by dissolving a leaf extract in a hydrocarbon solvent. He added the extract to the top of a column of powdered chalk and then passed more pure solvent through the column to complete the separation. Running solvent through the column to separate the mixture is called "elution" and the liquid which flows out of the bottom it the eluate.

This is an example of *adsorption* chromatography. Each compound in the mixture has its own equilibrium between adsorption on the solid and solution in the solvent. Compounds which are strongly adsorbed by the stationary phase move slower; compounds which are more soluble in the solvent move faster.

The liquid leaving the bottom of the column is collected in a series of tubes. Tubes containing parts of the mixture can be detected by their color or some other method such as their appearance in UV light. Components of the mixture can be recovered by evaporating the solvent from these tubes.

liquid crystal: a state of matter which is more ordered than a liquid but less ordered than a solid. Although the liquid crystal state was first noticed as long ago as 1888, it is only since the early 1970s that such crystals have been developed for use in digital watches, calculators and portable computer screens.

lithium (Li): a soft, shiny metal which turns dark grey in air. It is the first member of *group I* with the electron configuration [He]$2s^1$.

Like other group I metals, lithium:

- is stored in oil

- floats on water and reacts, but quite slowly, forming hydrogen and LiOH which is soluble and strongly alkaline

- burns in air with a colored flame (bright red) forming an oxide (Li_2O)

- forms an ionic, crystalline chloride.

The small size of the lithium ion means that some features of lithium chemistry are not typical of the group as a whole:

- hydration – the Li^+ ion is heavily hydrated in water and lithium has the most negative hydration energy in group I; as a result the standard electrode potential for $Li^+(aq)|Li(s)$ is the most negative for the group

- solubilities – lithium carbonate and lithium fluoride are only slightly soluble while most other simple compounds of group I metals are soluble

- thermal stability – lithium carbonate decomposes on heating while all the other group I carbonates are stable; lithium nitrate decomposes to the oxide on heating unlike the nitrates of sodium and potassium

- covalent bonding – lithium forms a wide range of covalent molecular compounds such as ethyl lithium, C_2H_5Li.

lithium aluminum hydride (LiAlH$_4$): a powerful reducing agent used to reduce *aldehydes*, *ketones*, *esters* and *carboxylic acids* to alcohols.

In its reactions the AlH_4^- ion can be regarded as a source of hydride ions (H^-). The reagent is rapidly hydrolyzed by water so it has to be used in an anhydrous solvent such as dry ether. Where possible *sodium borohydride* is preferred because it is easier and safer to use.

lithium cells provide the power in many mobile phones and notebook computers. These rechargeable cells are relatively light because lithium has a low density. The reactivity of lithium means that these *electrochemical cells* have a relatively large *emf*.

The electrodes consist of lithium atoms or ions trapped within a layer lattice. The electrolyte is a polymeric material containing lithium ions.

lithosphere: the rocks, weathered rocks and soils of the Earth's crust which make up less than 0.0001% of the volume of the planet. The crust consists largely of oxygen (46.6% by mass) mainly combined with silicon (27.7%) in *silicate* minerals. Only a few other elements are abundant: aluminum (8.1%), iron (5%), calcium (3.6%), sodium (2.8%), potassium (2.6%) and magnesium (2.1%). All other elements together make up the remaining 1.4%.

liter (symbol L): a unit of *volume* equal to 1000 cm^3. A liter is the same as a decimeter cubed (dm^3). 1 dm^3 = 10 cm × 10 cm × 10 cm = 1000 cm^3.

logarithms in chemistry are of two kinds, logarithms to base 10 (lg) and natural logarithms to base e (ln).

Chemists use logarithms to base 10 to handle values which range over several orders of magnitude. Logs to base 10 are defined such that:

- lg 1000 = 3
- lg 100 = 2
- lg 10 = 1
- lg 1 = 0
- lg 0.1 = −1
- lg 0.01 = −2

In general lg 10^x = x

The hydrogen ion concentration in aqueous solutions typically ranges from 1 to 10^{-14} mol/L. The definition of pH ($-$lg [H^+(aq)]) covers this range in a scale running from 0 to 14.

Also by definition: lg xy = lg x + lg y and lg x^n = n lg x

so lg $1/x$ = lg x^{-1} = $-$lg x.

Natural logarithms appear in relationships in thermochemistry and chemical kinetics. They follow similar mathematical rules as logarithms to base 10. In general ln e^x = x.

London forces: weak *intermolecular forces* due to transient dipoles of very short duration. The existence of these forces was proposed by the chemist Fitz London (1900–1954) to account for the attraction between nonpolar molecules.

lone pair of electrons: a pair of electrons in the outer shell of one of the atoms in a molecule or ion which is not involved in bonding.

Examples of molecules and ions with lone pairs of electrons

Lone pairs of electrons:

- affect the *shapes of molecules*
- form *coordinate bonds*
- are important in the chemical reactions of compounds such as *water* and *ammonia*
- are important in *nucleophilic addition* and *nucleophilic substitution* reactions.

(See also *Lewis acids/base theory*.)

lubricants are used to reduce friction wherever one surface slides over another. Lubricating oils for engines are usually based on oil fractions with molecules in the size range C_{20} to C_{34}. Synthetic lubricants for specialist purposes include *silicones*, esters and polymers made from *epoxyethane*.

luminescence: a term which includes all the examples of substances emitting light at low temperatures, in contrast to incandescence, which is the emission of light by objects when they are hot, such as the filament of an electric light bulb.

Examples of luminescence are:

- **fluorescence** – the immediate emission of light by a material exposed to other radiation such as UV radiation, X-rays or an electron beam

- **phosphorescence** – delayed and slow emission of light from a material after exposure to other forms of radiation,

- **chemiluminescence** – light given out during chemical changes (such as from light sticks, fireflies or glowworms)

- **triboluminescence** – light given out when crushing or breaking a material (such as the flashes of light seen in a darkened room when crushing sugar or opening a self-adhesive envelope).

macromolecular substances: materials made up of very large molecules. They are usually called *polymers*. Macromolecules, like small molecules, have strong *covalent bonds* holding the atoms together within the molecules but relatively weak attractions between the molecules (*intermolecular forces*).

magnesium (Mg): a reactive metallic element in *group II* of the periodic table. Its electron configuration is [Ne]3s^2. Samples of the silvery-white metal usually look grey because they are covered with a layer of magnesium oxide.

The source of magnesium metal is electrolysis of molten magnesium chloride obtained either from seawater or from salt deposits. The low density of the metal helps to make light alloys especially with aluminum. These alloys, which are strong for their weight, are especially valuable for cars and aircraft. The reactivity of magnesium makes it suitable as the "sacrificial" metal in anodes used to protect pipelines and ships from *corrosion* (see *cathodic protection*).

Magnesium burns very brightly in air with an intense white flame forming the white solid magnesium oxide (MgO). It is used in *fireworks* and flares.

Magnesium reacts very slowly with cold water but much more rapidly on heating in steam.

$$Mg(s) + H_2O(g) \longrightarrow MgO(s) + H_2(g)$$

Magnesium forms ionic compounds with nonmetals in which the metal is in the +2 oxidation state as Mg^{2+}. Mg^{2+} is the central metal ion in a chlorophyll molecule.

$$H_2SO_4(aq) \quad \begin{array}{c} + Mg(s) \\ \longrightarrow \\ + MgO(s) \\ \longrightarrow \\ + MgCO_3(s) \\ \longrightarrow \end{array} \quad \begin{array}{l} MgSO_4(aq) + H_2(g) \\ \\ MgSO_4(aq) + H_2O(l) \\ \\ MgSO_4(aq) + CO_2(g) + H_2O(l) \end{array}$$

Three reactions which produce magnesium sulfate

Magnesium oxide is a white solid made by heating magnesium carbonate. It is a *basic oxide*. In water it turns to magnesium hydroxide which is slightly soluble. The *antacid* milk of magnesia is a suspension of magnesium hydroxide in water. Magnesium oxide has a high melting point and is used as a *refractory ceramic* to line furnaces.

The reactions of the metal, its oxide or its carbonate with dilute acids provide a direct method for making soluble magnesium salts.

Epsom salt is hydrated magnesium sulfate, $MgSO_4.7H_2O$, which is a laxative.

main group elements: a term used by some chemists to describe the elements in the *s-block* and *p-block* of the periodic table but excluding the transition elements in the *d-block* and *f-block*.

malleability: metals are malleable if they can be hammered into shape without breaking. It is possible to make very thin sheets of a malleable metal such as gold by hammering.

manganese (Mn): a hard, grey brittle *d-block* metal with the electron configuration $[Ar]3d^5 4s^2$.

The main oxidation states of manganese are:

- +7, MnO_4^- – the purple permanganate ion, which is a strong oxidizing agent especially in acid solution

- +4, MnO_2 – an insoluble, black compound which is an oxidizing agent in acid solution

- +2, Mn^{2+} – the pink manganese(II) ion in salts such as manganese(II) sulfate.

manometer: an instrument for measuring *pressure*. A manometer consists of a U-tube containing a liquid such as water or mercury.

Typically one arm of a manometer is open to the atmosphere, so the instrument measures the extent to which the pressure inside the apparatus is higher or lower than atmospheric pressure.

margarine, and other non-dairy spreads, are solid emulsions consisting of water finely dispersed in *vegetable oils*, with a high enough proportion of hardened oils to make the product a spreadable solid. *Hydrogenation* is used industrially to add hydrogen to double bonds in oils. This produces saturated fats which are solid at room temperature. So the process is sometimes called "hardening." Manufacturers bubble hydrogen though the oil in the presence of finely divided nickel, which is the *catalyst*.

Markovnikov rule: a rule which predicts the main product when a compound HX (such as H — Br, H — OSO_3H or H — OH) adds to an unsymmetrical *alkene* (such as propene). The rule is that the hydrogen atom adds to the carbon atom that already has more hydrogen atoms attached to it. This pattern was first reported by the Russian chemist Vladimir Markovnikov who studied a great many alkene *addition reactions* during the 1860s.

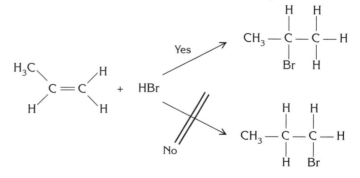

Markovnikov's rule

mass (symbol *m*): a fundamental physical quantity. The *SI unit* of mass is the kilogram (kg). 1 kg = 1000 g.

Mass can be determined by measuring the pull of gravity on a specimen. This is how chemists usually measure the masses of chemicals with a chemical balance.

Alternatively masses can be found by seeing how much an object accelerates when a force acts on it. This is how chemists measure the masses of ionized atoms and molecules in *mass spectrometry.*

mass number, *A*: the total number of protons and neutrons in the nucleus of an atom. *Isotopes* have the same atomic number but different mass numbers. The alternative term is nucleon number.

mass spectrometry: an accurate instrumental technique for determining *relative atomic masses* and *relative molecular masses*. Mass spectrometry can also help to determine molecular structures and to identify unknown compounds.

The instrument – inside a mass spectrometer there is a high vacuum so that it is possible to produce and study ionized atoms and molecules including fragments of molecules which do not otherwise exist.

Diagram of a mass spectrometer

The stages in producing a mass spectrum:

* inject a small sample into the instrument where it vaporizes

* bombard the sample with a beam of high-energy electrons which turns the atoms or molecules into positive ions by knocking out electrons

* accelerate the positive ions in an electric field

* deflect the moving stream of ions with a magnetic field to focus ions with a particular mass onto the detector

* feed the signal from the detector to a computer which prints out a mass spectrum as the magnetic field steadily changes over a range of values to focus the series of ions with different masses one by one onto the detector.

The instrument is calibrated using a reference compound of known structure and molecular mass so that the computer can print a scale on the mass spectrum.

Isotopes of elements – the mass spectrum for an element shows the relative abundance of different *isotopes* of the element. This makes it possible to calculate the *relative atomic mass* for the element.

Mass spectrum of magnesium   Mass of the positive ions detected

Molecular masses and structures – when molecular compounds are being analyzed, the peak of the ion with the highest mass is usually the whole molecule, ionized. So the mass of this "parent ion" is the *relative molecular mass* of the compound.

Mass spectrum of butane, C_4H_{10} showing the parent ion with a value 58. This corresponds to the relative molecular mass of 58. The pattern of fragments is characteristic of this compound. The very small peak at 59 is the "M + 1" peak, which is present because some molecules of butane include carbon-13 atoms. Carbon-13 makes up 1.1% of natural carbon so in a molecule with four carbon atoms the "M + 1" peak is 4.4% of the molecular ion peak.

Bombarding molecules with high-energy electrons splits them into fragments so the mass spectrum is a "fragmentation pattern." The computer has a database of mass spectra so it can identify an unknown by matching its spectrum with one in its database. The fragments which show up in the spectrum are positive ions. The highest peaks correspond to positive ions which are relatively more stable, such as tertiary *carbocations* or ions such as RCO⁺ (the acylium ion).

Chemists who synthesize new compounds can study their fragmentation patterns. They identify the fragments from their masses and then piece together likely structures with the help of evidence from other methods of analysis such as *infrared spectroscopy* and *NMR spectroscopy*.

The combination of *gas–liquid chromatography* (GLC) with mass spectrometry is of great importance in modern chemical analysis. First, GLC separates the chemicals in an unknown mixture, such as a sample of polluted water; then mass spectrometry detects and identifies the components.

matches use phosphorus compounds to make a flame. In a match that will strike anywhere, the head contains phosphorus sulfide (P_4S_3) and potassium chlorate as an oxidizing agent. The head contains the fuel and the oxidant and needs only friction to heat the match and start a fire.

Safety matches have sulfur and potassium chlorate in the head and red phosphorus in the striking strip on the side of the box.

Maxwell–Boltzmann distribution: the distribution of molecular kinetic energies for a gas at a particular temperature.

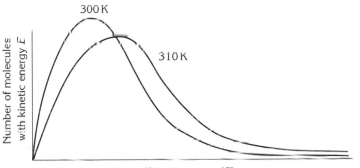

Boltzmann distribution of molecular kinetic energies in a gas at two temperatures. The modal speed gets higher as the temperature rises. The area under the curve gives the total number of molecules. This does not change as the temperature rises so the peak height falls as the curve widens.

The Boltzmann distribution is important in the *collision theory* of reaction rates and helps to account for the effects of temperature changes and *catalysts* on *rates of reaction*.

medicine: a substance or mixture of substances used to treat a disease or to give relief from the symptoms of disease. A medicine normally contains one or more active *drugs* dissolved in water or mixed with an inert solid. Mixing the active ingredient with an inert material makes it easier to give an accurate dose.

Doctors must prescribe some medicines – these are "prescription only medicines" (POMs). New medicines may start in this category but later be made available "over the counter" once they have been proved safe.

Medicines which people can buy from a pharmacist without a prescription are called over-the-counter (OTC) drugs. The sale of some of these medicines has to be supervised by a pharmacist. People can buy other medicines such as mild painkillers with no supervision.

For a medicine taken by mouth, added flavoring may make the medicine more palatable and coloring can help to identify a drug. Other drugs are taken by injection or drawn into the lungs by a deep breath.

The *pharmaceutical industry* employs many chemists in *drug development* for medicines.

mechanism of a reaction: a description of how a reaction takes place showing step by step the bonds which break and the new bonds which form. Some mechanisms involve *homolytic bond breaking* with *radical* intermediates. Other mechanisms involve *heterolytic bond breaking* and ionic intermediates.

Mechanisms with ionic intermediates can be classified according to the type of attacking molecules or ions involved. *Nucleophiles* attack atoms at the δ+ end of *polar covalent bonds*. *Electrophiles* attack regions which are electron rich, especially double bonds and the delocalized electrons in arenes.

Evidence used to support a proposed mechanism include the:

- *rate equation* for the reaction

- identification of *intermediates* using *spectroscopy* or other methods

- use of *isotopic labelling* to track what happens to particular atoms during a reaction.

melting: a change of state from a solid to a liquid. Another word for melting is fusion. The melting point for a pure substance is the temperature at which the solid and liquid are in equilibrium. It is the same temperature as the *freezing point*. Melting points vary with pressure but only very slightly.

Pure compounds have sharp melting points. Measuring melting points is a way of checking the purity and identity of compounds. This technique is especially important in *organic chemistry*.

Molecular substances melt at relatively low temperatures because the intermolecular forces between molecules are weak. *Network solids*, with strong bonding between the atoms, have high melting points. Energy is needed to overcome the bonds between particles as a substance melts. (See *enthalpy change of melting*.)

meniscus: the curved surface of a liquid in a tube. Water forms a concave meniscus in a glass tube because water molecules can form *hydrogen bonds* with oxygen atoms at a clean glass surface. This allows water to *wet* the glass. Mercury atoms bond strongly with themselves but not with atoms in glass, so mercury has no tendency to wet glass and forms a convex meniscus.

Chemists have to allow for the shape of the meniscus when calibrating and reading graduated glassware such as pipets and graduated flasks. The correct procedure is to adjust the level until the bottom of the meniscus just touches the mark as seen from eyes on a level with the mark.

mercury (Hg): the only liquid metal at room temperature. As a dense liquid, mercury (density 13.6 g/cm^3) has long been used to measure pressure in *manometers* and barometers. A *pressure* of 1 atmosphere is the pressure at the bottom of a column of mercury 760 mm high.

Mercury forms *amalgams* with metals. In the chlor-alkali industry, a mercury cell for the *electrolysis of brine* has a flowing mercury cathode. The product of electrolysis at the cathode is an amalgam of sodium in mercury. The amalgam flows to a separate vessel where the sodium in the amalgam reacts with water to make sodium hydroxide and hydrogen. This process produces very pure sodium hydroxide but it is being phased out because of the pollution problems caused by the traces of mercury which escape into the environment. Mercury vapor and mercury compounds are very *toxic*.

metal extraction involves two main types of process to reduce metal compounds to metals:

- **pyrometallurgy** – reactions at high temperature often above 1000°C;

    - electrolysis of a molten compound (for example *aluminum extraction*)

    - chemical reduction by coke in a blast furnace (for example *iron extraction*)

    - chemical reduction by a more reactive metal (for example *titanium extraction*)

- **hydrometallurgy** – reactions at low temperature in solution in water;

    - electrowinning using *electrolysis* of an aqueous solution (for example *zinc extraction*)

    - cementation using a *displacement reaction* (for example using iron to displace copper from a solution of copper(II) sulfate).

metallic bonds: the strong bonding between atoms in metal crystals. Each metal atom in a metal crystal contributes electrons from its outer shell to a "sea" of delocalized electrons. Only elements with relatively low first *ionization energies* form metallic crystals.

By losing electrons the metal atoms become positively charged. So a metal crystal consists of positively charged metal atoms held together by a "sea" of shared electrons.

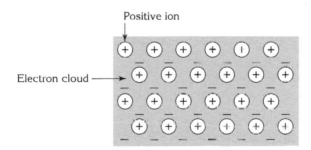

Simplified picture of metallic bonding

Metals conduct because the shared bonding electrons can drift through the crystal structure from atom to atom when there is an electric potential difference.

Metals can bend without breaking because metallic bonding is not highly directional. Lines or layers of metal atoms can shift their position in a crystal without the bonds breaking.

metalloid: an element with properties which are in between those of metals and non-metals. One example of such intermediate behavior is that metalloids are *semiconductors*. Their electrical conductivities are between those of metals and nonmetal insulators. The metalloids are boron, silicon, *germanium*, *arsenic*, antimony and tellurium.

metals: elements on the left-hand side of the *periodic table* with one, two or three electrons in the outer shell which take part in bonding and chemical reactions.

Physically, metals are:

- shiny when freshly polished and free of corrosion

- good conductors of electricity and thermal energy

- *malleable* and *ductile*

- (usually) solids with high melting and boiling points (only six metals melt below 100°C – mercury which is a liquid, gallium and four group I metals: sodium, potassium, rubidium and cesium).

Chemically, metals tend to:

- lose electrons forming positive ions (they are *reducing agents*)

- form *basic* or *amphoteric oxides* and hydroxides (which are *alkalis* if soluble in water)

- form solid, ionic chlorides.

The more an element shows these properties the greater its "metallic character."

methanol, CH₃OH, is a liquid alcohol which is becoming more important as a fuel or *gasoline* additive and as a raw material for the chemical industry. Industry can manufacture methanol from the mixture of *carbon monoxide* and hydrogen produced by *steam reforming*. Oxidation of methanol by air, in the presence of an iron or silver catalyst, produces the methanal required to make thermosetting resins.

Methanol reacts with carbon monoxide to make acetic acid in the presence of a catalyst made from rhodium or iridium with iodide ions. This process happens in the liquid phase and is an example of *homogeneous catalysis*.

$$CH_3OH(l) + CO(g) \rightleftharpoons CH_3COOH(l)$$

micelle: a cluster of *surfactant* molecules in a solution. A micelle has the "water-hating" (hydrophobic) hydrocarbon chains tangled together inside with the "water-loving" (hydrophilic) ends of the molecules on the outside.

The inside of a micelle is nonpolar. Unlike water, a solution of a surfactant in water can dissolve oil and grease. The molecules of oily dirt diffuse to the inside of the micelles.

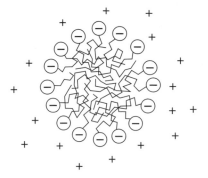

Cross section of a spherical micelle formed when a surfactant dissolves in water. A micelle is likely to consist of about 50 to a 100 molecules. The diameter of a micelle is in the range 5–50 nm.

microwaves: *electromagnetic radiation* with wavelengths in the range 1 mm to 30 cm and frequencies in the range 10^8 to 10^9 Hz. A polar molecule can absorb microwave radiation.

The quanta of microwave radiation correspond to the energy jumps when molecules gain energy and rotate faster (see *quantum theory*).

A microwave oven is "tuned" so that the radiation is absorbed by water molecules to make them spin. The microwaves can pass through food where they are absorbed by liquid water molecules. As the water molecules gain energy they bump into neighboring molecules and the energy spreads through the food as the molecules spin, vibrate and move around more, so the food gets hotter.

milliliter (symbol mL): a unit of *volume* often used on chemical glassware, in medicine and domestically. A milliliter is one thousandth of a *liter*.

$$1000 \text{ mL} = 1000 \text{ cm}^3 = 1 \text{ dm}^3 = 1 \text{ L}$$

minerals: the naturally occurring elements and compounds in the crust of the Earth. Most rocks are mixtures of minerals. There are about 3000 minerals of which 300 or so are important to the mineral and chemical industries.

Many *ores* contain only a small percentage of the useful mineral. The purpose of mineral processing is to separate the valuable mineral ("values") from the waste rock ("gangue"). Processing generally starts by crushing and grinding the ore to break it up and separate the mineral grains. The methods used to separate the "values" include:

● mechanical separation, often based on differences in *density*

● *froth flotation*

● chemical extraction by *leaching*.

mirror image molecules: see *chiral compounds*.

miscible liquids: liquids which mix with each other. The general rule is that "like dissolves like." This means that:

● nonpolar liquids mix freely with other nonpolar liquids (for example, *hydrocarbons* mix freely, as they do in *gasoline*)

● polar liquids mix with polar liquids (for example ethanol mixes with water)

● but polar and nonpolar liquids do not mix (gasoline floats on water).

mixed oxides: oxides which consist of a lattice of oxide ions (O^{2-}) with two or more types of metal ions, M^{n+}, in the spaces between the negative ions. So they are not mixtures of oxides but a distinct compound with a mixture of metal ions.

Examples of mixed oxides are:

● **red lead** (Pb_3O_4), which has a crystal lattice made up of ions in the proportions $2Pb^{2+}: Pb^{4+}: 4O^{2-}$

● **magnetite** (Fe_3O_4), the magnetic ore of iron originally called lodestone which contains $2Fe^{3+}$ ions for each Fe^{2+} ion in the lattice, and other magnetic *ferrites* such as $CaFe_2O_4$ in which Ca^{2+} ions take the place of Fe^{2+}

● **spinel** ($MgAl_2O_4$), made by heating aluminum oxide with magnesium oxide. Replacing some Al_2O_3 with 1% Cr_2O_3 produces a ruby-like gemstone.

mixtures are many and varied but all consist of two or more substances mixed together. Unlike compounds which have a definite formula, the composition of mixtures can vary widely.

The study of mixtures is very important because chemists have to know how to:

- separate mixtures – by *chromatography*, *distillation*, *filtration*, *froth flotation*, and *solvent extraction*

- formulate new mixtures for special purposes – when making *medicines*, *paints*, *detergents*, *dyes*, *pesticides*, *fertilizers* and many other products.

mobile phase: the liquid or gas which moves over the stationary *phase* during *chromatography*.

Mohs' scale of hardness: a scale for comparing the *hardness* of materials and minerals. The scale is based on scratching one substance with another. A harder material scratches a less hard material. The scale ranges from 1 for talc which is very soft to 10 for diamond which is very hard.

molality: a measure of concentration used when investigating *colligative* properties of solutions. The molality of a solution is the amount of solute in moles per kilogram of solvent.

molar concentration: see *concentration of solutions*. The term "molarity" is used for the molar concentration. Chemists write 2.0 M to describe a solution with a molar concentration of 2.0 mol/L.

molarity: the amount of solute in moles per liter of solution.

molar mass: the mass per mole of a chemical (see *amount of substance*). The symbol for the molar mass is M. The unit is g/mol. As always with molar amounts the chemical *entity* must be specified. The molar mass of hydrogen atoms, $M(H) = 1$ g/mol. The molar mass of hydrogen molecules $M(H_2) = 2$ g/mol.

The molar mass for the atoms of an element is numerically equal to the *relative atomic mass* of the element. The relative atomic mass of carbon is 12. the molar mass of carbon atoms $= 12$ g/mol.

The molar mass of a substance is calculated by adding up the molar masses of the atoms in the formula.

$$M_r(H_2O) = [(2 \times 1 \text{ g/mol}) + 16 \text{ g/mol}] = 18 \text{ g/mol}$$

$$M_r(CuSO_4) = [63.5 \text{ g/mol} + 32 \text{ g/mol} + (4 \times 16 \text{ g/mol}] = 159.5 \text{ g/mol}$$

$$M_r(CuSO_4.5H_2O) = [159.5 \text{ g/mol} + (5 \times 18 \text{ g/mol})]$$
$$= 249.5 \text{ g/mol}$$

Molar masses can be measured experimentally by *mass spectrometry*.

molar volume: the volume of a one mole *amount of substance*. The molar volume of any gas is the same under given conditions of temperature and pressure. This follows from *Avogadro's law* or from the *ideal gas* equation.

Rearranging the ideal gas equation, $pV = nRT$ gives: $V = \dfrac{nRT}{p}$

This shows that, for one mole of gas ($n = 1$), the volume (the molar volume) depends only on the temperature and pressure since R is a constant. The molar volume of an ideal gas at stp (273 K and 101.3×10^{-3} N/m^2) $= 22\,400$ cm^3.

mole: the chemist's term for an amount of stuff containing a particular number of atoms, molecules ions or any other type of particles. The word entered the language of chemistry at the end of the nineteenth century based on the Latin word for a heap or pile.

The mole is the *SI unit* for *amount of substance*. One mole is the amount of substance that contains as many specified *entities* (atoms, molecules, ions, electrons and so on) as there are atoms in exactly 12 g of the carbon-12 *isotope*. The number of entities per mole is the *Avogadro constant*.

mole fraction: a way of measuring the proportion of a substance in a mixture where it is the amount in moles which is important and not the chemical nature of the components.

In a mixture of n_A moles of A with n_B moles of B and n_C moles of C, the mole fractions (symbol X) are given by the following:

$$X_A = \frac{n_A}{n_A + n_B + n_C} \qquad X_B = \frac{n_B}{n_A + n_B + n_C} \qquad X_C = \frac{n_C}{n_A + n_B + n_C}$$

So the mole fraction of B is the fraction of all the moles which are moles of B.

The sum of all the mole fractions is 1, so $X_A + X_B + X_C = 1$.

Mole fractions are used by chemists when studying:

● gas mixtures (see *partial pressures*)

● the vapor pressure of solutions and mixtures of liquids (see *Raoult's law*).

mole of electrons: see *Faraday constant*.

molecular formula: a formula which shows the number of atoms of each element in a molecule. The molecular formula of chlorine is Cl_2, of ammonia is NH_3 and of ethanol is C_2H_5OH. The term molecular formula applies only to substances which consist of *molecules*.

A molecular formula is always a simple multiple of the *empirical formula*. Analysis shows that the empirical formula of hexane is C_3H_7. The mass spectrum shows that the relative molecular mass of hexane is 86. The relative mass of the empirical formula, $M_r(C_3H_7) = (3 \times 12) + (7 \times 1) = 43$. So the molecular formula is twice the empirical formula. The molecular formula of hexane is C_6H_{14}.

molecular models: physical and computer models used to show the atoms in molecules, the bonding between them and the shapes of molecules in three dimensions. (See *ball and stick models* and *space-filling models*.)

molecular orbitals: the orbitals in molecules which form as *atomic orbitals* overlap and interact when atoms bond together. Molecular orbitals fill with electrons according to the same rules as atomic orbitals. Molecular orbitals show the relative energies of the electrons in molecules. Their shapes also show the regions in space where there is a high probability of finding electrons. (See *sigma bonds* and *pi bonds*.)

molecular sieve: a solid with very fine pores which can separate molecules according to size. Small molecules can fit into the holes or channels in the crystal structure while larger molecules cannot. (See *zeolites*.)

molecularity of reaction steps: the number of reacting particles taking place in one step in the *mechanism of a reaction*. Often the term refers specifically to the rate determining step in the mechanism. The S_N1 mechanism for *nucleophilic substitution* is unimolecular. The S_N2 mechanism is bimolecular.

molecule: a group of atoms held together by covalent bonding. Most nonmetals are molecular, such as H_2, N_2, P_4, S_8, Cl_2, Br_2 and I_2. Most compounds of nonmetals with other nonmetals are also molecular, such as H_2O, CO_2, CH_4, NH_3, HBr, $SiCl_4$, PCl_3.

The covalent bonds in molecules are strong. The forces between molecules (*intermolecular forces*) are weak. Molecular substances have low melting and boiling points because of the weakness of intermolecular forces.

monochromatic radiation: radiation with a single wavelength. Spectrometers contain a diffraction grating or prism to provide monochromatic radiation. Rotating the grating or prism allows the instrument to scan across the range of wavelengths provided by the source of radiation.

A filter in a *colorimeter* is a cheap way of selecting a narrow range of wavelengths typically with a bandwidth of 20 to 50 nm. The disadvantage of a filter is that the operator has to change the filter to change from one band of wavelengths to another.

Monochromatic light is also important when measuring the rotation of *polarized light* by optically active compounds.

monodentate ligands: *ligands* which use a single *lone pair of electrons* to form one bond with the central metal atom in a *complex* ion. Examples of monodentate ligands are ammonia molecules, water molecules, chloride ions and cyanide ions.

monomers: the small molecules which join together in long chains to make *polymers*.

monoprotic acid: an *acid* which can give away (donate) only one *proton* per molecule. Examples of monoprotic acids are: hydrogen chloride, HCl; acetic acid, CH_3COOH; and nitric acid, HNO_3.

monosaccharide: see *carbohydrates.*

mordant: a chemical which fixes a dye to the fibers of a textile. Alums and other aluminum salts are used as mordants. *Dative bonds* form both between —OH groups of cotton and aluminum ions and between oxygen atoms on the dye and aluminum ions.

Aluminum ions acting as a mordant for the dye alizarin

Alizarin is a natural red dye once extracted from the roots of the madder plant but now made synthetically. Alizarin only sticks fast to cloth in the presence of a mordant and the color of the dye varies depending on the metal ions in the mordant.

multiple bonds: *double bonds* or *triple bonds* between atoms in molecules. In a double bond there are two shared pairs of electrons. In a triple bond there are three shared pairs.

mutagens: agents that cause mutations. A mutation is a change in the genetic material of living cells. Some chemicals are mutagenic and so are the forms of *ionizing radiation.*

names of carbon compounds: modern *IUPAC* names make it possible to determine the name from the formula and the formula from the name.

The name of an organic compound is based on the longest straight chain or main ring of carbon atoms in the skeleton of carbon atoms. If the main part is a straight chain the name is based on the corresponding *alkane*. For ring compounds the name is based on the corresponding *alicyclic* compound or *arene*.

Numbering the carbon atoms identifies the positions of side chains and *functional groups*, with the number repeated if there are two side groups on the same carbon atom. At all times chains are numbered from the end that gives the lowest possible numbers in the name. Numbers are omitted when there is no doubt about the position of the side chains or functional groups as in ethanol.

Prefixes (in front) and suffixes (following) in the hydrocarbon name identify the side chains and functional groups.

Where there are two or more of the same side chain or functional group the number is stated as di, tri, tetra and so on.

Prefixes are used for:

- *alkyl groups*, such as 2,3-dimethylbutane or 1,3,5-trimethylbenzene
- *haloalkanes*, such as 2-iodopropane or tetrachloromethane.

2,3-dimethylbutane

1,3,5-trimethylbenzene

2-iodopropane

tetrachloromethane
(carbon tetrachloride)

Structures of the examples

Suffixes are used for:

- double bonds in *alkenes*, such as 2-butene (with a double bond between the second and third carbon atoms)
- *alcohols*, such as 1,2,3-propanetriol
- *aldehydes*, such as propanal
- *ketones*, such as 3-pentanone
- *carboxylic acids*, such as hexanoic acid and ethanedioic acid
- *amines*, such as ethylamine and *phenylamine* (aniline).

$$CH_3 - CH_2 = CH_2 - CH_3$$

2-butene

1,2,3-propanetriol
(glycerol)

propanal

$$CH_3 - CH_2 - \overset{\overset{\displaystyle O}{\|}}{C} - CH_2 - CH_3$$

3-pentanone

hexanoic acid

ethanedioic acid
(oxalic acid)

$$CH_3 - CH_2 - NH_2$$

ethylamine

phenylamine
(aniline)

Structures of the examples

For other examples see entries for the various series of organic compounds.

The systematic names of complex molecules can be hard to pronounce and cumbersome. Chemists still often use older, traditional names which were often based on the Latin name for the source of the compound or where it was discovered. The organic acid 2,3-dihydroxybutanedioic acid, for example, occurs in grape juice. The traditional name is tartaric acid from "tartar" the name of the recrystallized deposits of tartaric acid salts collected from inside wine containers.

names of complex ions: the systematic names of complex ions show:

- first the number of *ligands*, "di", "tri", "tetra", "penta", "hexa"

- then the type of ligands (in alphabetical order if there is more than one type of ligand), such as "aqua" for water molecules, "ammine" for ammonia, "chloro" for chloride ions and "cyano" for cyanide ions

- next the identity of the central metal atom in a form which shows whether or not the ion is a cation or an anion:
  - for cations (and uncharged complexes) the metal name is normal, such as silver, iron or copper
  - for anions the metal name ends in "ate" and often has an old-fashioned style such as argentate for silver, ferrate for iron and cuprate for copper

- finally the oxidation number of the metal.

Examples:

- $[Ag(NH_3)_2]^+$ is the diamminesilver(I) ion,

- $[Cu(H_2O)_6]^{2+}$ is the hexaaquacopper(II) ion,

- $[CuCl_4]^{2-}$ is the tetrachlorocuprate(II) ion,

- $[Fe(CN)_6]^{4-}$ is the hexacyanoferrate(II) ion,

- $[Fe(CN)_6]^{3-}$ is the hexacyanoferrate(III) ion.

Note that the overall charge on the complex ion is the sum of the charges on the metal ion and the ligands.

names of inorganic compounds are becoming increasingly systematic but chemists still use a mixture of names. Most chemists prefer to call $CuSO_4.5H_2O$ hydrated copper(II) sulfate, or perhaps copper(II) sulfate-5-water but not the fully systematic name tetraaquo-copper(II) tetraoxosulfate(VI)-1-water. The systematic names has much more to say about the arrangement of atoms, molecules and ions in the blue crystals but it is too cumbersome for normal use.

These are some of the basic rules for common inorganic names:

- the ending "ide" shows that a compound contains just two elements mentioned in the name. The more *electronegative* element comes second, for example, sodium sulfide (Na_2S) carbon dioxide (CO_2) and phosphorus trichloride (PCl_3)

- the small Roman numerals in names are the oxidation numbers of the elements, for example iron(II) sulfate ($FeSO_4$) and iron(III) sulfate, $Fe_2(SO_4)_3$

- the names of *oxoacids* end "ic" or "ous" as in sulfuric (H_2SO_4) and sulfurous (H_2SO_3) acids and nitric (HNO_3) and nitrous (HNO_2) acids, where the "ic" ending is for the acid in which the central atom has the higher oxidation number

- the corresponding endings for the salts of oxoacids are "ate" and "ite" as in sulfate (SO_4^{2-}) and sulfite (SO_3^{2-}) and in nitrate (NO_3^-) and nitrite (NO_2^-).

To avoid misunderstandings chemists give the name and formula. If necessary they may also give two names: the systematic name and the traditional name.

nanometer: a unit of length used for the sizes of atoms, molecules and ions. One nanometer (1 nm) is one thousand millionths of a meter (10^{-9} m). One nanometer (1 nm) is one-thousand *picometers* (1000 pm). There are one million nanometers in a millimeter (1000 nm = 1 mm). One nanometer is roughly four times the diameter of an iron atom.

nanotubes: single-walled cylinders of carbon atoms created as graphite-like sheets of atoms which curl up into tubes, similar to rolls of chicken wire. They take their name from the *nanometer* which indicates their scale. Nanotubes form under similar conditions to *buckminsterfullerene* but with a few percent of nickel or cobalt atoms added to the mixture of carbon vapor and helium. Research into the properties of nanotubes is exploring the possibility that they could be adapted to make minute diodes, transistors and other electronic devices on an atomic scale – much smaller than existing devices on silicon chips. Nanotubes are also being incorporated into plastics to make them conduct electricity. Nanotube fibers are very strong and tough. In time they may become key ingredients of new composite materials.

naphtha: a mixture of hydrocarbons from the *fractional distillation of oil*. The naphtha fraction is an important feedstock for the *petrochemical industry*. Naphtha contains *hydrocarbons* with 6 to 10 carbon atoms in their molecules.

narcotics: powerful painkillers (*analgesics*) which also lessen anxiety and give a feeling of well-being. Regular use of narcotic analgesics can lead to drug dependence. Morphine is the oldest and most well known narcotic which is obtained from the dried juice of opium poppies (so it is an opiate). Morphine is used medically to treat patients in severe pain.

Diamorphine, otherwise known as heroin, is made from morphine. It is an even more powerful analgesic than morphine. It is also highly addictive so its medical use is limited mainly to the treatment of pain in patients who are going to die.

Codeine is another opiate. It is a much less powerful painkiller than morphine but it does not cause dependence. Unlike morphine, codeine can be taken by mouth. Codeine is an ingredient of some cough medicines because it helps to stop people coughing.

natural gas: a *fossil fuel* used for domestic heating, for raising steam in power stations and as a feedstock for the *petrochemical industry*. The composition of natural gas varies from one gas field to the next. It consists mainly of methane mixed with other *hydrocarbons* such as ethane, propane and butane together with variable amounts of carbon dioxide, nitrogen, helium and hydrogen sulfide.

Natural gas is processed before being supplied to homes. Hydrocarbons other than methane are separated for use in the petrochemical industry. Sulfur compounds are removed and a trace of a smelly chemical is added so that people notice gas leaks.

natural product chemistry: the study of chemicals produced by living things. Natural products include *carbohydrates*, *proteins*, *nucleic acids* and *lipids*. Before the start of the modern organic chemical industry, natural products from plants were the important ingredients of perfumes, dyes, oils, food flavors and drugs.

Plants may become important again as a large scale source of chemicals now that genetic engineering makes it possible to transfer genes from one species to another.

neon (Ne): the second member of the family of *noble gases,* coming below helium, with the *electron configuration* $[He]2s^22p^6$. Neon is best known for the red glow of neon lamps and tubes. It is separated from air as the other gases are liquefied. Neon boils at 27 K so it does not condense at the temperatures used to liquefy oxygen, nitrogen and argon.

network solids: crystal structures in which all the atoms are strongly linked by a network of covalent bonds extending throughout the crystal. Network solids have high melting and boiling points.

A few *nonmetal* elements consist of network structures of atoms held together by *covalent bonding*. Examples are carbon (as *graphite* or *diamond*) and silicon. Typically these elements do not conduct electricity because the electrons in covalent bonds are localized between pairs of atoms. Graphite is the exception with electrons *delocalized* over the layers of carbon atoms.

Some compounds of nonmetals with nonmetals, such as silicon dioxide and boron nitride, are network solids. These compounds are nonconductors because each bonding pair of electrons is localized between two atoms.

Many silicates are based on networks of silicon and oxygen atoms as in the *clay minerals*. Chains, sheets and three dimensional arrays of silicate ions combined with metal ions give rise to many different minerals.

neutralization reaction: a reaction in which an acid reacts with a base to form a *salt*.

$$HCl(aq) + NaOH(aq) \longrightarrow NaCl(aq) + H_2O(l)$$

Mixing equal amounts (in moles) of hydrochloric acid with sodium hydroxide produces a *neutral solution* of sodium chloride.

Strong acids, such as hydrochloric acid, and *strong bases*, such as sodium hydroxide, are fully ionized in solution, as is the salt formed, sodium chloride. Writing ionic equations for these examples shows that neutralization is essentially a reaction between aqueous hydrogen ions and hydroxide ions. This is supported by the values for the *enthalpy changes* of neutralization.

$$H_3O^+(aq) + OH^-(aq) \longrightarrow 2H_2O(l)$$

The surprise is that neutralization reactions do not always produce neutral solutions. Neutralizing a weak acid such as acetic acid with an equal amount in moles of a strong base sodium hydroxide produces a solution of sodium acetate which is alkaline.

Neutralizing a weak base, such as ammonia, with an equal amount of the strong acid hydrochloric acid produces a solution of ammonium chloride which is acidic.

Where a salt has either a "parent acid" or a "parent base" which is weak it dissolves to give a solution which is not neutral (see *hydrolysis of salts*). The "strong parent" in the partnership "wins":

• weak acid/strong base – the salt is alkaline in solution

• strong acid/weak base – the salt is acidic in solution.

neutral oxides: nonmetal oxides which do not react with water to form acids. The common examples are carbon monoxide (CO), nitrogen monoxide (NO) and dinitrogen oxide (N_2O). All three of these oxides are insoluble in water.

neutral solution: a solution with pH 7. Since pH = $-lg\ [H_3O^+]$, this means that the concentration of aqueous hydrogen ions in a neutral solution is 10^{-7} mol/L at 298 K.

neutrons: the uncharged particles in the nuclei of *atoms*. The number of neutrons in an atom can vary without changing its chemical properties. This accounts for the existence of *isotopes*.

nickel (Ni): a hard, greyish but shiny *d-block* metal with the electron configuration [Ar]$3d^84s^2$. Nickel is relatively unreactive so it is used to make spatulas and crucibles.

Nickel is a constituent of many alloys including some alloy *steels* and the ferromagnetic alloy Alnico in permanent magnets.

The common oxidation state of nickel is +2. Nickel(II) salts, such as the sulfate $NiSO_4$, are green.

Finely divided nickel metal is a good catalyst for *hydrogenation* reactions. It is a *heterogeneous catalyst*. It is used to "harden" unsaturated *vegetable oils* by adding hydrogen across the double bonds.

After extraction from its ores, nickel is purified by *electrolysis*.

nickel–cadmium cells are now widely used as rechargeable "NiCad" cells. As in a *lead–acid cells*, when a current flows the products formed at the electrodes are insoluble solids which stay put instead of dissolving in the electrolyte. This means that the electrode processes can be reversed as the cell is recharged. The electrode processes as a NiCad cell supplies a current are:

anode (oxidation): $Cd(s) + 2OH^-(aq) \longrightarrow Cd(OH)_2(s) + 2e^-$

cathode (reduction): $NiO(OH)(s) + H_2O(l) + e^- \longrightarrow Ni(OH)_2(s) + OH^-(aq)$

Unlike lead–acid cells, "NiCad" cells must be regularly discharged fully and then recharged if they are to retain their full capacity.

nitrates: *salts* of nitric acid (HNO_3) which all contain the nitrate ion (NO_3^-). Nitrates are common in laboratory chemistry because they are all soluble in water. Since nitric acid is a *strong acid*, the nitrate ion is a very weak base and does not change the pH when dissolved in water.

The reason that the nitrate ion is a weak base is that the negative charge is *delocalized* over the planar ion. Delocalization stabilizes the ion (see *oxoacids*).

Heating decomposes nitrates. Hydrated nitrates first give off steam, then the usual products are the metal oxide, oxygen and the brown gas nitrogen dioxide.

$$2Mg(NO_3)_2(s) \longrightarrow 2MgO(s) + O_2(g) + NO_2(g)$$

The exceptions are the nitrates of sodium and potassium which are hard to decompose and give only the nitrite and oxygen.

$$2NaNO_3(s) \longrightarrow 2NaNO_2(s) + O_2(g)$$

Nitrates are *oxidizing agents*. In alkaline conditions they are reduced to ammonia by aluminum metal on heating. This reaction is used as a test for nitrates since the basic ammonia gas can be detected by turning litmus paper blue (see *anion tests*).

$$NO_3^-(aq) + 6H_2O(l) + 8e^- \longrightarrow NH_3(g) + 9OH^-(aq)$$

Plants take up nitrogen from the soil in the form of nitrate ions. So a diet rich in vegetables is rich in nitrate too. Nitrates have had a bad press because they are associated with the environmental problem *eutrophication*.

nitration of benzene: an *electrophilic substitution* reaction which takes place in the presence of a *nitrating mixture* of concentrated nitric and sulfuric acids. The product is nitrobenzene.

The purpose of the concentrated sulfuric acid is to produce an electrophile which is the nitronium ion, NO_2^+. The concentrated sulfuric acid gives a proton to the —OH group in nitric acid which then loses a molecule of water, forming the nitronium ion.

$$HO-NO_2 + H_2SO_4 \longrightarrow HO^+_H-NO_2 + HSO_4^-$$

$$\downarrow$$

$$H_2O + NO_2^+$$

Mechanism for the nitration of benzene

Nitration of arenes is important because it produces a range of useful products including the explosive TNT (trinitrotoluene). Nitro compounds are also intermediates in the synthesis of chemicals used to make *polyurethanes*. Nitro groups are easily reduced to amine groups. Aryl amines, such as *aniline*, are important intermediates in the production of dyes, such as *azo dyes*.

nitric acid (HNO_3): pure nitric acid is a colorless, fuming liquid but it gradually turns yellow as it decomposes forming nitrogen dioxide. It is a highly corrosive but important chemical reagent because it can act as a:

- **strong acid** – nitric acid ionizes fully as it dissolves in water:

  $$HNO_3(l) + H_2O(l) \longrightarrow H_3O^+(aq) + NO_3^-(aq).$$

  Dilute nitric acid neutralizes bases producing soluble salts called *nitrates*.

- **powerful oxidizing agent** – in this role nitric acid oxidizes:
  - most metals including metals such as copper which do not react with non-oxidizing acids
  - nonmetals, forming oxides in the highest oxidation state of the element, giving I_2O_5 with iodine, for example
  - ions in solution such as iodide ions to iodine and iron(II) ions to iron(III)

- **nitrating agent** – in this role it is used for the *nitration of benzene* and other arenes.

nitric acid manufacture: a process for converting ammonia to nitric acid in two stages:

- oxidation of ammonia by oxygen on the surface of a catalyst gauze made of an alloy of platinum and rhodium:

  $$4NH_3(g) + 5O_2 \rightleftharpoons 4NO(g) + 6H_2O(g) \qquad \Delta H = -909 \text{ kJ/mol}$$

- absorption in water in the presence of oxygen to make nitric acid:

  $$2NO(g) + O_2(g) \rightleftharpoons 2NO_2(g)$$

  $$4NO_2(g) + O_2(g) + 2H_2O(l) \longrightarrow 4HNO_3(g)$$

Flow diagram for the manufacture of nitric acid

243

nitriles

Nitric acid is commonly converted to ammonium nitrate. Ammonium nitrate is largely used as a nitrogen *fertilizer*. Ammonium nitrate is also widely used as part of most *explosives* for mining and quarrying. Nitric acid is used to make other explosives such as nitrocellulose and nitroglycerine.

Nitric acid is an important reagent in the production of intermediates for making *plastics*, especially *polyamides* (such as nylon) and *polyurethanes*.

nitriles: organic compounds with the functional group — C≡N. Their general formula is R — CN where R is an *alkyl* or an *aryl group*.

Two ways of making nitriles

Hydrolysis converts a nitrile to a *carboxylic acid*. Acids or alkalis can catalyze this reaction. Reduction with *lithium aluminum hydride* converts nitriles to amines.

Reactions of nitriles

Nitriles are useful intermediates in organic synthesis, especially when there is a need to add to the carbon skeleton of the molecule.

Lengthening the chain of a carbon skeleton

nitrites: salts of *nitrous acid*. The two common examples are sodium nitrite ($NaNO_2$) and potassium nitrite (KNO_2). Sodium nitrite and potassium nitrite are preservatives and curing agents used in cooked meats and sausages.

nitrogen (N_2): an element which is chemically quite *inert* but which plays a vital part in the environment and forms many compounds of natural and economic importance. Nitrogen is the first element in group V with the electron configuration $[He]2s^22p^3$. Nitrogen gas (N_2) is unreactive because the atoms are joined by a very strong triple bond and the molecule is nonpolar.

The bonding in a nitrogen molecule. The bond energy for the triple bond is 945 kJ/mol.

$$:N::N: \qquad N≡N$$

The inertness of nitrogen makes it useful wherever oxygen must be excluded, as in food packaging, the *float glass process*, some chemical processing and the production of *semiconductor* chips.

Nitrogen makes up 79% of the air and can be separated by *fractional distillation* of liquid *air*. Alternatively, when only nitrogen is needed, the gas can be separated using:

- *semipermeable membranes* which let though all the gases except nitrogen which is retained
- molecular sieves which absorb oxygen but let nitrogen pass through.

nitrogen cycle: the ways by which the element nitrogen circulates through the environment. The nitrogen cycle is chemically complex because nitrogen crops up in the environment in a range of oxidation states. Some of the nitrogen is inorganic as NH_3 and NH_4^+, N_2, N_2O, NO_2^- or NO_3^-. Some of it is organic as *amino acids*, *proteins* and *nucleic acids* in living things or their remains.

There are three main reservoirs of nitrogen:

- the air (about 3×10^{20} mol N),
- the crust of the Earth (about 2×10^{16} mol N),
- the oceans (about 6×10^{16} mol N).

The main types of chemical change in the cycle are:

- *nitrogen fixation* – converting nitrogen from the air into inorganic and then organic nitrogen
- **denitrification** – converting inorganic nitrogen in the soil or oceans to nitrogen gas in the air
- **assimilation** – converting inorganic nitrates to organic nitrogen compounds
- **mineralization** – converting organic nitrogen in living things into inorganic nitrites and nitrates.

Diagram of the nitrogen cycle. Note that this version of the cycle leaves out the oceans which are major reservoirs of nitrogen compounds.

nitrogen fixation: any process which converts *nitrogen* gas in the *air* to nitrogen compounds which can be taken up by plants and converted to amino acids, proteins and nucleic acids.

Nitrogen fixation happens naturally:

- during thunderstorms the energy in lightning allows nitrogen and oxygen to combine to form oxides of nitrogen which are washed into the soil by rain and taken up by plants as nitrates

- some soil bacteria have an enzyme nitrogenase which can harness the energy of *ATP* to convert nitrogen from the air into ammonium compounds

- bacteria in the nodules of leguminous plants such as peas, beans and clover can also make ammonium salts from nitrogen.

As agriculture became more intensive the demand for nitrogen by crops outstripped the rate it could be produced by natural processes. Manures could not meet the demand either, so the challenge for chemists was to find a practicable and economic way to fix nitrogen industrially. This is the problem which was solved by Fritz Haber, leading to *ammonia manufacture* by the Haber process. Some of the ammonia is used for *nitric acid manufacture* and the two chemicals combine to make nitrogen *fertilizers*.

nitrogen oxides: there are six compounds of nitrogen with oxygen. All the oxides have positive enthalpies of formation. They are unstable relative to the elements and are easily decomposed by heating. Two of the oxides, N_2O_3 and N_2O_5, are particularly unstable and of little practical importance. Oxides of nitrogen contribute to air pollution. NO_x is a general formula used in accounts of air pollution to stand for any of the various nitrogen oxides when fuels burn at high temperatures in engines and furnaces. NO_x contributes to *acid rain* and the formation of *photochemical smog*.

Oxidation state	Formula	Name	Type	ΔH_f° kJ/mol	Notes
+4	$2NO_2 \rightleftharpoons N_2O_4$	nitrogen dioxide and dinitrogen tetroxide	acidic	+33.2 (for NO_2)	NO_2 is the brown gas formed on heating nitrates. It is used as an oxidant in rocket fuels.
					Lowering the temperature makes the gas paler as the equilibrium favors the colorless N_2O_4.
+2	NO	nitrogen monoxide	neutral	+90.2	Colorless but reacts with oxygen in the air to make NO_2; an intermediate in the manufacture of nitric acid from ammonia.
+1	N_2O	dinitrogen oxide	neutral	+82.0	Colorless; also called laughing gas.

Laughing gas is the traditional name for dinitrogen oxide (N_2O). The gas can be used as an *anesthetic* for dentistry and minor surgery. Dinitrogen oxide is soluble in fats. It is tasteless and nontoxic, and is used as the foaming agent and propellant in cans of whipped cream.

In recent years scientists have discovered that NO is a vital messenger molecule in the human body. When tiny quantities of NO are released from cells lining blood vessels the muscles around the vessels relax so that they can carry more blood. This prevents the blood pressure from rising during exercise and avoids strain on the heart.

nitrous acid (HNO_2): an unstable, *weak acid* which is an important reagent for producing the *diazonium salts* needed to make *azo dyes*. Nitrous acid is too unstable to be stored so it is made in solution when needed by adding a strong acid, such as hydrochloric acid, to a solution of sodium nitrite:

$$NO_2^-(aq) + H_3O^+(aq) \longrightarrow HNO_2(aq) + H_2O(l)$$

The solution of the acid is blue. It starts to decompose at room temperature, giving off nitrogen monoxide (NO) which turns brown as it meets the air and turns into nitrogen dioxide (NO_2).

In nitrous acid the nitrogen is in the +3 oxidation state. When it decomposes it *disproportionates* to the +5 and the +2 states.

$$3HNO_2(aq) \longrightarrow HNO_3(aq) + 2NO(g) + H_2O(l)$$

$\quad$ +3 $\qquad\qquad\qquad$ +5 $\qquad$ +2 $\qquad\qquad\qquad$ *Oxidation states of nitrogen*

NMR: see *nuclear magnetic resonance spectroscopy*.

noble gases: the unreactive gases in *group VIII* of the periodic table. They were called the "inert gases" until the discovery that *krypton* and *xenon* can form compounds. Even so the gases keep apart from most chemical changes and react only with highly reactive chemicals such as fluorine and oxygen.

The term "noble" has been used for a long time to describe metals such as gold and silver. These metals were *inert* to the reagents used by alchemists and early chemists. Like the nobility in society, these elements were seen to be special by standing apart from everyday changes and ordinary events.

nomenclature: see *names of carbon compounds, names of complex ions* and *names of inorganic compounds*.

non-aqueous solvent: any solvent other than water. A non-aqueous solvent has to be used if the *solute* will not dissolve in water. For this reason a *hydrocarbon* solvent is used to make oil paint while other organic solvents are used as nail-varnish removers, stain removers and dry-cleaning fluids. A non-aqueous dry cleaning fluid removes dirt without causing shrinkage or other damage to fabrics harmed by water.

A non-aqueous solvent has to be used for a reaction in solution if one of the reactants or products is rapidly hydrolyzed by water. For this reason *lithium aluminum hydride* reductions take place in ether solution.

The choice of solvent can affect the outcome of a reaction. A solution of potassium hydroxide or sodium hydroxide in water hydrolyzes *haloalkanes* to alcohols. If the solvent is ethanol it is much more likely that an *elimination* reaction will happen and the product will be an alkene.

nonmetals: the elements towards the right-hand side of the *periodic table* which do not show the characteristic properties of metals. In general nonmetals:

A
B
C
D
E
F
G
H
I
J
K
L
M
N
O
P
Q
R
S
T
U
V
W
X
Y
Z

- consist of small molecules with the atoms linked by *covalent bonding* (H_2, S_8, P_4 and Cl_2); the exceptions include carbon, silicon and red phosphorus in which covalent bonding is continuous throughout the network of atoms

- are mostly gases at room temperature because of the weak *intermolecular forces* (for example hydrogen, oxygen, nitrogen, chlorine and all the noble gases)

- if solid, are not shiny like metals; they are brittle rather than bendable and do not normally conduct electricity or thermal energy

- tend to form negative ions by gaining electrons when they react with metals

- form *acidic oxides* which react with water to produce *oxoacids*

- form covalent molecular chlorides which are liquids and usually rapidly hydrolyzed by water, such as PCl_3 and $SiCl_4$

- form covalent molecular hydrides which are gases at room temperature, such as CH_4, SiH_4, NH_3, HCl, HBr and HI – the notable exception is water which is a liquid because of *hydrogen bonding*.

nonpolar solvent: a solvent in which the molecules are not *polar*. Examples of nonpolar solvents are the many liquids *hydrocarbons* and the mixtures of hydrocarbons obtained by refining crude oil. These are also *non-aqueous* solvents.

Following the general rule that "like dissolves like," nonpolar solvents dissolve many compounds which consist of small molecules. Chlorine, bromine and iodine, for example, dissolve freely in hexane. The stain removers and dry cleaning fluids intended to remove oily grease from clothes consist of nonpolar solvents.

Nonpolar solvents do not dissolve ionic crystals. The interaction between nonpolar molecules and ions is much too weak to surround ions and pull them away from the crystal against the *electrostatic forces* between ions with the opposite charge.

nonstoichiometric compounds: compounds with formulas which do not have simple whole number ratios of atoms. The formulas of nonstoichiometric compounds vary. Many oxides and sulfides of *d-block* elements have variable formulas for instance iron(II) sulfide ($Fe_{1.1}S$ to $FeS_{1.1}$) and iron(II) oxide ($Fe_{1.06-1.19}O$). This variation is a result of variable oxidation states and irregularities in the crystal lattice. Other examples are the *interstitial hydrides* of *d*-block elements such as vanadium hydride, $VH_{0.6}$.

NO$_x$: see *nitrogen oxides.*

nuclear fusion: the process which produces the energy of the Sun and other stars. It is also the process which accounts for the origin of all the elements.

In the Sun, at a temperature of ten million degrees or so, hydrogen atoms fuse to make helium atoms releasing about 10^9 kJ/mol of helium formed.

$$4\ {}_1^1H \longrightarrow {}_2^4He + 2 \text{ positrons}$$

The pressure and temperature at the center of very large stars is high enough for helium nuclei to fuse to make heavier elements such as carbon, silicon and iron. Since helium has an even number of *nucleons*, it turns out that elements with even *atomic numbers* are more abundant than other elements.

Iron is the final product of the series of *exothermic* fusion reactions. Energy is needed to make heavier elements. Heavier elements form during the highly energetic explosion when massive stars run out of nuclear fuel and start to collapse and then burst apart. The massive explosion is a supernova which scatters dust and gas through the universe where it mixes with hydrogen and helium.

In time the remains of "dead" stars start to coalesce into new stars and the process starts all over again. The Sun is an example of a second generation star; it and the planets of the solar system formed from the remains of supernovae. This explains the variety of elements on Earth in a universe where over 90% of all atoms are hydrogen.

nuclear magnetic resonance spectroscopy (NMR): a powerful analytical technique for finding the structures of carbon compounds. The technique is used to identify unknown compounds, to check for impurities and to study the shapes of molecules.

In medicine, magnetic resonance imaging uses NMR to detect the hydrogen nuclei in the human body, especially in water and lipids. A computer translates the information from a body scan into 3-D images of the soft tissue and internal organs which are normally transparent to X-rays.

The name of the technique summarizes its key features:

- **nuclear** – the technique detects nuclei of atoms such as hydrogen-1 (protons)

- **magnetic** – the nuclei detected by the technique are the ones which act like tiny magnets which can line up either in the same direction or in the opposite direction to an external magnetic field

- **resonance** – the absorption of energy is from radiowaves with the frequency corresponding to the size of the energy jump as the nuclei flip from one alignment in a magnetic field to the other.

Diagram of an NMR spectrometer

Samples for analysis are dissolved in a solvent with no hydrogen-1 atoms. Also in the solution is some tetramethyl silane (TMS) which is a standard reference compound which produces an absorption peak well away from the sample peaks.

The tube with the sample is supported in a strong magnetic field. The operator turns on an oscillator which produces radiation at radio frequencies (RF). The RF detector records the intensity of the signal as the oscillator scans across a range of wavelengths.

The recorder prints out a spectrum with peaks wherever the sample absorbs radiation strongly. The zero on the scale is fixed by the absorption of hydrogen atoms in the reference chemical TMS. The distances of the sample peaks from zero are called their "chemical shifts" (δ).

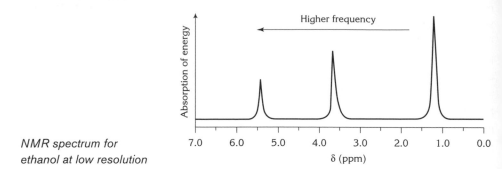

NMR spectrum for ethanol at low resolution

Each peak corresponds to a hydrogen nucleus in a different chemical situation. The area under a peak is proportional to the number of nuclei in each situation. In ethanol (CH_3CH_2OH) there are hydrogen nuclei in three environments which have different chemical shifts.

At high resolution it is possible to produce NMR spectra with more detail, which provide even more information about molecular structures.

High resolution NMR spectrum for ethanol. Note the extra peaks compared with the low resolution spectrum. This illustrates the "n + 1" rule. A peak from protons bonded to an atom which is next to an atom with two protons splits into three lines. A peak from protons bonded to an atom which is next to an atom with three protons splits into four lines.

Protons connected to neighboring atoms interact with each other. Chemists call this interaction "coupling" and they find that the effect is to split the peaks into a number of lines.

nuclear reactions: changes affecting the nuclei of atoms. Nuclear reactions are brought about by bombarding materials with high energy particles in particle accelerators or with beams of neutrons from nuclear reactors.

Nuclear reactions can be used to make elements which do not occur naturally, with higher proton numbers than uranium. Bombarding uranium-238 with neutrons produces uranium-239 which decays by *beta decay* to neptunium. Neptunium in turn decays to plutonium.

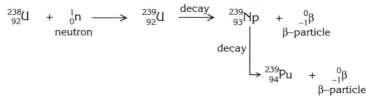

Nuclear reactions producing the transuranium elements neptunium and plutonium. Note that the mass numbers and atomic numbers balance.

Nuclear reactions are used to make radioactive isotopes for use as tracers.

nucleic acids: the molecules which carry genetic information and control protein synthesis in cells. There are two main types of nucleic acid, *DNA* and *RNA*. Nucleic acids are polynucleotides formed as *nucleotides* link together in long chains.

nucleon: a particle in the nucleus of an atom – either a *proton* or a *neutron*.

nucleon number: the nucleon number for an atom is an alternative term to *mass number*.

nucleophiles: molecules or ions with a lone pair of electrons which can form a new covalent bond. Nucleophiles are reagents which attack molecules where there is a partial positive charge, δ+. So they seek out positive charges – they are "nucleus-loving."

Some examples of nucleophiles are.

Examples of nucleophiles

Nucleophiles take part in *nucleophilic addition* and *nucleophilic substitution reactions*.

nucleophilic addition reaction: the attack of a nucleophile on a *carbonyl compound* leading to an *addition reaction*. The *electronegative* oxygen draws electrons away from carbon so that the $C=O$ bond in the aldehyde or ketone is *polar*. The incoming nucleophile uses its lone pair to form a new bond with the carbon atom. This displaces one pair of electrons in the double bond onto oxygen. Oxygen has thus gained one electron from carbon and now has a negative charge.

intermediate

First step of nucleophilic addition of hydrogen cyanide to acetaldehyde

To complete the reaction, the negatively charged oxygen acts as a *base* and gains a proton.

intermediate

Second step of nucleophilic addition of hydrogen cyanide to acetaldehyde

nucleophilic substitution in derivatives of carboxylic acids: nucleophilic attack on the carbon atom of the $C=O$ bond in the functional group which leads to a *substitution reaction*. The first step is similar to *nucleophilic addition*.

The *electronegative* oxygen draws electrons away from carbon so that the $C=O$ bond in the aldehyde or ketone is *polar*. The incoming nucleophile uses its lone pair to form a new bond with the carbon atom. This displaces one pair of electrons in the double bond onto oxygen. Oxygen has thus gained one electron from carbon and now has a negative charge.

The first step in the attack of a nucleophile on a derivative of a carboxylic acid. If $X = OC_2H_5$ the compound is an ethyl ester, if $X = Cl$ it is an acyl chloride, if $X = NH_2$ it is an amide.

At this point the negative oxygen does not gain a proton; instead it uses a pair of electrons to reform the double bond and displace X.

The second step which eliminates X as an X⁻ ion

Nucleophiles which react with esters, acyl chlorides and amides in this way are hydroxide ions and hydride ions from the aluminum hydride ion.

nucleophilic substitution in haloalkanes: *nucleophilic attack* on the carbon atom of the $C-X$ bond in a *haloalkane*, RX, which leads to a *substitution reaction*, where $X = Cl$, Br or I. Study of the *rate equations* suggests that there are two different mechanisms.

Hydrolysis of primary haloalkanes such as bromobutane, is overall *second order*. The rate equation has the form: rate $= k[C_4H_9Br][OH^-]$.

The suggested *mechanism* shows the $C-Br$ bond breaking as the nucleophile, OH^-, forms a new bond with carbon.

The S$_N$2 mechanism. Substitution–Nucleophilic–2 (for bimolecular – that is two molecules or ions involved in the rate determining step).

Hydrolysis of tertiary haloalkanes such as 2-bromo-2-methylpropane, is *first order*. The rate equation has the form: rate = $k[C_4H_9Br]$.

The suggested mechanism shows the C — Br bond breaking first to form a *carbocation intermediate*. Then the nucleophile OH⁻ forms a new bond with carbon.

The S$_N$1 mechanism. Substitution–nucleophilic–1 (for unimolecular – that is one molecule or ion involved in the rate determining step).

nucleotide: the monomer which polymerizes to make *nucleic acids*, DNA and RNA. A nucleotide consists of three parts:

- a **five-carbon sugar** – ribose in RNA and deoxyribose in DNA
- a **phosphate group**
- a **nucleotide base** – there are five different bases, adenine, cytosine, guanine, thymine and uracil often abbreviated as A, C, G, T and U.

In the nucleotides which make up DNA the base is one of A, C, G or T. In RNA, thymine, T, is replaced by uracil, U.

octet rule: this was first suggested as a guide by the American chemist Gilbert Lewis in 1916. The rule says that atoms tend to gain, lose or share electrons when they combine with other atoms to acquire a stable octet of electrons. The "stable octet" is the eight $s^2 p^6$ electrons, corresponding to the outer *electron configuration* of the nearest noble gas in the periodic table.

Lewis electron dot formulas for molecules and ions showing outer shells with eight electrons as in the nearest noble gas

There are many exceptions to the octet rule, so it is not a safe guide. The rule works pretty well for the elements Li to F in period 2 because there are only four orbitals in the second shell (one s and three p orbitals). The octet rule also works for ionic compounds of s-*block* elements with the halogens, oxygen and sulfur. Exceptions arise in period 3 and beyond, because *d-orbitals* in the third shell can become involved in bonding.

Born–Haber cycles can help to account for the octet rule in ionic compounds. When magnesium, for example, forms ionic compounds with the Mg^{2+} ion the extra *ionization energy* needed to remove two rather than one electron from the outer shell is more than compensated for by the big increase in the energy released when the solid forms from gaseous ions (– *lattice energy*). The lattice energy is much larger for two reasons which increase the attraction between the positive and negative ions. An Mg^{2+} ion has a larger charge than an Mg^+ ion. Also an Mg^{2+} ion is much smaller than an Mg^+ ion because it has no electrons in the outer shell, so the ions can pack closer together.

Electron configurations of Mg^+, Mg^{2+} and Mg^{3+} ions

Removing a third electron from a magnesium atom needs much more energy because it involves taking an electron from the next shell where the electrons are more strongly held by the nucleus. If it could form, an Mg^{3+} ion would be little smaller than an Mg^{2+} ion and the greater energy released when the solid forms from gaseous ions is not enough to compensate for the very large third ionization energy.

odd-electron molecules: molecules with an odd number of electrons in an outer shell. Examples are two *nitrogen oxides*: nitrogen oxide and nitrogen dioxide. Both are exceptions to the *octet rule*.

Lewis electron dot diagrams showing the odd-electrons in molecules

optical isomers: *isomers* which have opposite effects on *polarized light*. One isomer rotates the plane of polarized light clockwise (the + isomer). The other isomer rotates the plane of polarized light in the opposite direction (the − isomer). Rotations are measured with *monochromatic light* in an instrument called a polarimeter. Optical activity is characteristic of *chiral compounds*.

Optical isomerism is shown by molecules with the same structure but different three-dimensional shapes, so it is a type of *stereoisomerism*.

orbitals: see *atomic orbitals* and *molecular orbitals*.

order of reaction: see *rate equation*.

ore: a *mineral* mass which can be profitably mined and processed to produce a metal. Prospectors seek ore bodies which are local concentrations of ores. After mining or quarrying the first stages of processing separate the valuable minerals from waste rock.

The work of the mineral industry would be impossible if minerals were evenly distributed throughout the *lithosphere*. Fortunately, natural processes have produced concentrations of valuable minerals with much higher percentages of rare metals than in the Earth's crust.

organic acids: see *carboxylic acids*.

organic analysis consists of chemical methods for *qualitative* and *quantitative analysis* to identify organic compounds and determine their composition and structure.

In a modern laboratory organic analysis is based on a range of automated and instrumental techniques including *combustion analysis*, *mass spectrometry*, *infrared spectroscopy*, *ultraviolet spectroscopy* and *nuclear magnetic resonance spectroscopy*.

Melting points and boiling points also provide a check on the identity and purity of compounds.

Traditionally, chemical tests helped to identify *functional groups* in organic molecules. Preliminary tests typically involve observing:

- the state and appearance of the compound
- its solubility in water and the pH of any solution
- the type of flame when a small sample burns.

(See the table on page 256.)

organic chemistry: the study of carbon compounds (except the very simplest such as the oxides and carbonates). There are more carbon compounds than the compounds of any other element because carbon atoms have a remarkable ability to join up in stable chains,

Organic analysis: the functional groups (see page 255)

Functional group	Test	Observations
$\diagup C = C \diagdown$ in an *alkene*	• Shake with a dilute solution of bromine • Shake with a very dilute, acidic solution of potassium permanganate	• Orange color of the solution decolorized • Purple color fades and the solution turns colorless
$-\overset{\mid}{\underset{\mid}{C}} - X$ where X = Cl, Br or I	• Warm with a solution of sodium hydroxide, cool, acidify with nitric acid, then add silver nitrate solution	• Precipitate is white from a chloro compound, creamy yellow from a bromo compound and yellow from an iodo compound (*hydrolysis* produces ions from the covalent molecules)
$-\overset{\mid}{\underset{\mid}{C}} - OH$ in a primary alcohol	• Add a solution of sodium carbonate • Add PCl_5 to the **anhydrous** compound • Warm with an acidic solution of potassium dichromate	• No reaction; unlike acids, alcohols do not react with carbonates • Colorless, fuming gas forms (hydrogen chloride) • Orange solution turns green and gives a fruity smelling vapor. Alcohols are oxidized by dichromate but not by *Fehling's solution*
$\diagup C = O$ in *aldehydes* and *ketones*	• Add 2,4-dinitrophenyl-hydrazine solution • Warm with fresh *Fehling's solution* • Warm with fresh *Tollen's reagent*	• Aldehydes and ketones both give yellow or red precipitates • Solution turns greenish and then an orange–red precipitate forms with aldehydes, but not ketones • Silver mirror forms with aldehydes but not with ketones
$-COOH$ in *carboxylic acids*	• Measure the pH of a solution • Add a solution of sodium carbonate • Add a neutral solution of iron(III) chloride	• Acidic solution with pH about 3–4 • Mixture fizzes, giving carbon dioxide • Methanoic and acetic acids give a red color

branched chains and rings. Organic chemists have devised ways to organize the mass of information about millions of compounds and reactions so that it makes sense.

One approach is to group compounds into series each with a distinctive *functional group*. Examples are the *alkanes, alkenes, arenes, carbonyl compounds* and *carboxylic acids*.

Another way of making sense of carbon chemistry is to classify reactions. One classification describes what happens to the molecules: *addition reactions, substitution reactions, elimination reactions* and *rearrangement reactions*. An alternative classification of reactions is based on the *mechanisms of the reaction* showing the types of reagents, the method of bond breaking and the nature of any *intermediates*. This includes:

- *homolytic bond breaking* with *radical intermediates*
- *heterolytic bond breaking* with *electrophiles* and *nucleophiles* as the attacking reagents leading to ionic intermediates.

Biochemistry and *polymer chemistry* are specialist aspects of organic chemistry.

Organic chemists have proved their understanding of organic structures and reactions by devising ways to synthesize increasingly complex molecules.

organic preparations take place in four main stages to make pure organic compounds.

Stage	Activities	Commonly used techniques
Planning	• Deciding on the reaction • Working out reacting quantities from the equation • Deciding on the conditions for reaction	• Consulting reference books • *Risk* analysis
Carrying out the reaction	• Measuring out the reactants • Mixing them in a suitable apparatus • Controlling the temperature	• Heating in a flask with a *reflux condenser* • Shaking *immiscible* reactants in a stoppered container
Separating the product from the reaction mixture	• "Working up" the reaction mixture to obtain a crude product	• *Distillation* • *Steam distillation* • *Vacuum filtration*
Purifying the product and testing for purity	• Purifying liquids by shaking with reagents to extract impurities in a separating funnel, followed by drying and distillation • Purifying solids by recrystallization and filtration • Using tests to check on the identity and purity of the product • Calculating the percentage yield (see *yield calculations*)	• *Fractional distillation* and measuring boiling points • *Solvent extraction* • *Recrystallization* • Measuring *melting* points • *Thin-layer chromatography* • *Infrared spectroscopy* • *Organic analysis*

organic routes: pathways from reactants to the required product in one or several steps. Organic chemists often start by examining the "target molecule" and then work back through a series of steps to find suitable starting chemicals which are cheap enough and available. In recent years chemists have developed computer programs to help with the process of working back from the target molecule to a range of possible starting points in a systematic way.

All the reactions in organic chemistry convert one compound to another, but there are some reactions which are particularly useful for developing a synthetic route, such as:

- addition of hydrogen halides to *alkenes*
- substitution reactions to replace halogen atoms with other functional groups such as — OH or — NH$_2$
- substitution of a chlorine atom for the — OH group in an alcohol or *carboxylic acid*
- elimination of a hydrogen halide from a *haloalkane* to introduce a double bond
- oxidation of primary *alcohols* to *aldehydes* and then carboxylic acids
- reduction of *carbonyl* compounds to alcohols
- lengthening of a carbon chain by forming a *nitrile* or using a *Grignard reagent*.

organometallic compounds: organic compounds which contain a metal atom. *Grignard reagents* are organometallic compounds formed with magnesium.

osmosis: the movement of a solvent by diffusion through a *semipermeable membrane*. The membrane allows small solvent molecules through but not dissolved molecules or ions.

The concentration of solvent molecules is higher in the pure solvent than in the solution. Solvent molecules diffuse in both direction but overall they move from the solvent into the solution.

The membranes which surround living cells are selectively permeable. This makes osmosis important in biology. Solutions for medical treatment of the eyes and other delicate parts of the body are formulated to have the same osmotic pressure as body fluids, as are iso-tonic drinks.

osmotic pressure: the pressure needed to stop *osmosis* when the solution is separated from pure solvent by a semipermeable membrane. Osmotic pressure depends on the con-centration of solute particles and not on their chemical nature. There is an equation relating osmotic pressure to concentration at a given temperature:

$$\Pi = \frac{n}{\upsilon} RT$$

Where Π is the osmotic pressure, n/υ the concentration in moles per liter, R the *gas con-stant* and T the absolute temperature in *Kelvin*.

With the help of this equation osmotic pressure measurements can be used to determine the molar masses of large molecules such as proteins and polymers.

If the pressure applied to the solution is greater than the osmotic pressure, the direction of net flow reverses and pure solvent moves out of the solution. This is reverse osmosis which can make drinking water from salty water.

oxidant: an alternative term to *oxidizing agent* which is convenient when describing half-equations for *redox reactions* and *standard electrode potentials*. By convention, when electrode potential values are being assigned the half-equation takes the form:

$$\text{oxidant} + ne^- \rightleftharpoons \text{reductant}$$

Every half-equation involves an oxidant and a reductant.

oxidation: originally this meant combination with oxygen but the term now covers all reactions in which atoms, molecules or ions lose electrons. The definition is further extended to cover molecules as well as ions by defining oxidation as a change which makes the *oxidation number* of an element more positive, or less negative.

Oxidation and reduction always go together in *redox reactions*.

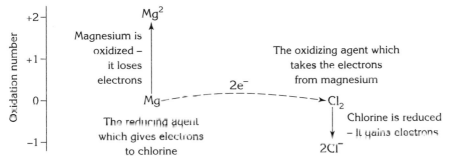

Chlorine oxidizes magnesium by taking two electrons from each magnesium atom

Oxidation number rules apply in principle in organic chemistry but it is often easier to use the older definitions. Oxidation is either addition of oxygen to a molecule or removal of hydrogen.

primary alcohol   aldehyde   carboxylic acid

Two-stage oxidation of an organic compound

oxidation numbers are used by chemists to define *oxidation* and *reduction* and to identify *redox reactions*. The advantage of oxidation numbers is that they apply to molecules and complex ions as well as to simple atoms and ions. The sum of the oxidation numbers in an uncharged compound is zero.

Chemists have agreed upon a set of rules for deciding on the oxidation numbers of elements:

- the oxidation numbers of uncombined elements are zero

- in a simple ion the oxidation number is the charge on the ion

- the sum of oxidation numbers in an uncharged compound is zero

- the sum of the oxidation numbers in an ion made of several atoms is equal to the charge on the ion

259

- some elements have fixed oxidation numbers (see the Table below).

Metals		Nonmetals	
group 1 (Li, Na, K)	+1	hydrogen	+1
group 2 (Mg, Ca, Ba)	+2	(except in metal hydrides)	
group 3 (Al)	+3	fluorine	−1
		oxygen	−2
		(except in peroxides and compounds with fluorine)	
		chlorine	−1
		(except in compounds with fluorine and oxygen)	

Oxidation numbers appear in the *names of inorganic compounds* and the *names of complex ions*. They also help with writing balanced redox equations.

Oxidation number rules are a formal way of keeping track of electrons in redox reactions. For the s-block elements the oxidation numbers are the same as the charges on the ions. Magnesium has an oxidation state of +2 and the metal atoms turn into Mg^{2+} ions in its compounds. The higher oxidation states of *d-block elements*, however, are certainly not ionic. Manganese in the +7 state cannot form an Mn^{7+} ion. The bonding between manganese and oxygen in the MnO_4^- ion is covalent.

MgO	NH_4^+	CCl_4	MnO_4^-
+2 −2	−3 +1	+4 −1	+7 −2

Al_2O_3	SO_4^{2-}	H_2SO_4	$Cr_2O_7^{2-}$
+3 −2	+6 −2	+1 +6 −2	+6 −2

Examples of the application of oxidation number rules. Note that each element in a compound is assigned an oxidation number

oxidation states: the states of oxidation or reduction shown by an element in its chemistry. The states are labelled with the oxidation numbers of the element in each state.

Describing the chemistry of an element according to oxidation state is a useful way of making sense of a large number of compounds and reactions. See for example the entries for *chlorine, manganese, hydrogen peroxide* and *nitrogen oxides*.

oxides: compounds of elements with oxygen. The compounds of *metals* with oxygen are *basic oxides* or *amphoteric oxides*. The compounds of *nonmetals* with oxygen are either *acidic oxides* or *neutral oxides*.

oxidizing agents: chemical reagents which can oxidize other atoms, molecules or ions by taking away electrons from them. Common oxidizing agents are *oxygen, chlorine, bromine, hydrogen peroxide*, the permanganate ion in *potassium permanganate*, and the dichromate ion.

Some reagents change color when they are oxidized which makes them useful for detecting oxidizing agents. In particular, a colorless solution of iodide ions is oxidized to iodine which turns the solution to a yellow–brown color.

$$2I^-(aq) \longrightarrow I_2(aq) + 2e^-$$

electrons taken by the oxidizing agent

This can be a very sensitive test if *starch* is also present because starch forms an intense blue–black color with iodine. Moistened starch–iodide paper is a version of this test which can detect oxidizing gases such as chlorine and bromine.

Chlorine acting as an oxidizing agent by taking electrons from iron(II) ions.
An oxidizing agent is itself reduced when it reacts. Oxidation and reduction always go together; hence the term redox reactions.

oxidised

$$2Fe^{2+}(aq) \longrightarrow 2Fe^{3+}(aq) + 2e^-$$

$$Cl_2(aq) + 2e^- \longrightarrow 2Cl^-(aq)$$

reduced

oxidizing substances: substances which are hazardous if they come into contact with materials that burn – especially if they are flammable liquids. The danger is that oxidizing substances will start a fire in contact with materials that can burn. (See *hazard symbols*.)

oxoacids: acids which form when acidic nonmetal *oxides* react with water.

Oxoacid	Structure	Oxoanion	Formula
sulfuric acid H_2SO_4	H—O, O / S \ H—O O	sulfate ion hydrogensulfate ion	SO_4^{2-} HSO_4
sulfurous acid H_2SO_3	H—O, S / H—O O	sulfite ion	SO_3^{2-}
nitric acid HNO_3	H, O / O—N \ O	nitrate ion	NO_3^-
nitrous acid HNO_2	H, O—N / O	nitrite ion	NO_2^-

Oxoacids ionize in solution by giving hydrogen ions to water molecules.

Ionization of nitric acid in water showing delocalization of the charge on the nitrate ion

Delocalization spreads the charge on the oxoanion, thus stabilizing the ion and favoring ionization. The more oxygen atoms doubly-bonded to the central atom, the stronger the acid. As a result, sulfuric acid is a stronger acid than sulfurous acid and nitric acid is a stronger acid than nitrous acid. So where an element forms two oxoacids, the one with the element in the higher oxidation state is the stronger.

oxygen (O): a colorless, odorless and highly reactive gas which makes up about a fifth of the *air* by volume and is essential to life. It is the first element in group VI with the electron configuration [He]$2s^22p^4$. Most oxygen occurs in the form of O_2 molecules but *ozone*, O_3, is a second *allotrope*.

All living organisms need oxygen for respiration. Green plants produce oxygen by *photosynthesis* in sunlight. Many biochemical molecules include oxygen atoms, for example *carbohydrates*, *fats*, *proteins* and *nucleic acids*.

Oxygen is the most abundant element in the *lithosphere*, mainly in the silicon–oxygen network solids of *silicates*.

Oxygen combines with most other elements to form *oxides*. Particularly important is *water*, the oxide of hydrogen.

Air separation plants use *fractional distillation* to separate oxygen and other gases. Alternatively, where only oxygen is required, the nitrogen from the air can be absorbed in a bed of a molecular sieve such as a *zeolite*.

Oxygen is commonly used in making *steel*. The manufacture of chemicals is another major use. Smaller quantities are used with a fuel for high-temperature metal cutting and welding. Oxygen is required medically by people with lung disease and is also used for *water treatment* to improve the quality of water contaminated with organic pollutants.

ozone (O_3): a colorless, reactive and unstable gas which is an *allotrope* of oxygen.

Ozone is an *oxidizing agent* and, like oxidizing bleaches, it will destroy microorganisms. Ozone is increasingly used to treat drinking water and swimming pools as an alternative to chlorine.

Ozone often features in stories about *environmental issues*. The ozone layer in the upper *atmosphere* is beneficial. The ozone high in the stratosphere protects living things by absorbing harmful UV radiation from the Sun. *CFCs* and other pollutants, such as the pesticide methyl bromide, tend to destroy the protective ozone layer.

In the lower atmosphere ozone is detrimental because it is harmful to living things and helps to cause *photochemical smog*. Ozone attacks and splits $C═C$ bonds. This is ozonolysis which can damage and destroy materials such as natural rubber.

paint: a product formulated to protect and decorate surfaces. A paint has three ingredients:

- pigments which scatter and absorb light so that the paint covers up the surface underneath and decorates it with color

- *polymers* which hold the pigment to the surface by forming a smooth plastic film as the paint dries and sets – in gloss paints the film-forming polymers are alkyd resins (see *polyesters*); in emulsion paints they are a latex polymer

- a solvent which is a liquid in which the other ingredients are dissolved or dispersed – in gloss paint the vehicle is traditionally a *hydrocarbon* solvent such as white spirit; in emulsion paints it is water.

paper chromatography: a type of chromatography in which the stationary phase is water in the fibers of paper and the moving phase is another solvent such as ethanol or acetone. The technique is used to analyze mixtures such as inks, food colors, dyes and amino acids.

During paper chromatography the chemicals in a mixture *partition* themselves between two solvents: water and the moving solvent. Each component has a different equilibrium constant (partition constant) which decides whether it has a greater tendency to dissolve in the stationary phase or in the mobile phase. As a result the mixture separates as the chemicals move at different speeds.

If the conditions are kept the same, each chemical in a mixture will move a fixed fraction of the distance moved by the solvent. The R_f value for the substance is a measure of this fraction.

If the chemicals in a mixture are colorless they are invisible on the paper. If so, the analyst has to "develop" the plates with a suitable "locating agent." After chromatography of amino acids, for example, the paper can be sprayed with ninhydrin solution and warmed. The amino acids then show up as purple spots.

paramagnetism: a type of magnetism shown by substances with unpaired electrons. Paramagnetic substances are weakly attracted into a magnetic field. The effect is very much weaker than the *ferromagnetism* of iron. Most compounds have orbitals filled by pairs of electrons with opposite *spins* so that their magnetic properties cancel out. These substances are very, very weakly repelled by a magnetic field. This is diamagnetism.

partial pressures are a useful alternative to concentrations when studying mixtures of gases and gas reactions. In a mixture of gases A, B and C the sum of the three partial pressures equals the total pressure.

$$P_{total} = P_A + P_B + P_B$$

According to the *ideal gas* equation: $pV = nRT$ and this equation is obeyed pretty well by real gases. In a mixture of gases the gas molecules move around independently. It is the number of molecules which matters and not their chemical nature. So the partial pressure for each gas is the pressure it would exert if it was the only gas in the container under the same conditions; thus:

$$p_A = \frac{n_A RT}{V}, \quad p_B = \frac{n_B RT}{V} \quad \text{and} \quad p_C = \frac{n_C RT}{V}$$

Since R is a constant, these equations show that at constant temperature:

$$p_A \propto \frac{n_A}{V}, \quad p_B \propto \frac{n_B}{V} \quad \text{and} \quad p_C \propto \frac{n_C}{V}.$$

So the partial pressures are proportional to the concentrations of the gases, since $n \div V$ is the concentration in moles per unit volume. This is the justification for working in partial pressures when applying the equilibrium law to gas reactions (see K_p).

The total pressure in a mixture can be "shared out" between the gases according their *mole fractions*, X. So:

$$p_A = X_A P_{total}, \quad p_B = X_B P_{total} \quad \text{and} \quad p_C = X_C P_{total}.$$

Worked example

An experimental study of the equilibrium mixture of $N_2O_4(g)$ and $NO_2(g)$ found that one equilibrium mixture contained 9.01×10^{-3} mol of $N_2O_4(g)$ and 7.16×10^{-3} mol of $NO_2(g)$ at 298 K and 3.99×10^4 Pa pressure. Calculate the partial pressures of the two gases.

Notes on the method

First determine the mole fractions of the two gases.

Multiply the total pressure by the mole fractions to get the partial pressures.

Check that the sum of the partial pressures equals the total pressure.

Answer

Total number of moles $= 9.01 \times 10^{-3}$ mol $+ 7.16 \times 10^{-3}$ mol
$= 16.17 \times 10^{-3}$ mol

Mole fraction of $N_2O_4(g) = \dfrac{9.01 \times 10^{-3} \text{ mol}}{16.17 \times 10^{-3} \text{ mol}} = 0.557$

Mole fraction of $NO_2(g) = \dfrac{7.16 \times 10^{-3} \text{ mol}}{16.17 \times 10^{-3} \text{ mol}} = 0.443$

Partial pressure of $N_2O_4(g) = 0.557 \times 3.99 \times 10^4$ Pa $= 2.22 \times 10^4$ Pa

Partial pressure of $NO_2(g) = 0.443 \times 3.99 \times 10^4$ Pa $= 1.77 \times 10^4$ Pa

particles: extremely small pieces of matter ranging from protons to grains of sand. Chemists use the term "particle" in several ways. At the smallest end of the scale they refer to protons, neutrons and electrons as *fundamental particles*. But they also use the term as a general word to cover atoms, molecules and ions when discussing the movements of particles in solids, liquids and gases. In addition, chemists use the term particle when describing much larger solid specks of matter finely dispersed in *colloids* such as paint, muddy water and smoke.

particulates: solid *particles* suspended in the air. In rural areas the main sources of particulates are natural. They include soil and sand swept up by winds, as well as volcanic dust and salt from sea spray.

In urban areas pollution comes mainly from the incomplete combustion of *fuels* in engines and power stations. The solid particles consist of carbon and unburnt hydrocarbons. *Diesel* engines are prone to producing particulates in their exhausts.

PM10 particles are of particular concern. These are very small particles which are less than 10 µm across. PM10 particles are small enough to penetrate deep into the lungs. Fine particles in the lungs can cause inflammation and be a threat to the people with heart and lung diseases. They may also carry *carcinogens* into the lungs.

partition: the distribution of a dissolved substance between two *solvents* which do not mix (they are *immiscible*). Partition is the basis for *solvent extraction* and *paper chromatography*.

Shaking a solid with two immiscible solvents produces an equilibrium system to which the *equilibrium law* applies. The ratio of the concentrations in the two layers is a constant, at a given temperature:

$$K = \frac{[X]_{solvent\ 1}}{[X]_{solvent\ 2}}$$

This special case of an equilibrium constant is called the partition constant or partition coefficient. (Some writers prefer the term "distribution constant.")

Note that partition obeys the equilibrium law only so long as the dissolved chemical has the same molecular structure in both solvents.

parts per million, ppm: a unit of *concentration* often used for low concentrations of pollutants in water. The density of water is 1 g/cm^3 and dilute solutions have almost the same density. So the mass of 1 L (= 1000 cm^3) of water is 1000 g. For dilute solutions:

1 ppm = 1 mg substance dissolved in 1 L water

Pauli exclusion principle: the fundamental principle which states that no two electrons in an atom can have the same value for all four *quantum numbers*. This rule accounts for the fact that no orbital can hold more than two electrons. The two electrons have opposite *spins*.

p-block elements: the elements in groups III, IV, V, VI, VII and VIII in the *periodic table*. For these elements the last electron added to the atomic structure goes into one of the three p-orbitals in the outer shell.

PCBs (polychlorinated biphenyls): a group of compounds formed from two linked benzene rings. They are liquids with valuable properties which mean that they have been widely used in transformers, in hydraulic equipment, as *plasticizers* and lubricants. PCBs are suited to these applications because they are very stable, nonflammable, involatile and particularly suited for use in some electrical equipment (because they have high dielectric constants). Some PCBs are also toxic pesticides.

Structure of a PCB

The problem with PCBs is that they escape into the environment where they are very persistent. PCBs can be a hazard too if they are mixed with waste burnt in an incinerator at too low a temperature. The danger is that incomplete combustion produces highly toxic polychlorobenzodioxins and polychlorobenzofurans. Modern incinerators are designed to operate at a high enough temperature to break $C-Cl$ bonds and destroy PCBs.

peptides: compounds made up of chains of *amino acids*. The simplest example is a dipeptide with just two amino acids linked. For chemists this is an example of an *amide* bond but the tradition in biochemistry is to call it a "peptide bond." The formation of a peptide bond is an example of a *condensation reaction*.

Formation of a peptide bond between two amino acids

Polypeptides are long-chain peptides. There is no precise dividing line between a peptide and a polypeptide. A *protein* molecule consists of one or more polypeptide chains.

Note that some writers make a distinction between polypeptides and the longer amino acid chains in proteins. They restrict the definition of polypeptides to chains with 10 to 50 or so amino acids.

Heating a peptide with dilute sulfuric acid leads to hydrolysis of the peptide bonds. In time this process splits the peptide into separate amino acids.

percent composition: the percent by mass of each of the elements in a compound. Percent composition is one way of expressing the results of chemical analysis, such as *combustion analysis*. The *empirical formula* of the compound can be calculated from these results.

Worked example

What is the empirical formula of copper pyrites which has the analysis 34.6% copper, 30.5% iron and 34.9% sulfur?

Notes on the method

Follow the procedure in the worked example for finding an *empirical formula*.

The percentages show the combining masses in a 100 g sample.

Answer

	copper	iron	sulfur
Combining masses	34.6 g	30.5 g	34.9 g
Molar masses of elements	64 g/mol	56 g/mol	32 g/mol
Amounts combined	$\dfrac{34.6 \text{ g}}{64 \text{ g/mol}}$	$\dfrac{30.5 \text{ g}}{56 \text{ g/mol}}$	$\dfrac{34.9 \text{ g}}{32 \text{ g/mol}}$
	= 0.54 mol	= 0.54 mol	= 1.09 mol
Simplest ratio of amounts	1	1	2

The formula is $CuFeS_2$.

The percent composition of a compound can be determined from its chemical formula. This is a guide to people who formulate products such as fertilizers, medicines and cleaning agents.

Worked example

Two common nitrogen fertilizers are urea $(H_2N)_2CO$, and ammonium nitrate, NH_4NO_3. Compare the percentage of nitrogen in the two compounds.

Notes on the masses

Look up the relative atomic masses of the elements and use them to determine the two *relative formula* masses. (Note that when part of a formula is in brackets the number outside refers to all the atoms in the bracket.)

Answer

Relative formula mass of urea $= 2 \times (2 + 14) + 12 + 16 = 60$

of which $(2 \times 14) = 28$ is nitrogen

Percentage of nitrogen in urea $= \dfrac{28}{60} \times 100\% = 46.7\%$

Relative formula mass of ammonium nitrate $= 14 + 4 + 14 + (3 \times 16) = 80$
of which $(2 \times 14) = 28$ is nitrogen

Percentage of nitrogen in ammonium nitrate $= \dfrac{28}{80} \times 100\% = 35\%$

Urea contains the higher percentage of nitrogen.

percent yield: see *yield calculations*.

perfumes: fragrant chemicals used not only for expensive cosmetics but also to give a smell to everyday household products. Many natural perfumes are essential oils separated from plants by *steam distillation* or *solvent extraction*. Increasingly, however, perfumes now contain synthetic chemicals, some of which are identical to those extracted from natural sources.

Experts use an analogy with music to describe the odors from complex perfumes. "Top" notes are the most volatile chemicals, the main effect depends on the "middle notes" and the most long lasting are the "end notes." So a perfume chemist has to understand the volatility of perfume chemicals.

undecanal
top note; boiling point 117°C

indane musk
bottom note; melting point 53°C

geraniol
middle note;
boiling point 146°C

Skeletal structures *of some perfume chemicals showing top, middle and end notes*

The connection between odor and chemical structure is little understood so there is no easy way to predict smells by examining molecular models.

period 3: the third horizontal row of elements in the *periodic table* which starts with sodium and ends with argon. In each main group of the periodic table the element in period 3 is often regarded as the typical element in the group. Chemists then generalize about trends down the group by comparing the elements in periods 4, 5 and 6 with the period 3 element. The first member of each group (usually an element in period 2) is often exceptional in some way.

The changes in properties from Na to Ar (across period 3) show periodic patterns clearly (see *periodicity of physical properties* and *periodicity of chemical properties*).

periodic table: a table of the elements arranged in order of *atomic number* (proton number). *Periods* are the horizontal rows in the table. Each period ends with a *noble gas*. The vertical columns are *groups* arranged in blocks – the *s-block*, *p-block*, *d-block* and *f-block* based on the electron configurations of the elements.

The periodic table was a triumph for nineteenth-century chemistry. In 1869 Dmitri Mendeleev published a version of the table based on his periodic law that when elements are arranged in order of atomic mass, similar properties recur at intervals. He left gaps in his table for undiscovered elements and predicted their properties. His success with these predictions helped to persuade other scientists of the merits of his ideas when the missing elements were discovered.

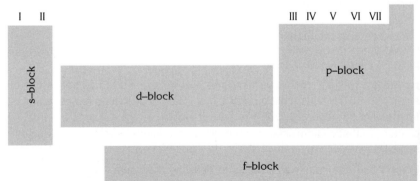

Outline periodic table showing the main blocks. Note that IUPAC now rules that the groups should be numbered from 1 to 18 across the table. In this system the old group IV becomes group 14 and the old group VII becomes group 17.

periodicity: a pattern which repeats itself. The most obvious repetition in the periodic table is from metals on the left of each period to *nonmetals* on the right. Another repeating pattern from one period to the next is shown by the main *oxidation states* of the elements in their common compounds.

Periodicity of oxidation states of the period 2 and 3 elements in their oxides. Note that the pattern repeats, but with variations.

periodicity of chemical properties: the repeating patterns from one period to the next in the chemical properties of the elements and their compounds.

The *oxides* of reactive metals are ionic and basic. The oxides of nonmetals are covalently bonded and acidic with *amphoteric oxides* in between.

Ionic crystals		Covalent network structures		Covalent molecular gases and solids		
$Li_2O(s)$	$BeO(s)$	$B_2O_3(s)$	$CO_2(g)$	$N_2O_5(s)$	$O_2(g)$	$F_2O(g)$
$Na_2O(s)$	$MgO(s)$	$Al_2O_3(s)$	$SiO_2(s)$	$P_4O_{10}(s)$ $P_4O_6(s)$	$SO_3(s)$ $SO_2(g)$	$Cl_2O(g)$
	Basic		Amphoteric	Acidic		

Formulas, structures, bonding and acid–base character of oxides in periods 2 and 3

The chlorides of reactive metals are ionic. They dissolve in water to give neutral solutions. The chlorides of *nonmetals* consist of covalent molecules and they are normally *hydrolyzed* by water.

Ionic crystals	Polar covalent network structures		Covalent molecular gases and solids			
$LiCl(s)$	$BeCl_2(s)$	$BCl_3(g)$	$CCl_4(l)$	$NCl_3(l)$	$Cl_2O(g)$	$ClF(g)$
$NaCl(s)$	$MgCl_2(s)$	$AlCl_3(s)$	$SiCl_4(s)$	$PCl_5(s)$ $PCl_3(l)$	$S_2Cl_2(s)$	$Cl_2(g)$
Dissolve in water giving a neutral solution	Partially hydrolyzed to give an acidic solution		Hydrolyzed by water giving an acidic solution (unless inert as is CCl_4)			

Periodicity: formulas, structures, bonding and behavior in water of chlorides in periods 2 and 3

periodicity of physical properties: the repeating patterns in the *physical properties* of the elements and their compounds from one period to the next. Metals on the left of a period conduct electricity well. *Nonmetals* to the right of a period are nonconductors. The in-between elements are the *metalloids*, many of which are *semiconductors*.

Plots of the melting points and boiling points against atomic number for periods 2 and 3. Note the repeating pattern. This is an example of periodicity. A plot of first ionization energy against proton number is another example of a periodic pattern.

The periodicity in the physical properties of the elements in periods 2 and 3 reflects the underlying patterns of structure and bonding.

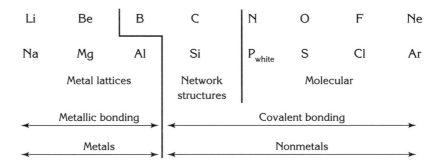

Li	Be	B	C	N	O	F	Ne
Na	Mg	Al	Si	P$_{white}$	S	Cl	Ar

Periodicity: structure and properties of elements in periods 2 and 3

permanent dipole: see *polar molecules* and *intermolecular forces*.

peroxides: compounds related to *hydrogen peroxide*, H_2O_2. The inorganic peroxides of *s-block elements* are ionic and contain the O_2^{2-} ion. When sodium burns in air it forms sodium peroxide Na_2O_2. Barium similarly produces barium peroxide, BaO_2.

Inorganic peroxides are powerful *oxidizing agents*. The peroxide ion is a *strong base* so it also reacts with water and dilute acids to make hydrogen peroxide.

Organic peroxides can act as a source of *radicals* to initiate *addition polymerization* of unsaturated compounds.

A benzoyl peroxide molecule splitting to form two radicals. The O — O bond is relatively weak.

pesticides: agrochemicals which kill insects or vermin that cause damage to crops, food, wood, fabrics and other materials. Examples are: insecticides including organochlorines such as DDT, organophosphorus compounds such as malathion, and pyrethroids.

Research chemists make and test many new compounds in search of pesticides which are:

• fast-acting

• specific so that they kill only the pests without harming other organisms

• effective at low doses

• biodegradable so that they quickly break down in the environment.

petrochemical industry: the part of the *chemical industry* based on crude oil and natural gas. The industry has developed and expanded massively since the late 1930s.

petroleum: a name for crude oil which is also used for some products from the *fractional distillation* and refining of oil. The *naphtha* fraction which is an important feedstock for the chemical industry is sometimes called the "light petroleum fraction."

pH changes during acid-base titrations: the variation in *pH* as a solution of a base flows from a buret into a flask containing a measured volume of acid. Plotting a graph of pH against volume of alkali added gives a shape determined by the nature of the acid and the base. (See diagram on page 272.)

The indicator chosen to detect the end point must change color completely in the pH range of the near vertical part of the curve. Note that at the equivalence point, when exactly equal amounts of acid and base have been added, the pH is not always neutral. (See also *neutralization reactions*.)

It is possible to read off the value of K_a for a weak acid from the corresponding titration curve (see *Henderson–Hasselbalch equation*).

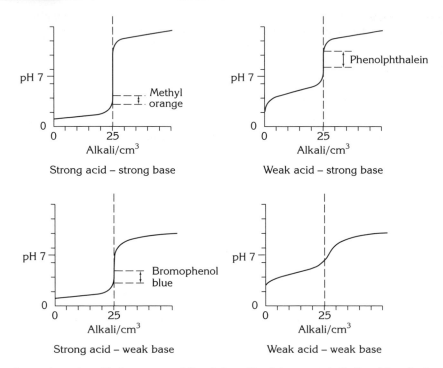

Graphs to show the pH change on adding 0.1 mol/L of the named alkali to 25 cm³ of a 0.1 mol/L of the named acid. Note the sharp change of pH around the equivalence point except when both the acid and the base are weak. The graphs show the pH range for the color change of an acid–base indicator *for detecting the* end point *of the titration.*

pH scale: a *logarithmic* scale for measuring the concentration of aqueous hydrogen ions in solutions.

$$pH = -\lg [H_3O^+(aq)]$$

pH	0	1	2	3	4	5	6	7	8	9	10	11	12	13	14
$[H_3O^+(aq)]$, mol/L	10^0	10^{-1}	10^{-2}	10^{-3}	10^{-4}	10^{-5}	10^{-6}	10^{-7}	10^{-8}	10^{-9}	10^{-10}	10^{-11}	10^{-12}	10^{-13}	10^{-14}

increasingly acidic   ←   neutral   →   increasingly alkaline

Worked example

What is the pH of 0.02 mol/L hydrochloric acid?

Notes on the method

Hydrochloric acid is a strong acid so it is fully ionized. Note that 1 mol HCl gives 1 mol $H_3O^+(aq)$.

Enter the value of $[H_3O^+(aq)]$ in your calculator, then press the log button, finally press +/– to reverse the sign.

Answer

$[H_3O^+(aq)]$ = 0.02 mol/L

$pH = -\log (0.02) = 1.7$

Worked example

What is the aqueous hydrogen ion concentration in a cola drink with pH 2.3?

Notes on the method

$pH = -\log [H_3O^+(aq)]$

From the definition of *logarithms* this rearranges to $[H_3O^+(aq)] = 10^{-pH}$

Enter the pH value in your calculator, press +/- to reverse the sign and then the inverse log button (10^x). (The order of pressing the buttons matters.)

Answer

pH = 2.3

$[H_3O^+(aq)] = 10^{-2.3} = 5 \times 10^{-3}$ mol/L

pharmaceuticals: chemicals used in *medicines*. (See also *drug development*.)

phase: the three states of matter – solid, liquid or gas. A phase diagram is a map showing the states of a compound over a range of temperatures and pressures.

The "liquid" area on the phase diagram for water shows all the conditions of temperature and pressure when water is a liquid. Similarly the "solid" area shows the conditions for water to be solid. The "vapor" region shows the conditions under which water is a gas.

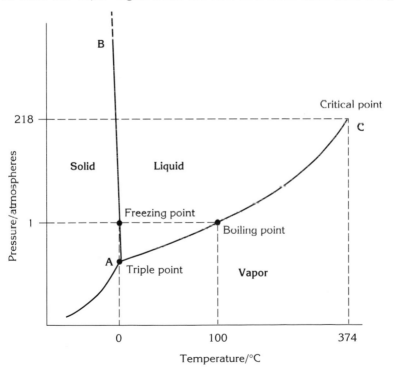

Phase diagram for water. Line AB shows how the melting point of ice varies with pressure. Line AC shows how the boiling point varies with pressure. Point A is the triple point which defines the only conditions of temperature and pressure at which ice, water and water vapor can be in equilibrium.

A B C D E F G H I J K L M N O P Q R S T U V W X Y Z

The lines show the conditions when two phases can be together in equilibrium. At 1 atmosphere pressure, ice and liquid water are in equilibrium at the melting point, 0°C; while liquid water and steam are in equilibrium at the boiling point, 100°C.

Chemical systems often have more than one phase. Each phase is distinct but need not be pure:

● a solid in equilibrium with its saturated solution is a two-phase system

● in the reactor for *ammonia manufacture* the mixture of nitrogen, hydrogen and ammonia gases is one phase with the iron catalyst being a separate solid phase

● *chromatography* separates mixtures by letting the components in a mixture come to equilibrium between two phases – the stationary phase and the mobile phase.

The same material broken up into little pieces still counts as one phase. In a *colloid* one phase is finely dispersed as specks, droplets or bubbles in a continuous phase.

phenols: a series of compounds with one or more —OH groups directly attached to a benzene ring. The simplest example is phenol, C_6H_5—OH.

The main uses of phenols are to make:

● *thermosetting polymers*

● *epoxy resins* for adhesives

● *intermediates* for the manufacture of nylon.

Structure of phenol and other phenols

The hydroxyl group and the benzene ring in phenol interact. The benzene ring makes the hydroxyl group more acidic in phenol than it is in *alcohols*. Phenol is acidic enough to form salts with alkalis. So phenol dissolves easily in a solution of sodium hydroxide.

Phenol reacting with alkali. Note that the phenoxide ion is stabilized because the negative charge can be spread by delocalization over the whole molecule. This cannot happen in alcohols.

Phenol is not as acidic as *carboxylic acids*. It does not ionize significantly in water and it does not react with carbonates to produce carbon dioxide. (See also *acid strength of organic hydroxy compounds*.)

Unlike alcohols, phenols do not react directly with carboxylic acids to form esters. It is, however, possible to convert a phenol to an ester with an *acyl chloride*.

The hydroxyl group makes the benzene ring more reactive. *Electrophilic substitution* takes place under much milder conditions with phenol than with benzene. Adding aqueous bromine to a solution of phenol produces an immediate white precipitate of 2,4,6-tribromophenol as the bromine color fades.

Reaction of phenol with bromine. The reaction is rapid at room temperature and there is no need for a catalyst.

Many phenols form colored complexes with $Fe^{3+}(aq)$ ions in neutral iron(III) chloride solution. Phenol itself forms a violet complex.

Halogenated phenols are used as *antiseptics* and *disinfectants*.

phenyl group: the group C_6H_5 — which is present in many compounds where one of the hydrogen atoms in benzene has been replaced by another atom or group.

| phenyl group | phenylamine (aniline) | phenylethene (styrene) |

Structure of the phenyl group and compounds derived from benzene

The use of phenyl in this way dates back to the first studies of benzene when "phene" was suggested as an alternative name for the compound, based on a Greek word for "giving light." Benzene was found in the tar formed on heating coal to produce gas for lighting.

Note that "benzyl" refers to aryl compounds with a carbon atom attached directly to the benzene ring.

| benzyl alcohol | benzaldehyde | benzoic acid |

Benzyl alcohol and other compounds related to $C_6H_5.CH_2$—

phosphorescence: see *luminescence*.

phosphoric(v) acid (H_3PO_4) is manufactured on a large scale from phosphate rock to make *fertilizers, detergent* phosphates and phosphates for food and drink. The pH of a typical cola drink is 2.3 because it contains 0.05% phosphoric acid. Another use of phosphoric acid is to make iron more resistant to corrosion.

The salts of phosphoric acid have many uses. These are a few examples:

- NaH_2PO_4 – used as a source of phosphate to make drugs and for *pH* control in tooth-paste

- Na_2HPO_4 – used to make processed cheese (by a method patented by J. L. Kraft in 1916), also a *corrosion* inhibitor and an ingredient of enamels and glazes

- Na_3PO_4 – the most basic of the sodium salts used in heavy duty cleaners, paint strippers and as a water softener

- $Ca(H_2PO_4)_2$ – the acid in baking powders

- $CaHPO_4$ – the abrasive in toothpaste.

Organic phosphates play a vital part in biochemistry (see *ATP, nucleotides* and *nucleic acids*).

phosphorus (P): a highly reactive element and important *nonmetal*. It is the second element in group V of the periodic table, $[Ne]3s^22p^3$.

There are two common *allotropes*:

- **white phosphorus** which is molecular (P_4) and has to be stored under water otherwise it catches fire in air

- **red phosphorus,** which is made up of long chains of P_4 units, is more stable and does not catch fire in air at room temperature.

There is also a black allotrope.

White phosphorus condenses when phosphorus vapor cools. This happens in the manufacture of phosphorus from phosphate rock (fluorapatite). The red form is more stable and, at 540 K, white phosphorus changes rapidly to the red form without a catalyst.

Phosphorus forms two oxides. They are both molecular and both *acidic oxides*. The more important oxide is P_4O_{10} which forms as a white solid when white phosphorus burns in excess air (see *phosphoric acid*).

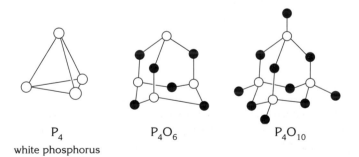

P_4
white phosphorus

P_4O_6

P_4O_{10}

Structures of white phosphorus, phosphorus(III) oxide, P_4O_6, and phosphorus(V) oxide, P_4O_{10}

As a laboratory reagent P_4O_{10} is used as a drying and dehydrating agent. It dehydrates *amides* to *nitriles*.

Phosphorus also forms two chlorides. Phosphorus trichloride (PCl_3) is made commercially from white phosphorus and chlorine. It is an important intermediate in the manufacture of medical drugs, insecticides and flame retardants. PCl_3 is a molecular compound. At room temperature it is a colorless fuming liquid which is rapidly *hydrolyzed* by water.

Phosphorus pentachloride (PCl_5) forms when PCl_3 reacts with excess chlorine. It is a volatile solid. It too is rapidly hydrolyzed by water and is a laboratory reagent for replacing hydroxyl groups with chlorine atoms in *alcohols* and *carboxylic acids*.

photochemical smog is produced by sunlight and the pollutants from the exhaust gases of motor vehicles. This type of *smog* forms on still, sunny days when there is no wind to blow away the gases. It is severe in cities such as Los Angeles where weather conditions and the local geography tend to trap pollutants in the city.

The primary pollutants are *nitrogen oxides* and unburnt *hydrocarbons* which are emitted in large amounts during the morning rush hour in a city. Bright sunlight during the middle of the day sets off photochemical reactions involving oxygen in the air. The products are the secondary pollutants which create smog.

The level of ozone in the air rises and oxidizing radicals form. The unburnt hydrocarbons are then oxidized to *aldehydes*, *ketones* and other chemicals such as organic nitrates which irritate the eyes and lungs.

Catalytic converters are designed to lessen this kind of pollution.

photochemistry: the study of reactions caused by light or other forms of *electromagnetic radiation*. Photons of ultraviolet light have enough energy to break covalent bonds and form *radicals*. These reactive intermediates can then start *radical chain reactions*. Photochemistry is important in the environment as well as in the laboratory (see *ozone* and *photochemical smog*).

photography uses chemical reactions to form images. Photography is based on the chemistry of silver and shows that, like other *d-block* elements, silver can act as a *catalyst*, form *complex ions* and take part in *redox reactions*. Photographic film consists of tiny silver bromide crystals in a film of gelatin on a strip of transparent plastic (see *ionic precipitation*). The stages of photography are as follows.

1   **Exposure** – in both black and white and color photography the first step is a photo-chemical reaction. Light produces a few atoms of silver on the surface of exposed silver bromide crystals:

$$2AgBr(s) \longrightarrow 2Ag(s) + Br_2(l)$$

At this stage there is no visible image.

2   **Development** – the atoms of silver are the catalyst for the next stage when the film is taken out of the camera and treated with a solution of the developer. In black and white photography the developer is a *reducing agent* which turns many more of the silver ions in exposed crystals to silver. Exposed parts of the film turn black. In color photography the developer is oxidized to an intermediate which then reacts with other chemicals in the gelatin layers of the film to produce colored dyes.

3   **Fixation** – this removes unexposed and undeveloped silver bromide to stop the whole film turning black when it is brought out into the light. The fixer is a solution of sodium thiosulfate. Thiosulfate ions form a complex with silver ions, turning insoluble silver bromide into a soluble complex:

$$AgBr(s) + 2S_2O_3^{2-}(aq) \longrightarrow [Ag(S_2O_3)_2]^{3-}(aq) + Br^-(aq)$$

4   **Bleaching** – this is needed only in color photography. In black and white photography the grains of silver formed by the developer stay to form the black areas of the negatives and prints. In color photography the silver has done its job once it has helped to make the colored dyes. So the silver is removed by treatment with a solution of iron(III) ions. The silver ions are thus removed from the film.

$$Ag(s) + Fe^{3+}(aq) \longrightarrow Ag^+(aq) + Fe^{2+}(aq)$$

photon: a quantum of light energy (see *quantum theory* and *electromagnetic radiation*).

photosynthesis: the process by which plants use energy from the Sun to convert carbon dioxide and water to carbohydrates and oxygen (see *carbon cycle*). It is the starting point for the synthesis of all the organic molecules in plants, including *proteins*, *lipids* and *nucleic acids*. The nitrogen, phosphorus and other elements which plants need are taken in from the soil as soluble salts.

Chlorophyll, the green pigment in plants, absorbs light and transfers the energy to electrons. With the help of other molecules the energy is used to split water into hydrogen and oxygen and to make *ATP*. All the oxygen released into the air during photosynthesis comes from water.

Through a series of biochemical steps, the hydrogen atoms from water produce a reducing agent which, with ATP, converts carbon dioxide to the basic building block for carbohydrates and other chemicals. So all the carbon and oxygen in carbohydrates comes from carbon dioxide.

Overall, photosynthesis is an *endothermic* process.

physical chemistry: the branch of chemistry which provides the theories that help to make sense of inorganic and organic chemistry. Physical chemistry explores the states of matter and uses *kinetic theory* to explain the properties of *solids*, liquids and gases. Physical chemistry also investigates the interactions between *electromagnetic radiation* and matter to explain structures and *bonding*. Physical chemists use the full range of the electromagnetic spectrum in the various types of *spectroscopy*.

The methods and theories of physical chemistry help to answer four important questions about all sorts of chemical reactions:

*   **How much?** – measuring *amounts* of chemicals makes it possible to find the *formulas* of compounds and the *balanced equations* for reactions which in turn makes it possible to calculate the quantities of reactants and products involved in reactions.

*   **How far and in which direction?** – the *equilibrium law*, *enthalpy changes*, *free energy changes* and *entropy changes* as well as *electrode potentials* help chemists to predict the direction and likely extent of chemical change.

*   **How fast?** – the study of *reaction rates* explores the factors which determine the speed of chemical change.

*   **How?** – *rate equations*, the use of *isotopes* as *tracers* and the various types of spectroscopy help chemists to determine the *mechanisms of reactions*.

physical properties: properties which describe how a substance behaves when it is chemically unchanged. Examples include:

- appearance (*color*, transparency)
- mechanical properties (*strength, hardness, ductility*)
- electrical properties (*electrical conductivity*)
- thermal properties (*melting and boiling* points, thermal conductivity, *specific heat capacity* and latent heats of melting and vaporization).

pi (π)-bonds: the type of bond found in molecules with double and triple bonds. The bonding electrons are in a π-orbital formed by sideways overlap of two atomic p-orbitals. In a π-bond the electron density is concentrated on either side of the line between the nuclei of the two atoms joined by the bond.

Pi bond in ethene   p-orbital

Pi bonds prevent rotation around the double bond and give rise to *geometrical isomerism*. Electrons in π-bonds are *delocalized* in molecules such as benzene with alternating double and single bonds (*conjugated systems*).

A π-bond is not as strong as a sigma (σ) bond and so the double bond in ethene (or any other alkene) is not twice as strong as a single bond.

picometer: a unit of length used for the sizes of atoms, molecules and ions. One picometer (1 pm) is a million millionths of a meter (10^{-12} m). The diameter of a hydrogen atom is about 75 pm. One *nanometer* (1 nm) equals a thousand picometers (1000 pm).

pigments: insoluble solids used in *paints*, cosmetics and printing inks which spread a thin layer of the pigment over a surface. Pigments may also color the bulk of a material such as a plastic. *Titanium(IV) oxide* is a bright white pigment. A famous blue pigment is monastral blue which is the very stable *coordination compound* copper phthalocyanine.

pipets are used to measure small volumes of liquid accurately. A volumetric pipet usually has a bulb for the main volume of the solution and then a single calibration line in the narrow tube above the bulb. This type of pipet can be used to deliver the same fixed volume of a solution again and again during titrations. (See *uncertainty of measurement*.)

pK: the logarithmic form of an equilibrium constant which is particularly useful for *pH* calculations. Taking *logarithms* produces a convenient small scale of values.

Examples:

- $pK_a = -\lg K_a$
- $pK_b = -\lg K_b$
- $pK_w = -\lg K_w$

The *Henderson–Hasselbalch equation* is one illustration of the advantage of working with pK_a instead of K_a. Another is the use of pK_w (the logarithmic version of the *autoionization constant of water*).

$$K_w = [H_3O^+][OH^-] = 10^{-14} \text{ at 298 K}$$

Taking logarithms gives: $\lg K_w = \lg [H_3O^+] + \lg[OH^-] = \lg 10^{-14} = -14$

Multiplying through by -1, reverses the signs:

$$-\lg K_w = -\log [H_3O^+] - \lg[OH^-] = 14$$

Hence: $pK_w = pH + pOH = 14$, where pOH is defined as $-\log[OH^-]$ by analogy with pH.

So: $pH = 14 - pOH$ which makes it easy to calculate the pH of alkaline solutions

Worked example

What is the pH of a 0.02 mol/L solution of sodium hydroxide?

Notes on the method

Sodium hydroxide (NaOH) is a *strong base* so it is fully ionized.

Find the values of logarithms by entering the value in a calculator and then pressing the lg button.

Answer

$[OH^-] = 0.02$ mol/L

$pOH = -\lg 0.02 = 1.7$

$pH = 14 - pOH = 14 - 1.7 = 12.3$

planar molecules: flat molecules with all the atoms in the same plane (see *shapes of molecules*).

Planck's constant: see *quantum theory.*

plaster of Paris: the main ingredient of building plasters and much is used to make plasterboard. The white powder is made by heating the mineral gypsum in kilns to remove most of the *water of crystallization.*

$$CaSO_4.2H_2O(s) \longrightarrow CaSO_4.\tfrac{1}{2}H_2O(s) + \tfrac{3}{2}H_2O(g)$$

Stirring plaster of Paris with water produces a paste which soon sets as it turns back into interlocking grains of gypsum. Plaster makes good molds because it expands slightly as it sets so that it fills every crevice.

plastic materials: materials which can be molded by gentle pressure. Examples are potter's clay and Plasticine. Once molded, a plastic material keeps its new shape, unlike an *elastomer* which tends to spring back to its original shape.

plasticizers: additives mixed with *polymer* materials to make them more flexible. A plasticizer is usually a liquid with a high boiling point, such as large *ester* molecules. The molecules of the plasticizer get between the polymer chains and reduce the intermolecular forces between the chains so that they can slide past each other. Rigid uPVC (unplasticized) is suitable for window frames and gutters. Adding a plasticizer to PVC changes the polymer to a material suitable for squeeze bottles, hose pipes and the insulation on electric cables.

plastics: materials made of long-chain molecules which at some stage can be easily molded into shape. *Thermoplastics* become *plastic materials* when they are hot and harden on cooling. Some thermoplastics are rigid and brittle at room temperature such as polystyrene and uPVC, others, such as polythene, are flexible. (See also *recycling.*)

platinum (Pt): a valuable metal which is the most abundant of the group of so-called "platinum metals" which occur together in sulfide ores. The platinum metals are the *d-block elements* ruthenium and osmium, rhodium and iridium plus palladium and platinum.

Platinum is about as abundant as gold. It is mined in South Africa where the yield of metal is about 30 g from 10 metric tons of rock.

Platinum is an attractive metal and it is very unreactive so it does not tarnish. The main use of platinum is for jewelery. As well as being expensive, it is a *malleable* and *ductile* metal so it can be worked into shape.

Chemists take advantage of the inertness of the metal in electrochemistry. Platinum electrodes are required to measure standard electrode potentials. The conductor in a *standard hydrogen electrode* is platinum covered with a thin layer of finely divided metal (platinum black) deposited by electrolysis.

Platinum is an effective *catalyst*. Its catalytic activity is enhanced by alloying with rhodium. The second main use of platinum is to make *catalytic converters* for cars. Platinum–rhodium gauzes catalyze the oxidation of ammonia in *nitric acid manufacture*.

Platinum forms a range of *complex ions* some of which, like *cisplatin*, are used to treat cancer.

polar covalent bonds: covalent bonds formed between atoms of different elements so that the shared electrons are drawn towards the more *electronegative* atom. One end of the bond has a slight excess of negative charge ($\delta-$). The other end of the bond has a slight deficit of electrons so that the charge cloud of electrons does not cancel the positive charge on the nucleus ($\delta+$).

Examples of molecules
with polar bonds

Molecules with polar bonds are generally more reactive than nonpolar molecules, especially with ionic reagents such as acids, alkalis and oxidizing agents. The presence of polar bonds makes molecules open to attack by *electrophiles* and *nucleophiles*.

polar molecules have polar bonds which do not cancel each other out, so that the whole molecule is polar.

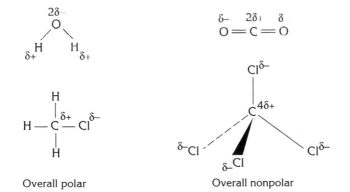

Overall polar   Overall nonpolar

Molecules with polar bonds. Note that in the examples on the left the net effect of all the bonds is a polar molecule. In the examples on the right the overall effect is a nonpolar molecule.

Polar molecules are little electrical dipoles (they have a positive pole and a negative pole). Dipoles tend to line up in an electric field. The bigger the dipole the bigger the twisting effect (dipole moment). By making measurements with a polar substance between two electrodes it is possible to calculate dipole moments. The units are Debye units named after the physical chemist Peter Debye (1884–1966).

The *intermolecular forces* are stronger between molecules with permanent dipoles.

Molecule	Dipole moment (in debye units)
HCl	1.08
H_2O	1.94
CH_3Cl	1.86
$CHCl_3$	1.02
CCl_4	0
CO_2	0

Measuring dipole moments can distinguish geometrical isomers.

Overall polar   Overall nonpolar

Polarity of geometric isomers

polar solvents: *solvents* made of *polar molecules*. Highly polar solvents such as water are needed to dissolve ionic compounds (see *hydration*). In organic chemistry the use of polar solvents favors *heterolytic bond breaking* and reactions with ionic intermediates.

polarizability: an indication of the extent to which the electron cloud in a molecule or ion is distorted by a nearby electric charge.

Larger molecules with more electrons are generally more polarizable. The *intermolecular forces* which account for weak intermolecular forces between nonpolar molecules are stronger between molecules which are more polarizable.

polarization of ions: the tendency of a positive metal ion to distort the charge cloud of a neighboring nonmetal ion. This gives rise to some degree of electron sharing (that is a degree of covalent bonding). Ionic bonding is favoured if:

● the charges on the ions are small (1+ or 2+, 1– or 2–)

● the radius of the positive ion is large and the radius of the negative ion small.

The smaller a positive ion and the larger its charge the greater the extent to which is tends to polarize a negative ion. So polarizing power increases along the series: Na^+, Mg^{2+}, Al^{3+}, Si^{4+}. Sodium chloride is an ionic, crystalline solid. The bonding in anhydrous *aluminum chloride* is largely covalent. Silicon chloride is a covalent, molecular liquid.

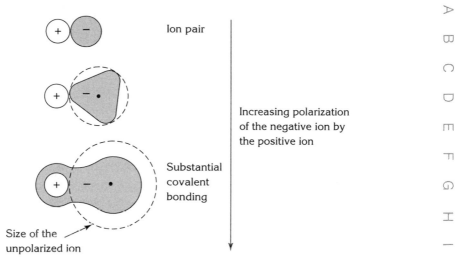

Ion pair

Increasing polarization
of the negative ion by
the positive ion

Substantial
covalent
bonding

Size of the
unpolarized ion

Ionic bonding with increasing degrees of electron sharing because the positive ion has polarized the neighboring negative ion. Dotted circles show the unpolarized ions.

The larger the negative ion and the larger its charge the more polarizable it becomes. So iodide ions are more polarizable than fluoride ions. Fluorine, which forms the small, singly charge fluoride ion, forms more ionic compounds than any other nonmetal.

Na^+	Mg^{2+}	Al^{3+}
1+	2+	3+
0.102 nm	0.072 nm	0.053 nm

Ionic radii for metal ions in the third period

polarized light: a light beam in which all the waves are vibrating in the same plane.

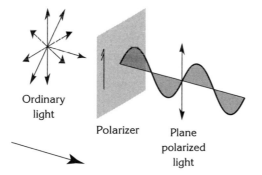

Ordinary
light

Polarizer

Plane
polarized
light

Ordinary light and polarized light

Light is plane-polarized after passing through a sheet of Polaroid. *Optical isomers* rotate the plane of polarized light.

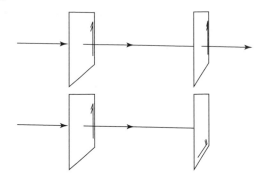

Effect of passing a light beam through two sheets of Polaroid. After the first sheet the light is polarized. The beam passes through the second sheet if it is aligned the same as the first. No light gets through if the second sheet is rotated through 90°.

polarizing power: an indication of the extent to which a positive ion is able to distort the electron cloud around a neighboring negative ion. The larger the charge on a positive ion and the smaller its size the greater its polarizing power (see *polarization of ions*).

pollution: contamination of the *atmosphere, hydrosphere* or *lithosphere* as a result of natural processes or events (such as excretion, the decay of dead organisms and volcanic eruptions) and human activities (such as agriculture, sewage discharges, combustion of fuels, extraction of resources, industrial processes and the disposal of *wastes*).

polyamides: polymers in which the monomer molecules are linked by amide bonds. The first important polyamides were formed by *condensation polymerization* between diamines and dicarboxylic acids. (See also *aramids*.)

Condensation polymerization to make Nylon-6,6. The numbers in the name indicate the numbers of carbon atoms in the two monomers.

polyesters: *polymers* formed by *condensation polymerization* between acids with two carboxylic acid groups and alcohols with two or more hydroxyl groups. Units in the polymer chains are linked by a series of *ester* bonds.

The monomers for making polyester fibers are **ter**ephthalic acid and eth**ylene** glycol (1,2-ethanediol), hence the commercial name Terylene for one brand of polyester. The name for the polymer is **poly**ethylene **t**erephthalate which gives rise to the name PET when the same polymer is used as a plastic to make bottles for carbonated drinks.

Alkyd *resins*, ingredients of gloss paint, also consist of long chains linked by ester bonds. The polymer chains have *unsaturated hydrocarbon* side-chains. When exposed to air, the *double bonds* in the side-chains of neighboring polymer molecules react so that the chains become *cross-linked*. As a result the resin hardens to a smooth film as the paint dries and sets.

Some polyesters are *biodegradable* because bacteria can break them down by hydrolyzing the ester links.

polymer chemistry: the study of the synthesis, structure and properties of *polymers*. It is a branch of chemistry which has developed in the twentieth century. Leo Baekland developed *Bakelite* in 1905 with little help from theory. The first *thermosetting polymers* were discovered before chemists understood the structure of big molecules.

In 1922 the German chemist Hermann Staudinger published his theory that natural rubber, cellulose and related substances consist of long chain molecules. He had to fight hard to persuade other chemists to accept his theory which is now taken for granted.

One of the people who was convinced by the new theory was the American industrial chemist Wallace Carothers. He wrote an article in 1931 in which he introduced the terms *addition polymerization* and *condensation polymerization*. His team at Du Pont invented the synthetic rubber Neoprene. They also discovered the first completely synthetic polymer (nylon) which went into production in 1939.

The 1930s were probably the most important years in the development of the plastics industry. It was the time when the addition polymers polythene, PVC, polystyrene and acrylics were all developed commercially. The high-pressure process for making polythene was discovered and developed in the mid-1930s as an unexpected offshoot of the study of high-pressure gas reactions by Eric Fawcett and Reginald Gibson working for the UK chemical company ICI.

Also in the 1930s, Otto Bayer was pursuing his interests in isocyanates in the laboratories of the IG Farben Industries in Germany. He too faced opposition in his pursuit of new polymeric materials. His breakthrough came in 1941 when he and his team discovered the possibility of producing *polyurethane* foams. It took ten more years of development work before large scale manufacturing could begin

It was in the early 1950s that Karl Ziegler in Germany discovered how to polymerize ethene at low temperatures using a new kind of catalyst. At the same time Guilio Natta in Italy discovered the benefits of producing *isotactic* polymers (see *Ziegler–Natta catalysts*).

Research and development since the 1950s has produced many new specialized polymer materials including the *aramids* and biodegradable polymers.

polymers: long chain molecules. Natural polymers include *proteins*, polysaccharides (see *carbohydrates*) and *nucleic acids*. Synthetic polymers include *polyesters*, *polyamides* and the many polymers formed by the *addition polymerization* of compounds with $C = C$ bonds.

The properties of polymers are very varied. Polymeric materials include *plastics*, *elastomers* and *fibers*.

As polymer science has developed, chemists and materials scientists have learned how to develop new materials with particular properties. Some of the ways of modifying polymeric materials include:

- altering the average length of the polymer chains
- changing the structure of the monomer perhaps by adding side groups which increase *intermolecular forces*
- varying the degree of *cross-linking* between chains
- *copolymerization*
- controlling the three-dimensional shape of the polymer (see *isotactic* polymers)
- changing the alignment of the polymer chains, for example by spinning the polymer into fibers and then stretching the fibers
- adding fillers, pigments and *plasticizers*
- making *composites*.

polymerization: a process in which many small molecules (*monomers*) join up in long chains (see *addition polymerization* and *condensation polymerization*).

polymorphism: the existence of two or more crystalline forms of a substance. Some ionic compounds are polymorphic. Zinc sulfide, for example, can crystallize either with the zinc blende structure or with the alternative wurtzite structure. Polymorphism in solid elements is an example of *allotropy*.

polyols: compounds with more than one hydroxyl group. They include *diols*, such as 1,2-ethanediol (ethylene glycol) and triols such as 1,2,3-propanetriol (better known as glycerine or glycerol). Polyols are used to make *polyurethanes*.

polysaccharides: see *carbohydrates.*

polyurethanes: a varied range of cross-linked polymers made from two liquids – a *polyol* and an isocyanate. Polymerization is exothermic and happens at room temperature. Adding a chemical to make a gas means that the polymer forms as a plastic foam. By choosing different isocyanates and polyols, manufacturers can vary the properties of the polyurethane. The products range through flexible foam for bedding and upholstery, rigid foam for wall panels and refrigerators, durable but bendable soles for shoes and ingredients for paints and adhesives.

One of the great advantages of polyurethanes is that they can be made where they are needed without any complex machinery. A furniture manufacturer, for example, can buy the two liquid ingredients and then mix them in a mold so that the polymer takes up the required shape for a chair as it forms.

p-orbitals: see *atomic orbitals.*

position of equilibrium: see *Le Chatelier's principle*, the *equilibrium law* and the *temperature effect on equilibria.*

potassium (K): a very soft, shiny metal which rapidly tarnishes in moist air. It is the third member of *group I* with the electron configuration [Ar]$4s^1$.

Like other group I metals, potassium:

- is stored in oil
- floats on water, melts and reacts violently forming hydrogen which catches fire and KOH which is soluble and strongly alkaline
- forms an ionic, crystalline chloride K^+Cl^-.

Unlike lithium and sodium, potassium produces a *superoxide* (KO_2) when it burns in oxygen. This is an ionic compound with the O_2^- ion. One of the main uses of potassium is to make this oxide for use in emergency breathing apparatus. The oxide removes carbon dioxide from moist exhaled air and replaces it with oxygen.

$$4KO_2(s) + 4CO_2(g) + 2H_2O(l) \longrightarrow 4KHCO_3(s) + 3O_2(g)$$

Potassium, as potassium ions, is an essential nutrient for plants and an ingredient of NPK *fertilizers*.

potassium permanganate ($KMnO_4$) consists of greyish-black crystals which dissolve in water to give a deep purple solution. It is used as a powerful *oxidizing agent*.

Potassium permanganate is an important reagent in *redox titrations* because it will oxidize many reducing agents in acid conditions, such as iron(II) ions. The reactions go according to their equations which makes them suitable for quantitative work.

$$MnO_4^-(aq) + 8H^+(aq) + 5e^- \longrightarrow Mn^{2+}(aq) + 4H_2O(l)$$

No indicator is required for a permanganate titration. When the solution is added from a buret the permanganate rapidly changes from purple to colorless (because the color of the Mn^{2+} ion is so pale). At the end point it takes only the slightest excess of permanganate to give a permanent red-purple color.

In organic chemistry potassium permanganate is a reagent used to oxidize the side-chains of *arenes*.

precipitation: a reaction which produces an insoluble product from soluble chemicals in solution. A common example is the formation of an insoluble salt on mixing solutions of two soluble salts (see *ionic precipitation*).

Ionic precipitation is the basis of many *anion tests* and *cation tests*.

In *organic analysis* the *Fehling's solution* or *Tollen's reagent* tests, used to distinguish aldehydes from ketones, both make precipitates by redox reactions.

precision of data: data is precise if repeat measurements have values which are close to each other. Precise measurements have a small random *error*.

Precise measurements may or may not be accurate. As a result of a systematic error a series of precise measurements may give values which are almost the same but are not the true value.

pressure: force per unit area. The *SI unit* of pressure is the pascal (Pa) which is a pressure of one newton per square meter (1 N/m^2). The pascal is a very small unit so pressures are often quoted in kilopascals (kPa).

When gases are being studied the standard pressure is *atmospheric pressure* which is:

$$101.3 \times 10^3 \text{ N/m}^2 = 101.3 \text{ kPa}$$

In accounts of chemical processes, multiples of atmospheric pressure give an indication of the extent to which gases are compressed (see for example *ammonia manufacture*).

The standard pressure for definitions in thermodynamics is now 1 bar which is:

$$100 \ 000 \text{ N/m}^2 = 100 \text{ kPa}$$

The *partial pressure* of a gas is a measure of its concentration in a mixture of gases.

primary, secondary and tertiary organic compounds: labels which distinguish different chemical situations for functional groups in *alcohols, haloalkanes, amines* and *carbocations*.

The labels have the same meaning for alcohols, haloalkanes and carbocations but a different meaning for amines.

primary standard: a chemical which can be weighed out accurately to make up a *standard solution* for *volumetric analysis*. A primary standard must:

- be very pure

- not gain or lose mass when exposed to the air (so it must not be *hygroscopic* or *deliquescent*)

- be soluble in water

- have a relatively high molar mass so weighing errors are minimized

- react exactly as described by the chemical equation.

A convenient primary standards for *acid–base titrations* is anhydrous sodium carbonate.

Primary standards for *redox titrations* include: potassium dichromate and potassium iodate.

promoters: chemicals which make catalysts more effective. The iron catalyst in the Haber process for *ammonia manufacture* contains potassium hydroxide as a promoter. The vanadium(v) oxide catalyst used in *sulfuric acid manufacture* has potassium sulfate on a silica support to act as a promoter.

propagation: a step in a *radical chain reaction*.

proportionality: a description of any relationship which takes the form:

$$y \propto x \text{ or } y = \text{constant} \times x$$

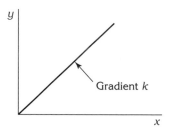

A graph showing that y is proportional to x. A straight line through the origin. y = kx. The gradient of the line gives the value of the constant k.

The volume of a fixed amount of an ideal gas, for example, is proportional to the temperature on the *Kelvin* scale at constant pressure. The rate of a *first order reaction* is proportional to the concentration of the reactant in the rate equation.

protecting group: a group introduced into an organic molecule during a synthesis to stop an unwanted reaction. Once the reaction is complete the protecting group is removed.

proteins consist of long chains of *amino acids* joined by *peptide* bonds. Some fibrous proteins are vital to the structure of animals, including the proteins in muscles, ligaments, tendons, skin and hair. Other proteins coil up into a globular shape and dissolve in body fluids where they act as oxygen carriers, *enzymes* and *hormones*.

Biochemists describe the structure at a series of levels:

- **the primary structure** is the sequence of amino acids in the polymer chains

- **the secondary structure** is the way in which the chains are arranged and held in place by *hydrogen bonding* within and between chains – this includes the coiling of chains into an α-helix in proteins such as keratin and the formation of layers of parallel chains as in the β-pleated sheets of silk

- **the tertiary structure** describes the three dimensional folding of protein chains which gives some proteins, such as enzymes, a definite three-dimensional shape held in place by hydrogen bonding, disulfide "bridges" and interactions between amino acid side-chains with surrounding water molecules

- **the quaternary structure** describes the linking of two or more amino acid chains as in the hormone *insulin* which consists of two chains linked by disulfide bonds, or in antibodies consisting of four chains.

| **Primary** (sequence of amino acids) | **Secondary** (coiling of the amino acid chain) | **Tertiary** (folding of coiled chain to create an active site) | **Quaternary** (association of two or more coiled chains) |

The four levels of protein structure

protium (${}^{1}_{1}H$): the abundant isotope of hydrogen. It is sometimes called *protium* to distinguish it from deuterium, ${}^{2}_{1}H$, and tritium, ${}^{3}_{1}H$.

proton: one of the two types of particle which make up the nucleus of an atom. The relative mass of a proton is 1 and its charge is +1. The number of protons in the nucleus of an atom is the *atomic number* (or proton number).

A hydrogen atom, H, normally consists of one proton in the nucleus and one electron in the first shell. So a hydrogen ion, H^+, is simply a proton. Chemists often use the terms "hydrogen ion" and "proton" interchangeably, especially when describing *acid–base reactions* as *proton transfer* reactions.

proton magnetic resonance spectroscopy: *nuclear magnetic resonance spectroscopy* (NMR) when the nuclei involved are hydrogen nuclei (or protons).

proton number: the number of protons in the nucleus of an atom. An alternative to *atomic number*.

proton transfer describes the movement of a hydrogen ion (or *proton*) from an acid to a base during an *acid–base reaction*. According to the *Brønsted–Lowry theory* an acid is a proton donor and a base is a proton acceptor.

purification: a process used by chemists to remove impurities from the products of reaction. Common methods of purification include:

* *distillation* or *fractional distillation* for liquids

* *steam distillation* to obtain a liquid with a high boiling point which does not mix with water

* *recrystallization* to separate solids from solid impurities

* *solvent extraction* for liquids or solids.

Some types of *chromatography* are also used not only for analysis but also to separate pure products.

Chemists use a range of techniques to test for purity. A chemical is pure if it:

* melts exactly at its melting point

* boils at its boiling point

* gives only one spot or peak when analyzed by chromatography

* gives a printout which matches the spectrum of a known pure sample when analyzed by *infrared spectroscopy*.

qualitative analysis: any method for identifying chemicals in a sample. Examples of qualitative analysis include:

- *gas tests* such as the use of limewater to detect carbon dioxide
- *anion tests* such as the use of silver nitrate solution to detect and distinguish chlorides, bromides and iodides
- the separation and identification of amino acids by *paper chromatography*
- the identifying of *functional groups* in an organic compound using its *infrared spectroscopy.*

quantitative analysis: any method for determining the amount of a chemical in a sample. Examples of quantitative analysis include:

- an *acid–base titration* to determine the concentration of a solution of hydrochloric acid
- the use of a *colorimeter* to determine the concentration of a colored *complex ion*
- testing the blood-alcohol concentration by *infrared spectroscopy*
- finding the relative abundances of the *isotopes* of chlorine with a *mass spectrometer.*

quantum numbers identify the *energy level* or *orbital* occupied by an electron in an atom. Theory shows that four quantum numbers uniquely identify each electron in an atom:

- the principal quantum number indicates the distance of the electron from the nucleus
- the second quantum number identifies the type of *atomic orbital*, s, p, d or f
- the third quantum number shows the direction in space of the orbital, p_x, p_y or p_z
- the fourth quantum number states the alignment of the *spin*, spin up or spin down.

The principal quantum number identifies the main shell occupied by an electron. The first shell ($n = 1$) can hold 2 electrons, the second shell ($n = 2$) up to eight electrons and the third shell ($n = 3$) up to 18 electrons. Thus the maximum number of electrons in a shell is given by $2n^2$.

quantum theory states that radiation is emitted or absorbed in discrete amounts called energy quanta. Max Planck, the German physicist, put forward the theory in a paper published in 1900. Quanta have energy $E = h\nu$ where h is Planck's constant and ν is the frequency of the radiation.

The Danish physicist Niels Bohr took up quantum theory in 1913 to explain the lines in hydrogen's *atomic emission spectra*. Bohr's theory could account very well for the frequencies of the lines in the spectrum of the hydrogen atom by making these assumptions:

291

- electrons in a hydrogen atom can only be at certain definite *energy levels*

- a photon of light is emitted or absorbed when an electron jumps from one energy level to another

- the energy of the photon, *E*, equals the difference between the two energy levels

- the frequency of the radiation emitted or absorbed is related to the energy of the photon by $E = h\nu$.

quartz: a crystalline form of *silica* (silicon dioxide). Quartz has a high melting point because it consists of a covalently bonded *network solid* in which each silicon atom is linked to four oxygen atoms and each oxygen atom to two silicon atoms.

quaternary ammonium cations: the cations formed when all the hydrogen atoms in an ammonium ion are replaced by *alkyl groups*.

$$R-\overset{\overset{\displaystyle R'}{|}}{\underset{\underset{\displaystyle R'''}{|}}{N}}{}^{+}-R''$$

General structure of a quarternary ammonium cation

Cationic *surfactants* used in fabric and hair conditioners are quaternary ammonium salts such as dodecyltrimethylammonium bromide, $CH_3(CH_2)_{11}N^+(CH_3)_3Br^-$.

quenching: the rapid cooling of a hot metal sample during heat treatment by plunging it into cold oil or water. Quenching a hot metal produces a different structure to slow cooling.

Similarly, chemists "quench" reactions by sudden cooling. This effectively stops (or "freezes") the reaction. In this way it is possible to find the composition of an equilibrium mixture or measure the concentration of a reactant or product during a study of rate of reaction.

R_f values: ratios which measure how far a particular chemical moves during *paper chromatography* and *thin-layer chromatography* (TLC).

$$R_f = \frac{\text{distance moved by solute}}{\text{distance moved by solvent}}$$

$$= \frac{x}{y}$$

The R_f value is the ratio of the distance moved by the chemical in the mixture to the distance moved by the solvent front

R_f values can help to identify components of mixtures so long as conditions are carefully controlled. The values vary with the type of paper or TLC plate and the nature of the solvent.

racemic mixture: a mixture of equal amounts of the two mirror image forms of a *chiral compound*. The mixture does not rotate *polarized light* because the two *optical isomers* have equal and opposite effects so they cancel each other out.

Lactic acid (2-hydroxypropanoic acid) from muscles is optically active because it consists of the (+) isomer. Lactic acid from sour milk is not optically active because it is a 50:50 racemic mixture of the two mirror image forms. Laboratory synthesis of lactic acid also produces the racemic mixture because there is an equal chance of the two isomers forming. (See diagram on page 294.)

radicals: reactive particles with unpaired electrons. Radicals form when covalent bonds break in the way which leaves one electron on each of the atoms joined by the bond. This is *homolytic fission*. The symbol for a radical generally shows the unpaired electron as a dot. Other paired electrons in the outer shells are generally not shown.

$$Br \!:\! Br \longrightarrow Br^\bullet + {}^\bullet Br$$

Formation of radicals

Radicals are *intermediates in reactions* taking place:

- in the gas phase at high temperature or in ultraviolet light

- in a non-polar solvent, either when irradiated by ultraviolet light or with an initiator.

Examples of radical processes include the *cracking* of hydrocarbons, burning of *gasoline* in an engine cylinder and the formation and destruction of the *ozone* layer. (See also *radical chain reactions*.)

Racemic mixture *A normal laboratory synthesis produces a racemic mixture of the mirror image forms of lactic acid (see page 293)*

radical chain reactions involve three stages:

- initiation – the step which produces radicals
- propagation – steps giving products and more radicals
- termination – steps which remove radicals by turning them into molecules.

Radical substitution – the reaction of an alkane with bromine in sunlight in a radical chain reaction. The main products are bromomethane and hydrogen bromide. The presence of some ethane in the mixture of products is evidence for the termination step.

Initiation: $Br-Br \longrightarrow Br\bullet + Br\bullet$

Propagation: $CH_4 + Br\bullet \longrightarrow CH_3\bullet + HBr$

            $CH_3\bullet + Br_2 \longrightarrow CH_3Br + Br\bullet$

Termination: $CH_3\bullet + CH_3\bullet \longrightarrow CH_3CH_3$

            $CH_3\bullet + Br\bullet \longrightarrow CH_3Br$

Radical addition polymerization – radical chain reactions can also be used to make *addition polymers* from alkenes and from epoxy compounds.

The initiator can be an organic *peroxide* such as benzoyl peroxide, which acts as a source of radicals, $R-O\bullet$.

$$R-O \diagdown O -R \longrightarrow 2R-O\bullet$$

This leads to the propagation steps which are repeated many times, producing a very long chain molecule.

$$2R - O \bullet \quad + \quad \overset{H}{\underset{H}{\diagdown}} C = C \overset{H}{\underset{H}{\diagup}} \longrightarrow R - O - \overset{\overset{H}{|}}{\underset{\underset{H}{|}}{C}} - \overset{\overset{H}{|}}{\underset{\underset{H}{|}}{C}} \bullet$$

Termination takes place when two radicals combine to make a molecule.

$$R - O - (CH_2)_p - CH_2 - CH_2 \bullet \quad \bullet CH_2 - CH_2 - (CH_2)_q - O - R$$

$$\downarrow$$

$$R - O - (CH_2)_p - CH_2 - CH_2 - CH_2 - CH_2 - (CH_2)_q - O - R$$

radioactive decay: the decay of a radioactive nucleus to form the nucleus of another element by *alpha decay* or *beta decay*. *Gamma rays* may also be emitted.

radioactive nucleus $\longrightarrow$ daughter nucleus + alpha or beta radiation

Radioactive decay is a *first order* process. The rate of decay is proportional to the number of radioactive atoms. This means that each radionuclide decays with a constant *half-life*.

Radionuclides with a short half-life decay more quickly and tend to give off more intense radiation. The unit of radioactivity is the *becquerel* (symbol Bq).

Half-lives for radioactive decay are unaffected by changes in temperature, pressure or chemical state. This is the principle of the technique used to find the age of the remains of living things or of rocks. Knowing the half lives it is possible to estimate ages by measuring the proportions of different nuclides in a sample.

Raoult's law: a law which predicts the *vapor pressure* of mixtures of liquids and solutions. The law states that in a mixture of two liquids A and B:

$$\begin{array}{l}\text{vapor pressure of A} \\ \text{over the mixture}\end{array} = \textit{mole fraction of A} \times \text{vapor pressure of pure A}$$

$$\begin{array}{l}\text{vapor pressure of B} \\ \text{over the mixture}\end{array} = \textit{mole fraction of B} \times \text{vapor pressure of pure B}$$

The total vapor pressure of the mixture is the sum of the vapor pressures of A and B over the mixture.

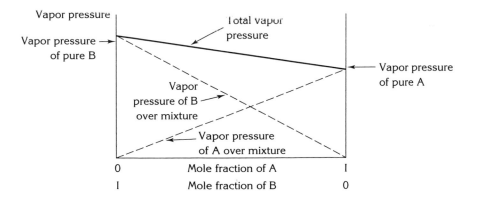

Graph to illustrate the meaning of Raoult's law

Only *ideal mixtures* obey the law. These are mixtures of very similar liquids so that the *inter-molecular forces* between molecules of A and B in the mixture are closely similar to the intermolecular forces in pure A and pure B.

If the intermolecular forces between A and B are overall stronger than in the pure liquids, then the mixture is less volatile than Raoult's law predicts and has a lower vapor pressure. This leads to a negative deviation from the law.

If the intermolecular forces between A and B are overall weaker than in the pure liquids, then the mixture is more volatile than Raoult's law predicts and has a higher vapor pressure. This leads to a positive deviation from the law.

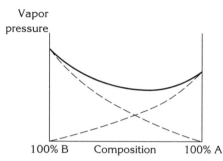

Negative deviation from Raoult's law because of stronger forces between the molecules in the mixture

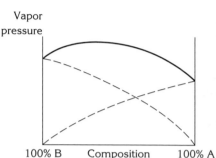

Positive deviation from Raoult's law because of overall weaker forces between the molecules in the mixture than in the pure liquids

rare earths: the traditional name of the *lanthanide elements*.

rate constants: the constants in *rate equations*. The units of a rate constant depend on the overall order of reaction.

Overall order	Units of the rate constant
zero	$mol \cdot L^{-1} \cdot s^{-1}$
first	s^{-1}
second	$mol^{-1} \cdot L \cdot s^{-1}$
third	$mol^{-2} \cdot L^{2} \cdot s^{-1}$

rate determining step: the slowest step in the mechanism of a reaction which therefore controls the overall rate of reaction. It is generally the molecules or ions involved (directly or indirectly) in the rate determining step which appear in the rate equation for the reaction.

Nucleophilic substitution in haloalkanes such as 2-bromo-2-methylpropane involves a two-step mechanism. The first step, which is the bond breaking step in this S_N1 mechanism, is the rate determining step. It involves only haloalkane molecules. Hydroxide ions are not involved. So the hydroxide ion concentration does not appear in the rate equation which has the form:

rate $= k[C_4H_9Br]$

The rate determining step in a multistep reaction is the one with the highest activation energy.

rate equations: equations which show how changes in concentration, temperature and catalysts affect the rate of a reaction.

For the reaction of hydrogen peroxide with hydriodic acid:

$$H_2O_2(aq) + 2I^-(aq) + 2H^+(aq) \longrightarrow 2H_2O(l) + I_2(aq)$$

rate $= k[H_2O_2(aq)][I^-(aq)]$

Note that the rate equation cannot be deduced from the balanced equation; it has to be found by experiment. The equation shows that this reaction is *first order* with respect to each reactant. Overall it is a *second order reaction*.

Methods for finding the order of reaction include:

* the *half-life* method
* the *initial rate method*.

The rate constant, k, is constant only for a particular temperature. The value of k varies with temperature. A rise in temperature of 10 K doubles the value of k and so doubles the rate of reaction if the *activation energy* is about 50 kJ/mol. Adding a *catalyst* increases the value of k and hence the rate of reaction because the catalyzed pathway has a lower activation energy.

The *Arrhenius equation* is the relationship between the rate constant, the activation energy and the temperature.

rates of reaction, in mol/L·s, are found by measuring the rate of formation of a product or the rate of removal of a reactant. The usual procedure for finding rates is to measure some property of the reaction mixture, such as its volume, and to see how this property varies with time.

$$\text{Rate} = \frac{\text{change in the measured property}}{\text{time}}$$

Rates are measured in practical units such as "cm per second" and then converted to mol/L·s if necessary.

Methods for studying rates include:

* collecting and measuring the volume of a gas formed
* removing measured samples of the mixture, stopping the reaction and then determining the concentration of one reactant or product by *titration*
* using a *colorimeter* to follow the formation of a colored product or the removal of a colored reactant
* using a conductivity cell and meter to measure the changes in electrical conductivity of the reaction mixture and the number or nature of the ions changes.

Factors which affect reaction rates are:

* the *concentration* of reactants in solution – in general the higher the concentration the faster the reaction (see also *collision theory*)
* the *pressure* of gaseous reactants – high pressures compress gases and increase their concentration
* the surface area of solids – breaking a solid into smaller pieces increases the surface area in contact with a liquid or gas

- the temperature – raising the temperature increases the proportion of atoms, molecules or ions with enough energy to react when they collide
- catalysts – adding a *homogeneous catalyst* or *heterogeneous catalyst* provides an alternative reaction mechanism with lower *activation energy*
- radiation – *electromagnetic radiation* can initiate *free-radical reactions* and the more intense the radiation the faster the reaction goes.

raw materials: see *chemical industry*.

reacting masses: the masses of elements and compounds which take part in a chemical reaction. Chemists work out *empirical formulas* and *balanced equations* by measuring the masses of reactants and products. They use equations to determine how much of each chemical they need for a chemical process and the theoretical yield of the products (see *yield calculations*).

Worked example

Carbon monoxide reduces iron(III) oxide to iron in a blast furnace. What mass of carbon monoxide is required to reduce 20 metric tons of the oxide?

Notes on the method

Start by writing the balanced equation for the reaction.

Work out the *molar masses*, M, of the reactants in g/mol.

The reacting masses are in the same ratio whether measured in grams or in tons.

Answer

$$Fe_2O_3(s) + 3CO(g) \longrightarrow 2Fe(s) + 3CO_2(g)$$

1 mol $Fe_2O_3(s)$ reacts with 3 mol $CO(g)$.

$M(Fe_2O_3) = (2 \times 56 \text{ g/mol}) + (3 \times 16 \text{ g/mol}) = 160 \text{ g/mol}$

$M(CO) = (12 + 16) \text{ g/mol} = 28 \text{ g/mol}$

So $(1 \text{ mol} \times 160 \text{ g/mol}) = 160$ g iron reacts with $(3 \text{ mol} \times 28 \text{ g/mol})$
$= 84$ g carbon monoxide,

or 160 tons iron reacts with 84 tons carbon monoxide

Hence mass of CO needed to react with 20 tons iron $= \dfrac{20}{160} \times 84$ tons

$= 10.5$ tons

reacting volumes of gases: see *gas-volume calculations*.

reaction kinetics: the study of the factors that affect the *rates of chemical reactions*. Rates are followed either by measuring the rate of formation of one of the products or the rate of removal of one of the reactants. The results of these studies are summed up in *rate equations* for reactions. *Collision theory* is one model which helps to explain the factors affecting rates. Studying the kinetics for a change can help to determine the *mechanism of a reaction*.

reaction mechanisms: see *mechanism of a reaction*.

reactors: reaction vessels used by the chemical industry for *synthesis*. There are two main types of reactor:

- batch reactors – for *batch processes* in which all the reactants are mixed in a vessel where they are heated or cooled until the process is complete

- continuous reactors – for *continuous processes* in which the reactants are fed continuously into a stirred tank or tube and products drawn out at an equal flow rate.

reagent: any pure chemical, mixture or solution supplied for chemical analysis or synthesis. "Reagent grade" chemicals match the minimum standards specified for normal laboratory use.

Primary standard grade reagents are chemicals which are much purer than standard laboratory grade reagents. They are intended for accurate analytical work where impurities in the reagents would affect the results.

real gases: gases which deviate to a greater or lesser extent from *ideal gas* behavior as predicted by the ideal gas equation. In practice, many gases at room temperature follow the gas laws sufficiently closely for it to be reasonable to treat them as ideal gases.

The assumptions of *kinetic theory* explain why real gases tend to deviate from ideal behavior at:

- low temperatures and moderate pressures – when gases are not far above their boiling points so that they are close to condensing because of *intermolecular forces*

- very high pressures – when the volume of the gas is so small that the volume of the molecules cannot be ignored and the gases are less compressible than the gas laws predict.

rearrangement reaction: a reaction which rearranges the atoms in a molecule to convert one *structural isomer* to another.

$$CH_3 - CH_2 - CH_2 - CH_2 - CH_3 \longrightarrow CH_3 - \overset{\overset{\displaystyle CH_3}{\displaystyle |}}{CH} - CH_2 - CH_3$$

pentane   2-methylbutane

Isomerization rearranges the atoms in pentane to make 2-methylbutane which has a higher octane number

One example of a rearrangement reaction is the isomerization of alkanes to make branched compounds to increase the octane number of *gasoline*.

recrystallization: a procedure often used to purify solid products of *organic preparations*. The procedure is based on a *solvent* which dissolves the product when hot but not when cold. The choice of solvent is usually made by trial and error. Use of *vacuum filtration* speeds filtering and makes it easier to recover the purified solid from the filter paper.

The procedure is as follows:

1 warm the impure solid with the hot solvent

2 if the solution is not clear, filter the hot solution though a heated funnel to remove insoluble impurities

3 cool the solution so that the product recrystallizes leaving the smaller amounts of soluble impurities in solution

4 filter to recover the purified product

5 wash the solid with small amounts of pure solvent to wash away the solution of impurities still clinging to the solid

6 allow the solvent to evaporate in a stream of air and then in a *desiccator*.

recycling: a method of dealing with *wastes*. The economics of recycling often depend on the energy required to process wastes compared to the energy needed to produce materials directly from raw materials, examples of which follow:

- steel, 32 MJ/kg (megajoules per kilogram)

- aluminum, 146 MJ/kg

- plastics such as polythene, 90 MJ/kg

- cardboard 56 MJ/kg.

The large amount of energy involved in *aluminum extraction* makes recycling particularly worthwhile.

There are other benefits to recycling such as reducing the environmental impact of extracting and processing raw materials.

Recycling metals – recycling *steel* can be done repeatedly because steel is as good as new after reprocessing. For every metric ton of steel recycled there is a saving of 1.5 metric tons of iron ore and half a metric ton of coal. There is also a big reduction in the total amount of water needed since large quantities of water are involved in mineral processing.

Worldwide, the steel industry recycles over 430 million metric tons of the metal each year which is a recycling rate of over 50%. The basic oxygen process uses a minimum of 25% scrap steel. Electric arc furnaces make steel plate, beams and bars by remelting nearly 100% scrap steel. Much of the recycled steel is waste from various stages of manufacturing steel objects. Another large source of scrap is from old motor vehicles.

Some steel is recovered from the cans in household waste. The advantage of steel is that it is magnetic so magnets can easily separate cans from other waste materials.

Aluminum scrap from manufacturing processes is always recycled. Less than 1% of the mass of household waste is aluminum, almost all of it in the form of soft drink cans. Aluminum is not magnetic but can be separated from a waste stream by a rapidly varying magnetic field which induces eddy currents in the metal of the can. The interaction of the external field and the magnetic effect of the eddy currents leads to a force which can push cans out of the waste stream. Aluminum is easy to recycle at low cost. The energy required for recycling the metal is a twentieth of the energy needed to extract the element from its ore.

Recycling plastics – plastics are much more difficult to recycle than iron or steel. One problem arises from the low density of plastic waste. It can take as many as 20,000 bottles to make up a metric ton of waste. Also, there are many types of plastic which have to be sorted. The industry has introduced codes for labeling plastic products to help consumers separate them for recycling.

Plastics vary greatly in their value. The material used to make sparkling drink bottles, PET, is generally the most valuable. PET can be remelted and spun into fibers for carpets, bedding and clothing. Another plastic worth recycling is PVC, which can be recycled to make sewage pipes, flooring and even the soles of shoes.

An alternative approach to recycling plastics is to use cracking to break the polymer molecules into small molecules which can be added to the *feedstock* for oil refineries and chemical plants.

redox potential: see *standard electrode potential.*

redox reactions: reactions which involve **red**uction and **ox**idation. In every redox reaction an atom, molecule or ion is reduced while another atom, molecule or ion is oxidized. Reduction and oxidation always go together.

Redox reactions involve the transfer of electrons from the *reducing agent* to the *oxidizing agent*. Writing *half-equations* for redox reactions help to show electron transfer.

Oxidation numbers help to identify redox reactions. In any redox reaction the oxidation number of one element becomes more positive (or less negative) while the oxidation number of the other element becomes less positive (or more negative).

Equations for redox reactions, like other balanced equations, show the *amounts* (in moles) of reactants and products involved. Oxidation numbers help to balance redox equations because the total decrease in oxidation number for the element reduced must equal the total increase in oxidation number for the element oxidized. This is illustrated here by the oxidation of iron(II) ions by permanganate ions in acid solutions.

Step 1 – Write down the formulas for the atoms, molecules and ions involved in the reaction:

$$MnO_4^- + H^+ + Fe^{2+} \longrightarrow Mn^{2+} + H_2O + Fe^{3+}$$

Step 2 – Identify the elements which change in oxidation number and the extent of change:

change of −5

$$MnO_4^- + H^+ + Fe^{2+} \longrightarrow Mn^{2+} + H_2O + Fe^{3+}$$

change of +1

Step 3 – Balance so that the decrease in oxidation number of one element equals the total increase of the other element.

In this example the decrease of −5 in the oxidation number of manganese is balanced by five iron(II) ions which each increase their oxidation number by +1.

$$MnO_4^- + H^+ + 5Fe^{2+} \longrightarrow Mn^{2+} + H_2O + 5Fe^{3+}$$

Step 4 – Balance for oxygen and hydrogen.

In this example the four oxygens from the permanganate ion join with eight hydrogen ions to form four water molecules:

$$MnO_4^- + 8H^+ + 5Fe^{2+} \longrightarrow Mn^{2+} + 4H_2O + 5Fe^{3+}$$

Step 5 – In an *ionic equation*, check that the positive and negative charges balance and add state symbols.

The net charge on the left is now 17+, which is the same as the net charge on the right.

$$MnO_4^-(aq) + 8H^+(aq) + 5Fe^{2+}(aq) \longrightarrow Mn^{2+}(aq) + 4H_2O(l) + 5Fe^{3+}(aq)$$

redox titration: a practical technique used to determine the *concentration of a solution* of an *oxidizing agent* or of a *reducing agent*. A *titration* measures the volume of a *standard solution* of oxidizing agent or reducing agent needed to react exactly with a measured volume of the unknown solution. The procedure gives accurate results only if the reaction is rapid and is exactly described by the chemical equation.

An acidic standard solution of the oxidizing dichromate ions, for example, can be used to estimate the concentration of iron(II) ions. The end point for a dichromate titration is detected using a redox indicator – a chemical which gives a clear color change when oxidized. The procedure and method of calculating results is similar to those for other titrations for example *iodine–thiosulfate titrations* and *potassium permanganate* titrations.

reducing agents: chemical reagents which can reduce other atoms, molecules or ions by giving electrons to them. Common inorganic reducing agents are metals such as zinc, iron and tin, often with acid; other common reducing agents are sulfur dioxide, iron(II) ions and iodide ions.

Iodide ions acting as a reducing agent by giving electrons to iron(III) ions. A reducing agent is itself oxidized when it reacts. Reduction and oxidation always go together in redox reactions.

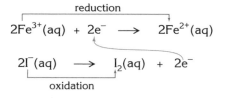

Some reagents change color when they are reduced which makes them useful for detecting reducing agents.

Test	Observations	Explanation
Add a solution of potassium permanganate acidified with dilute sulfuric acid	Purple solution turns colorless	Purple MnO_4^- ions are reduced to very pale pink Mn^{2+} ions
Add potassium dichromate solution acidified with dilute sulfuric acid	Orange solution turns green	Orange $Cr_2O_7^{2-}$ ions are reduced to green Cr^{3+} ions

Powerful reducing agents used in organic chemistry include *lithium aluminum hydride* and *sodium borohydride*.

reducing sugars: *sugars* such as glucose and fructose which give a positive result with *Fehling's solution* or *Benedict's solution*. These sugars reduce the blue copper(II) complex in the test solution to an orange–red precipitate of copper(I) oxide.

All monosaccharides and many disaccharides are reducing sugars, but the disaccharide sucrose is not a reducing sugar.

reductant: an alternative to *reducing agent* which is convenient when describing half-equations for *redox reactions* and *standard electrode potentials*. By convention, when electrode potential values are assigned the half-equation takes the form:

oxidant + ne⁻ $\rightleftharpoons$ reductant

Every half-equation involves an oxidant and a reductant.

reduction: originally this meant removal of oxygen or the addition of hydrogen but the term now covers all reactions in which atoms, molecules or ions gain electrons. The definition is further extended to cover molecules, as well as ions, by defining oxidation as a change which makes the *oxidation number* of an element more negative or less positive.

$$2Br^- + 2H^+ + H_2SO_4 \longrightarrow Br_2 + SO_2 + 2H_2O$$

Bromide ions reducing concentrated sulfuric acid to sulfur dioxide

Oxidation and reduction always go together in *redox reactions*.

Oxidation number rules apply in principle in organic chemistry but it is often easier to use the older definitions. Reduction is either removal of oxygen to a molecule or the addition of hydrogen.

Reduction of ketone to an alcohol by addition of hydrogen

reference electrodes: electrodes used to measure electrode potentials in place of the standard *hydrogen electrode*. A hydrogen electrode is difficult to use so it is much easier to use a secondary standard such as a silver/silver chloride electrode or a calomel electrode. These electrodes are available commercially and are reliable to use. They have been calibrated against a standard hydrogen electrode. (Calomel is an old-fashioned name for mercury(I) chloride.)

$$Hg_2Cl_2(s) + 2e^- \rightleftharpoons 2Hg(l) + 2Cl^-(aq) \qquad E^\circ = +0.27 \text{ V}$$

refining: the processes which separate, convert and purify chemicals in *crude oil* in an oil refinery, or the removal of impurities from metals (see *copper refining*).

reflux condenser: a condenser fitted to a chemical apparatus to prevent vapor from escaping while heating a liquid. Vapor from a boiling reaction mixture condenses and flows back into the flask.

reforming: a process in oil refining which converts *alkanes* to *arenes* such as benzene and toluene (methylbenzene). This helps to raise the octane number of mixtures used in *gasoline*.

An example of reforming

cyclohexane   benzene

Another example of reforming

refractive index: a *physical property* of chemicals and materials which can help to identify unknown samples.

Forensic scientists use an oil immersion method to find the reactive index of fragments of glass from the scene of an accident or crime. The procedure is to immerse a crushed fragment in silicone oil on a slide. The slide is slowly heated, making the refractive index of the oil change as the temperature rises. The fragments of glass seem to disappear when their refractive index exactly matches that of the oil, then reappear when the oil is heated beyond the point where the refractive indices match. The technique is extremely accurate. Repeat measurements lie within a range of ± 0.000 05 units.

refractories: materials with very high melting points used to line furnaces and make crucibles. *Ceramics* are refractory materials.

Fireclay is the raw material for making refractory bricks. Some industries need more specialized refractories. Molten *glass* and the *slags* formed during smelting are very corrosive when molten and would quickly destroy ordinary refractory bricks made of fireclay. These industries use refractories such as pure silicon dioxide (an *acidic oxide*) or pure magnesium oxide (a *basic oxide*).

relative atomic mass, A_r: the mean mass of the atoms of an element relative to the mass of atoms of the isotope carbon-12 for which A_r is defined as exactly 12. The values are relative so they do not have units.

Mass spectrometry is the technique for determining accurate relative atomic masses.

Amount of substance in chemistry is defined in such a way that the *molar mass* of an element is numerically equal to its relative atomic mass (but the latter is a ratio and has no units).

Relative atomic masses often do not have whole number values because elements have *isotopes*. Relative atomic masses are average values for the mixture of isotopes found naturally (see *isotopic abundance*).

relative formula mass, M_r: a term used for the relative mass of ionic compounds or ions to avoid the suggestion that their formulas represent molecules. The scale is the same as for relative atomic and relative molecular masses that is $^{12}C = 12$.

The relative formula mass of an ionic compound or ion is the sum of the *relative atomic masses* for the atoms in the formula.

For anhydrous magnesium nitrate, $M_r[Mg(NO_3)_2]$

$= 24 + (2 \times 14) + (6 \times 16) = 148$

    ↑   ↑   ↑

$A_r(Mg)$ $A_r(N)$ $A_r(O)$

Amount of substance in chemistry is defined in such a way that the *molar mass* of an ionic compound or an ion is numerically equal to its relative formula mass (but the relative mass has no units).

relative isotopic mass: the mass of the atoms of an isotope of an element relative to the mass of atoms of the isotope carbon-12. Chemists determine relative isotopic masses with a *mass spectrometer*. The values are always very close to the mass numbers of the elements. It is usually accurate enough to assume that the value of the relative atomic mass is the same as the *mass number*.

relative molecular mass, M_r: the relative mass of the molecules of an element or compound on the scale $^{12}C = 12$.

The relative molecular mass of the molecules of an element or compound is the sum of the *relative atomic masses* for the atoms in the *molecular formula*.

$$\text{For ethanol, } M_r(CH_3CH_2OH) = (2 \times 12) + (6 \times 1) + 16 = 46$$

$$A_r(C) \qquad A_r(H) \quad A_r(O)$$

Amount of substance in chemistry is defined in such a way that the *molar mass* of the molecules of an element or compound is numerically equal to its relative molecular mass but as with other relative masses, the relative value has no units.

reserves and resources describes the quantities of *minerals* and *fossil fuels* available in the Earth's crust. Estimates of reserves and resources include some which are proven because they have been identified and evaluated and others which it seems likely to exist by a mixture of inference and guesswork.

The proven reserves of a mineral have been identified and evaluated so that it is known that it is economical to extract and process them at the current market price of the product. The resources of a mineral include all the possible sources, including those that it is not economical to extract at the time.

Judgements about changes in technology, in the demand for materials and in the economy can all shift the balance of reserves and resources. Figures for reserves have to be interpreted with care because commercial companies do not find it economic to establish reserves for more than about 25 years ahead.

residence time: a measure of the time a chemical stays in any part of a system where chemicals are continuously flowing in and out of containers or reservoirs. In environmental systems the reservoirs are parts of the Earth system such as the *atmosphere*, oceans or land. In a *steady state system* the residence time equals the amount of the chemical in the reservoir divided by the rate of addition (or removal) of the chemical. The residence time of calcium ions in the oceans is a million years. The residence time of oxygen in the atmosphere is 7600 years. If the residence time of a pollutant is large it may stay in the environment for a long time before being inactivated.

residues: materials left over at the end of a chemical process. The term sometimes refers to a solid caught in a filter paper. Laboratories which use large quantities of valuable chemicals may have a "residues bottle" to collect wastes for recovery of the expensive substances. Residues worth collecting include silver compounds, iodine compounds and solvents.

A B C D E F G H I J K L M N O P Q **R** S T U V W X Y Z

resins: originally these were natural materials from the sap of trees or from insects. Examples are the rosin used for violin bows, shellac used in varnishes and myrrh which is added to perfumes. These gummy materials consist of long chain molecules and on warming they soften before they melt.

Now the term resin refers to synthetic materials made by *polymerization*. *Thermosetting polymers*, for example, are produced in two stages. The first stage produces a polymer resin. Compressing this resin in a hot mold shapes it and at the same time creates *cross-links* between the polymer chains so that the material sets to a hard plastic.

Similarly, one component of an *epoxy resin* adhesive is a polymer resin which sets on mixing with a hardener to start the chemical changes which cross-link the chains.

Synthetic *ion-exchange* resins consist of polymer beads which have been treated chemically so that they can hold cations or anions.

respiration: biochemical processes in cells which provide living organisms with the energy for growth, movement and warmth.

Aerobic respiration is respiration with oxygen. In the process *glucose* is *oxidized* to carbon dioxide and water. Overall:

$$C_6H_{12}O_6(aq) + O_2(g) \longrightarrow 6CO_2(g) + 6H_2O(l) \quad \Delta H = -2816 \text{ kJ/mol}$$

This is not a one-step reaction; it takes place in a complex series of biochemical reactions which harness the energy from the oxidation process to produce *ATP*.

Anerobic respiration is respiration without *oxygen*. Anerobic respiration in yeast converts *sugars* to carbon dioxide and ethanol. This is the process of *fermentation* used in baking, brewing and winemaking. Anerobic respiration in animal cells produces lactate ions (the anions of lactic acid). Anerobic respiration produces much less ATP than aerobic respiration and so is a much less efficient source of energy for processes.

restricted rotation: see *geometrical isomerism*.

retention times: see *gas-liquid chromatography*.

reversible changes: processes which can be reversed by altering the conditions.

Melting is an example of a reversible change of state. Changing the temperature alters the direction of change. Ice melts above 0°C. Water freezes below 0°C. Water and ice are in equilibrium at 0°C. This is an example of *dynamic equilibrium*.

$$H_2O(s) \rightleftharpoons H_2O(l)$$

Many chemical reactions are reversible, including the reactions used in *ammonia manufacture* and *sulfuric acid manufacture*. *Acid–base reactions* and *redox reactions* are generally reversible too. There are many important reversible processes in living things. *Hemoglobin*, for example, picks up oxygen from the air in the lungs and then releases the oxygen for *respiration* in body tissues.

The direction and extent of change for a reversible process may vary with the temperature or the concentration of chemicals. *Pressure* is an important variable affecting reactions of gases.

Chemists predict the direction and extent of change for reversible processes when they consider *feasibility*.

reverse osmosis: see *osmotic pressure*.

rhodium is one of the *platinum* metals which is largely used as a catalyst in the platinum–rhodium alloys found in *catalytic converters* and in chemical plants for *nitric acid manufacture*.

risk, in chemistry, is an estimate of the chance that a hazardous substance or process will cause harm.

There are three questions to consider when assessing the risk of a chemical process:

- what are the hazards?
- what is the likelihood that someone will come to harm because of the hazards?
- what can be done to control and reduce the risks?

When a chemical investigation is being planned the possibilities for controlling risks include the following:

- choosing a nonpractical approach using secondary sources, models and simulations
- adopting an alternative safer procedure with less hazardous chemicals
- modifying the experimental design with different apparatus or a different method of heating
- creating a barrier between the apparatus and people, using safety screens or a fume cupboard
- wearing personal protection such as eye protection and gloves.

(See also *hazard warning signs.*)

RNA: a type of *nucleic acid* in which the five-carbon sugar is ribose and the four nitrogenous bases are adenine, cytosine, guanine and uracil. Two types of RNA are important in *protein synthesis*:

- messenger RNA (m-RNA) which takes the genetic code from *DNA* molecules in the nucleus to the sites of protein synthesis in the cell
- transfer RNA (t-RNA) which helps to assemble amino acids in the right order during the translation of the genetic code into protein molecules.

In some viruses, such as HIV, the genetic material is RNA rather than DNA.

roasting converts sulfides to oxides during *metal extraction*. Oxides are much easier to reduce to metals than sulfides. Roasting is exothermic so once the reaction has begun it needs little fuel to keep the process going. Roasting produces hot metal oxides which may be fed directly to a furnace for reduction to the metal.

$$2PbS(s) + 3O_2(g) \longrightarrow 2PbO(s) + 2SO_2(g)$$

In the past roasting sulfide ores was a major cause of air pollution. The sulfur dioxide was simply released into the air where it caused acid rain. Now the sulfur dioxide is recovered and used for *sulfuric acid manufacture*. (See also *zinc extraction*.)

rotation around covalent bonds: rotation of one part of a molecule relative to another around a single covalent bond. This allow molecules to take up different *conformations*.

Double bonding stops free rotation and this accounts for the existence of *geometric isomers*.

rubber: an elastic material which is easily stretched or bent but which tends to spring back to its original shape when the force is removed.

Rubbers are **elastic** *polymers* often called *elastomers*.

Elastomer molecules tend to coil up in such a way that they can uncoil as they stretch and coil up again when unstretched. The molecules are *cross-linked* by strong covalent bonds so that they cannot permanently slide past each other unless pulled much harder.

Natural rubber comes from the sap of the rubber tree. This white latex coagulates to a very soft material. Vulcanizing converts natural rubber into a useful elastomer by cross-linking the polymer chains with sulfur atoms. This was the process discovered by Charles Goodyear in 1839.

The science of *polymer chemistry* has helped to develop a range of alternatives to natural rubber suited for specific purposes. Examples are:

- neoprene – used for rubber dinghies
- styrene-butadiene rubber – used for motor vehicle tires
- *polyurethanes* – used for furniture foams and stretch fabrics.

Tires are black because the rubber is mixed with carbon *filler* which makes the material more resistant to wear. Carbon also absorbs *ultraviolet radiation* helping to slow down the rate at which rubber degrades in sunlight.

rusting: the *corrosion* of iron.

sacrificial anode: see *cathodic protection.*

salt bridge: the ionic connection between the solutions of the two *half-cells* which make up an *electrochemical cell.* A salt bridge makes an electrical connection between the two halves of the cell by allowing ions to flow while preventing the two solutions from mixing. At its simplest a salt bridge consists of a strip of filter paper soaked in potassium nitrate solution. Potassium salts and nitrates are soluble so the salt bridge does not react with the ions in the half-cells.

In commercial *reference electrodes* the salt bridge is often a small plug of porous glass.

salt hydrates: see *hydration.*

salting out effect. the effect of making a chemical much less soluble in water by adding a high concentration of a salt. Adding common salt (sodium chloride) helps to separate *soap* from water after *hydrolysis* of fats or oils.

salts: ionic compounds formed when an *acid* reacts with a *base.* Salts therefore have two "parents." Salts are related to a parent acid and to a parent base.

Acid	Salts
hydrochloric acid, HCl	sodium chloride, NaCl
	calcium chloride, $CaCl_2$
	ammonium chloride, NH_4Cl
sulfuric acid, H_2SO_4	sodium sulfate, Na_2SO_4
	calcium sulfate, $CaSO_4$
	ammonium sulfate, $(NH_4)_2SO_4$
acetic acid, CH_3COOH	sodium acetate, CH_3CO_2Na
	calcium acetate, $(CH_3CO_2)_2Ca$
	ammonium acetate, $CH_3CO_2NH_4$

Base	Salts
sodium hydroxide, NaOH	sodium chloride, NaCl
	sodium sulfate, Na_2SO_4
	sodium acetate, CH_3CO_2Na
calcium oxide, $Ca(OH)_2$	calcium chloride, $CaCl_2$
	calcium sulfate, $CaSO_4$
	calcium acetate, $(CH_3CO_2)_2Ca$
ammonia, NH_3	ammonium chloride, NH_4Cl
	ammonium sulfate, $(NH_4)_2SO_4$
	ammonium acetate, $CH_3CO_2NH_4$

Neutralization is not the only way to make a salt. Some metal chlorides, for example, are made by heating metals in a stream of chlorine. This is useful for making anhydrous chlorides, such as *aluminum chloride* or iron(III) chloride, both of which are hydrolyzed by water so cannot be dehydrated by heating.

Insoluble salts are conveniently prepared by *ionic precipitation*.

salts of strong and weak acids: see *hydrolysis of salts* and *neutralization.*

sampling for analysis is a vital first step in any analysis because it is important to make sure that the sample is representative of the whole specimen. This is easy when analyzing solutions which are well mixed. It is more difficult when sampling an *ore* made up of lumps of rock each with variable but small amounts of a precious mineral. In these circumstances an analyst has the difficult task of preparing a sample with a mass of about 1 g from a batch of ore with a mass of several metric tons.

Analysts generally prepare multiple samples from the same specimen and work through the analysis with all of them to check on the *uncertainty* of the results.

saturated compounds: compounds containing only single bonds between the atoms in their molecules. Examples of saturated *hydrocarbons* are the *alkanes*.

The term "saturated" is also used for compounds with saturated hydrocarbon chains such as the saturated *fats* and *fatty acids*, even though these compounds include $C = O$ bonds. Saturated compounds do not undergo *addition reactions*.

saturated solutions: solutions which contain as much of the dissolved substance as possible at a particular temperature. A saturated solution is in equilibrium with undissolved excess of the substance in solution. The concentration of a saturated solution is the *solubility* of the substance at that particular temperature.

saturated vapor pressure: see *vapor pressure.*

s-block elements: the elements in groups I and II in the *periodic table*. For these elements the last electron added to the atomic structure goes into the *s-orbital* in the outer shell. All the elements in the s-block are reactive metals. (See also *atomic orbitals*.)

scaling up processes: converting a small scale synthesis in a laboratory or pilot plant to a process which can operate successfully and safely on a commercial scale (see table on page 311).

second law of thermodynamics: see *thermodynamics (laws of).*

second order reaction: a reaction is second order with respect to a reactant if the rate of reaction is proportional to the concentration of that reactant squared. The concentration term for this reactant is raised to the power two in the *rate equation*. At its simplest the rate equation for a second order reaction is:

$$Rate = k[X]^2$$

The rate of reaction of 1-bromopropane with hydroxide ions is overall second order. It is *first order* with respect to the haloalkane and first order with respect to hydroxide ions. The overall order is the sum of the powers in the rate equation.

$$Rate = k[CH_3CH_2CH_2Br][OH^-]$$

(See the diagram on page 311.)

Scaling up processes (see page 310)

On a small scale	On a big scale
• Laboratory preparations are normally *batch processes*.	• In industry, manufacturers use *continuous processes* where there is a steady demand for the product.
• A chemist in a laboratory stores chemicals in bottles and transfers them to a reaction vessel with a spatula or by pouring through a funnel.	• In industry, pumps and pipes transfer liquids to reactors. Solids may be handled by the bagful or ingot.
• Manual mixing or shaking is adequate in the laboratory.	• A large reactor typically has a stirrer driven by an electric motor to ensure thorough mixing.
• A laboratory synthesis may happen in a beaker or flask easily heated with a Bunsen flame or cooled by dipping in water.	• It is not to easy to heat or cool large quantities of material. Reactors often have pipes running through them to carry steam for heating or cold water for cooling. Energy from exothermic reactions can raise steam to generate electricity.
• To separate the product, laboratory apparatus can be rearranged for distillation or tipped to pour the products into a filter funnel.	• Solids can be separated as pumps force a liquid through large cloth filters.
• Chemists can pour products from one container to another, perhaps using handheld tap funnels to extract impurities. *Distillation* or *recrystallization* help to produce a pure product.	• One large-scale approach to crystallization and drying is to spray a solution into a stream of hot air.
• The product can be dried in a *desiccator* or small oven.	

Second order reaction (see page 310)
Variation of concentration of a reactant plotted against time for a second order reaction. The half-life for a second order reaction is not a constant. The time for the concentration to fall from c to c/2 is half the time for the concentration to fall from c/2 to c/4. The half-life is inversely proportional to the starting concentration.

Second order

Concentration of reactant

Time

Half-lives

secondary organic compounds: see *primary, secondary and tertiary organic compounds.*

seed crystal: a crystal added to a *supersaturated* solution to encourage it to crystallize. *Recrystallization* of an organic product sometimes produces a solution which is reluctant to crystallize. Scratching the sides of the container with a glass rod can encourage crystals to form. If this fails, adding a minute crystal of the product may be enough to start rapid crystallization.

semiconductors are not *electrical insulators* but they do not conduct electricity as well as metals. Examples are elements such as *silicon* and *germanium* and compounds such as gallium arsenide. Semiconductors consist of covalent *network solids* in which a few of the bonding electrons can break free and become *delocalized* in the structure. The conductivity of semiconductors is enhanced by "doping" with traces of impurity atoms. Doping silicon (in group IV) with an element such as arsenic (group V) increases the number of conducting electrons. An arsenic atom has one more electron in its outer shell than is needed for bonding in the network solid.

Extra electron
(free to move)

Silicon with an impurity atom from a group V element

semipermeable membrane: a membrane which allows solvent molecules and perhaps some other small molecules or ions to pass through but is impermeable to larger molecules and other ions. *Osmosis* and *dialysis* both depend on semipermeable membranes. So does the manufacture of chlorine by *electrolysis of brine* in a membrane cell.

separating funnel: a tap funnel used to separate liquids which do not mix (*immiscible liquids*). Separating funnels are used to:

- separate an organic product by *solvent extraction* with a solvent such as ether – the product is more soluble in the organic solvent but the ionic reagents and ionic by-products remain in the water

- purify an impure organic product by "washing" it with aqueous reagents (such as dilute acids or alkalis) which dissolve impurities.

Pressure can build up in a separating funnel when a mixture is being shaken if one liquid is very volatile or if a reaction produces a gas. Hence the technique of inverting the funnel while holding in the stopper and releasing excess gas or vapor through the tap.

sequestering agent: a ligand which forms such stable complexes with metal ions that it effectively makes them chemically inactive. Examples of sequestering agents are:

- *EDTA* used to treat people suffering from metal poisoning

- tripolyphosphates added to *detergents* to complex with calcium or magnesium ions in hard water so a scale or scum precipitate is not formed.

shapes of complex ions depend on the number of ligands around the central metal ion. There is no simple rule for predicting the shapes of complexes from their formulas. Typically complexes with:

- six smaller ligands, such as H_2O and NH_3, are octahedral, as in $[Mn(H_2O)_6]^{2+}$ and $[Fe(H_2O)_5(SCN)]^{2+}$

- four larger ligands, such as Cl^-, are usually tetrahedral, as in $[CuCl_4]^{2-}$, but may be planar, as in $[Pt(NH_3)_4]^{2+}$

- two ligands are linear, as in $[Ag(NH_3)_2]^+$, $[Ag(S_2O_3)_2]^{3-}$ and $[Ag(CN)_2]^-$.

Octahedral   Tetrahedral

$[H_3N — Ag — NH_3]^+$

Linear

Shapes of complex ions

shapes of molecules and ions can be predicted by examining the number of bonding and nonbonding *lone pairs of electrons* in the outer shell of the central atom. This is the Valence Shell Electron Pair Repulsion (VSEPR) theory. The expected shape for a molecule is the one which minimizes the repulsion between electron pairs by keeping them as far apart as possible in three dimensions.

Shapes of molecules with two and three electron pairs around the central atom

Cl—Be—Cl

Linear

Trigonal planar

Nonbonding lone pairs are held closer to the central atom. The result is that the order of repulsion between electron pairs is:

lone pair–lone pair > lone pair–bond pair > bond pair–bond pair

109.5°   107°   105°

Bond angles in molecules with four electron pairs around the central atom. The greater repulsion between lone pairs, and between lone pairs and bonding pairs, means that the angles between the covalent bonds decrease as the number of lone pairs increases.

A molecule with six electrons pairs around the central atom is octahedral. An octahedron has eight faces but six corners.

An octahedral molecule

A PCl_5 molecule has five electron pairs around the central phosphorus atom. In the vapor phase this molecule is a trigonal bipyramid.

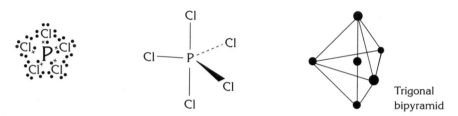

A trigonal bipyramid – a shape based on two pyramids with triangular bases

The two electron pairs in a double bond count as "one" when it comes to predicting molecular shapes.

Linear   Trigonal planar   Tetrahedral

Shapes of molecules with double bonds

The same principles apply to the shapes of ions.

Trigonal planar   Trigonal planar   Tetrahedral

Shapes of ions consisting of more than one atom

shells are the main *energy levels* in atoms and are numbered 1, 2, 3. These numbers are the principal *quantum numbers*.

shielding: an effect of inner electrons which reduces the pull of the nucleus on the electrons in the outer shell of an atom. Thanks to shielding, the electrons in the outer shell are attracted by an "effective nuclear charge" which is less than the full charge on the nucleus.

Shielding accounts for the fall in the first *ionization energy* for the elements down a group in the periodic table.

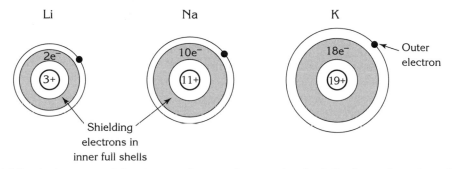

Li   Na   K

Outer electron

Shielding electrons in inner full shells

Shielding in the atoms of the elements in group I means that the "effective nuclear charge" is 1+. Down the group the outer electron is held less strongly, being further from the same effective nuclear charge.

SI units: the internationally agreed units for measurement in science. There are seven base units in the system. All other units are derived from the base units.

Every physical quantity in the system has a symbol. A physical quantity has a value and a unit. In calculations it is good practice to substitute both the value and the unit in formulas as shown in the worked examples in this book. The six units in this table are the base units used in chemistry. Note that in print the symbols of physical quantities appear in italics but the units do not.

Physical quantity	Symbol	Unit	Unit symbol
length	l	meter	m
mass	m	kilogram	kg
time	t	second	s
electric current	I	ampere	A
temperature	T	kelvin	K
amount of substance A	n_A	mole	mol

side reactions: unwanted reactions which reduce the yield of the product being formed by the main reaction. Side reactions create *by-products* which are wasteful because they have no use.

sievert (symbol Sv): the unit which measures the biological effect of radiation. Some types of radiation are more damaging to living cells than others. Some parts of the body are more at risk than others. This is allowed for by calculating an effective radiation dose measured in sieverts.

One sievert is a large dose, so effective doses are often quoted in millisieverts (mSv) or microsieverts (μSv). The dose for an individual depends on where the individual lives, how the individual travels and whether or not the individual has certain medical treatments. One dental X-ray may involve a dose of about 20 μSv. Radiotherapy can involve very large doses such as 40 Sv.

sigma (σ) bond: a single covalent bond formed by a pair of electrons in a *molecular orbital* with the electron density concentrated between the two nuclei. Free rotation is possible around single bonds unlike around *pi bonds*.

Sigma bonds can form by overlap of two *s-orbitals*, an s-orbital and a *p-orbital* or two p-orbitals.

1s 1s
hydrogen atoms   hydrogen molecule, H₂

Examples of sigma bonds in molecules

1s   3p
hydrogen   chlorine   hydrogen chloride, HCl
atom   atom

significant figures: the number of significant figures quoted for a measurement shows the degree of uncertainty. Measuring instruments vary in their *accuracy* and there is *uncertainty* associated with any measurement.

Putting values into *standard form* provides the clearest way of indicating the number of significant figures. This removes any doubt about any zeros, which sometimes just indicate the position of the decimal point, as in 0.0056 g which is a very small mass quoted to two significant figures. This is clearer when the value is in the form 5.6×10^{-3}. In other values the zeros are included to show the number of significant figures. When writing the distance of 3500 m as 3.50×10^3 m it become clear that this is a quantity quoted only to three significant figures.

Generally, when multiplying or dividing values the number of significant figures in the result is the same as the measurement with the **least** number of significant figures.

When values are being added or subtracted in a calculation the number of significant figures in the result is determined by the **least** precise measurement included in the calculation. In other words, the measurement with the fewest figures after the decimal point when the measurement is written out in nonstandard form.

silica (SiO_2): also known as silicon dioxide, this is the oxide of silicon which is abundant in rocks as quartz. Silica has a very high melting point at 1710°C, turning to a viscous liquid. On cooling the liquid becomes a *glass*.

Network structure of quartz. Each Si atom is at the center of a tetrahedron of oxygen atoms. The arrangement of silicon atoms is the same as the arrangement of carbon atoms in diamond but there is an oxygen atom between each silicon atom.

○ Si
● O

Amethyst is crystalline quartz colored purple due to the presence of iron(III) ions. Sandstone and sand consist mainly of silica. Flint is a noncrystalline form of silica.

silica gel: a noncrystalline, hydrated form of *silica*. Heating silica gel produces hard granules which absorb water strongly, so silica gel is used as a drying agent. Self-indicating silica gel contains enough of an anhydrous cobalt(II) salt to color the granules blue. After the gel has absorbed a certain quantity of moisture the cobalt ions become hydrated and turn pink as a warning that the gel is no longer as effective as a drying agent. Heating drives off the water so that the gel can be used again. Silica gel is also used as the stationary phase in *chromatography*.

silicates: the minerals that make up most of the Earth's crust. The basic building block for silicates is a SiO_4^{4-} tetrahedron. Zircon ($ZrSiO_4$) is an example of a simple silicate mineral with metal ions and silicate ions.

Silicate tetrahedra can join up in chains or strands as in asbestos. They can also join up in sheets as in mica, talc or *clay minerals*. In these minerals the negative charges on the silicate part is balanced by positive charges from metal ions such as Na^+, Ca^{2+}, Mg^{2+} or Al^{3+}.

A single SiO_4^{4-} tetrahedron

A fragment of a sheet of silicate tetrahedra

In *zeolites* the silicate tetrahedra are built up into three-dimensional networks.

A very wide range of silicate minerals is possible because some of the silicon atoms in the tetrahedra can be replaced by aluminum, boron or beryllium atoms.

silicon (Si): the second most abundant element in the Earth's crust. Combined with oxygen it forms many minerals including *silica* (SiO_2) and *silicates*. Silicon is the second element in *group IV* of the periodic table. Its electron configuration is $[Ne]2s^2 2p^2$.

Solid silicon has a diamond structure. It is a shiny grey material made by reducing silicon dioxide with carbon in an electric furnace. The silicon formed in this way is not pure enough for use in electronics as a *semiconductor*.

Silicon compounds are typical of the compounds of a *nonmetal*:

- the oxide SiO_2 is an *acidic oxide* though relatively unreactive because it has a *giant structure*
- the chloride ($SiCl_4$) is a molecular liquid which is rapidly *hydrolyzed* by water to hydrated silica and hydrogen chloride
- the hydrides, such as SiH_4 and Si_2H_6, are molecular gases (silanes).

silicones: *polymers* based on chains of alternating silicon and oxygen atoms. They are made by hydrolysis of compounds such as dimethylchlorosilane.

A silicone polymer

By controlling the chain length and degree of *cross-linking* it is possible to make a range of silicones for use as oils, greases and rubbery materials. Silicones are water repellent and can be more safely used at higher temperatures than polymers based on carbon atoms. They are electrical insulators and other materials do not stick to them. Silicones are colorless, have no smell and are inert.

Silicones are used:

- to waterproof fabrics

- as an ingredient of polishes

- to coat nonstick surfaces such as the paper backing for self-adhesive labels

- as *lubricants*, especially at high temperatures.

silver halides are used in qualitative analysis and in *photography*. The three insoluble silver salts form on mixing solutions of silver nitrate and a soluble chloride, bromide or iodide. Silver fluoride is soluble so there is no precipitate when silver nitrate is added to a solution of fluoride ions.

$$Ag^+(aq) + X^-(aq) \longrightarrow AgX(s) \text{ where } X = Cl, Br \text{ or } I$$

This precipitation reaction is used as a test for halide ions (see *anion tests*). The three silver compounds can be distinguished by their color and the ease with which they redissolve in ammonia solution. The values for the *solubility product constants* show the trend in solubility. Silver chloride is the most soluble.

Silver halide	Color	Effect of adding ammonia solution to a precipitate of the compound	Solubility, mol/L	K_{sp}
silver chloride, AgCl	white	redissolves readily in ammonia solution forming a complex ion	1.4×10^{-5}	2×10^{-10}
silver bromide, AgBr	cream	redissolves but only in concentrated ammonia	5.5×10^{-7}	3×10^{-13}
silver iodide, AgI	yellow	does not redissolve in ammonia solution	2.8×10^{-9}	8×10^{-18}

sizes of atoms and ions: see *atomic radius*, *covalent radius*, *ionic radius* and *van der Waals radius*.

slag: the unwanted waste material from *metal extraction* at high temperature (pyrometallurgy). Slag is tapped from a furnace, such as a *blast furnace*. The slag solidifies as it cools and can be dumped or crushed for use in construction or road building.

smelting: a process of *metal extraction* at high temperature (pyrometallurgy). Smelting involves melting the concentrates from metal ores to remove impurities and to reduce metal compounds to metals. Examples of smelting include the extraction of iron in a *blast furnace* and *aluminum extraction* by electrolysis of a melt.

smog: a smoky fog caused by air pollution. Burning coal in homes and industry created the dense city smogs of the nineteenth and first half of the twentieth centuries. The damaging components of these smogs were sulfur dioxide and soot. Many people affected by these acidic smogs died of lung disease. Smokeless zones and the shift of fuels from coal to oil and natural gas have largely eliminated this type of smog.

Today cities are affected by another type of smog caused by motor traffic. This is *photochemical smog*.

smoke: a *colloid* with specks of solid dispersed in a gas. Smokes are examples of aerosols.

S$_N$1 and S$_N$2 reactions: see *nucleophilic substitution in haloalkanes*.

soaps are made by the *hydrolysis* of *fats* or *vegetable oils* with alkali. Soaps are the sodium or potassium salts of *fatty acids*.

Soaps are *surfactants* which help to remove greasy dirt because they have an ionic (water-loving) head and a long (water-hating) hydrocarbon tail.

Most body soaps are made from a mixture of animal fat and coconut palm oil. Soaps from animal fat are less soluble and longer lasting. Soaps from palm oils are more soluble so that they lather quickly but wash away more quickly. A bar of soap also contains a dye and perfume together with an antioxidant to stop the soap and air combining to make irritant chemicals.

glycerol   sodium stearate, soap

Saponification − the hydrolysis of a fat or vegetable oil (triglyceride) with alkali to make soap

sodium (Na): a soft, shiny metal which rapidly tarnishes in moist air. It is the second member of *group I* with the electron configuration [Ne]$4s^1$.

Like other group I metals, sodium:

• is stored in oil

• floats on water, melts and reacts violently forming hydrogen which catches fire and NaOH which is soluble and strongly alkaline

• forms an ionic, crystalline chloride Na^+Cl^-.

Sodium produces a mixture of the oxide, Na_2O, and peroxide, Na_2O_2, when it burns in air.

Electrolysis of molten sodium chloride is the process used to manufacture sodium. The electrolyte also contains some calcium chloride to lower the melting point. Sodium forms at steel cathodes while chlorine bubbles off the graphite anodes. The cells are designed to keep these two reactive elements apart.

Sodium is a powerful reducing agent used for the extraction of some metals such as zirconium. Molten sodium is also the fluid which circulates through *heat exchangers* to transfer energy and raise steam in some nuclear power stations and other processes. Sodium is used in street lights.

sodium borohydride (NaBH₄): a *reducing agent* used in organic chemistry. It is a milder reducing agent than *lithium aluminum hydride* and has the advantage that it can be used in aqueous solution. NaBH₄ reduces *aldehydes* and *ketones* to alcohols.

sodium chloride structure: the cubic crystal structure of the ionic compound sodium chloride, NaCl. Each positive ion is surrounded by six nearest neighbors and each negative ion is surrounded by six positive ions, so the *coordination numbers* are 6 and 6.

Many other compounds have this structure, including: the chlorides, bromides and iodides of Li, Na and K; the oxides and sulfides of Mg, Ca, Sr, Ba as well as the fluoride, chloride and bromide of Ag.

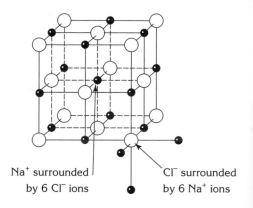

Structure of sodium chloride showing 6:6 coordination. The structure consists of a face-centered cubic array of negative ions with all the octahedral holes filled by positive ions.

Na^+ surrounded by 6 Cl^- ions

Cl^- surrounded by 6 Na^+ ions

sodium hydroxide: a white, translucent solid supplied as flakes or pellets. It is *deliquescent*. Sodium hydroxide is a *strong base*; it dissolves in water to form a highly alkaline solution. It is fully ionized both in the solid and in solution. Note that the solution is alkaline because of the hydroxide ions (not because of the sodium ions).

Sodium hydroxide is a useful test reagent (see *anion tests*, *cation tests* and *organic analysis*).

The traditional name for the alkali is caustic soda, which is a reminder that it is highly corrosive. Sodium hydroxide is more hazardous to the skin and eyes than many acids.

Sodium hydroxide is manufactured by the *electrolysis of brine*. Sodium hydroxide is widely used for manufacturing other chemicals including *soaps* and *detergents*, rayon *fibers* as well as *aluminum*, sodium cyanide and sodium peroxide.

sodium thiosulfate (Na₂S₂O₃) is used in *iodine–thiosulfate titrations*. In *photography* it is the fixer which removes unexposed silver salts after developing the image (see also *thio compounds*).

A milky precipitate of sulfur forms on adding dilute acid to sodium thiosulfate. This is an example of a *disproportionation* reaction.

$$S_2O_3^{2-}(aq) + 2H^+(aq) \longrightarrow SO_2(aq) + S(s) + H_2O(l)$$

solid: one of the *states of matter*. (See also *crystal structures of ionic compounds, crystal structures of nonmetals* and *crystal structures of metals*. See also *ceramics, glasses, metals* and *polymers*.)

solubility: the mass (g) or amount (mol) of a substance that will dissolve in 100 g water.

As a general rule, "like dissolves like." *Polar solvents*, such as water, dissolve polar or ionic compounds. *Nonpolar solvents*, such as hexane, dissolve other hydrocarbons and nonpolar elements or compounds. Solids generally become more soluble in water as the temperature rises. Gases become less soluble as the temperature rises. Boiling water, for example, removes gases dissolved from the air. Bubbles of gas appear around the edge of a saucepan before the water boils. The bubbles contain air coming out of solution as the gases become less soluble with the rise in temperature. Gases become more soluble as their pressure rises (see *Henry's law*).

No chemical is completely soluble and none is completely insoluble (see *solubility product constant*). Even so, chemists find it useful to use a rough classification of solubility based on what they see on shaking a little of the solid with water in a test tube:

• **very soluble,** like potassium nitrate, plenty of the solid quickly dissolves

• **soluble,** like copper(ii) sulfate; crystals visibly dissolve to a significant extent

• **sparingly or slightly soluble,** like calcium hydroxide; little solid seems to dissolve but the solution becomes quite strongly alkaline

• **insoluble,** like iron(iii) oxide; there is no sign that any of the material dissolves.

	Soluble in water	Insoluble in water
Acids	All common *acids* are soluble	
Bases	*Alkalis:* the hydroxides of sodium and potassium (calcium hydroxide which is slightly soluble), ammonia, plus the carbonates of sodium and potassium	All other metal oxides, hydroxides and carbonates
Salts	All nitrates	
	All chlorides ...	*except* silver and lead chlorides
	All sulfates ...	*except* barium sulfate, lead sulfate, and calcium sulfate which is slightly soluble
	All sodium and potassium salts	All other carbonates, chromates, sulfides and phosphates

A similar rough classification applies to gases dissolving in water. Ammonia and hydrogen chloride are very soluble. Sulfur dioxide is soluble. Carbon dioxide is slightly soluble. Nitrogen is insoluble.

Some generalizations about solubilities help to interpret observations during qualitative analysis. The generalizations in the table on the previous page apply to solutions in water at room temperature. Adding acid or alkali changes the patterns of solubility.

solubility product constants: equilibrium constants for almost insoluble salts in equilibrium with solutions of their own ions. Even salts which are insoluble for practical purposes do dissolve to a very slight extent. There is an equilibrium between the solid and its ions in solution.

$$AgCl(s) \rightleftharpoons Ag^+(aq) + Cl^-(aq)$$

The *equilibrium law* applies. As with other *heterogeneous equilibria*, the concentration of the solid silver chloride is constant and does not appear in the equilibrium law equation. In this context the equilibrium constant, K_{sp}, is the solubility product constant.

$$K_{sp} = [Ag^+(aq)][Cl^-(aq)] = 2 \times 10^{-10} \text{ for equilibrium concentrations}$$

K_{sp} values can be used to predict whether or not a precipitate will form on mixing two solutions.

solute: a substance which dissolves in a *solvent* to make a *solution*. In a sugar solution the solvent is water and the solute is sucrose.

solutions are formed when solids, liquids or gases dissolve in a *solvent*. Water is so abundant on Earth that solutions in water (*aqueous* solutions) are particularly important to the natural environment, to life and to chemistry in laboratories and in industry.

Most solutions are solids, liquids or gases dissolved in a liquid but there are also "solid solutions." Nickel–copper *alloys* are examples of solid solutions. In a solid solution atoms of one metal replace atoms of the other metal in the crystal lattice.

solvation takes places when solvent molecules bond to ions or molecules as they dissolve. The bonding may be through weak intermolecular forces, attraction between ions and polar molecules or via covalent bonds. *Hydration* describes solvation when the solvent is water.

Solvay process: a process for manufacturing sodium carbonate from salt and limestone.

solvent: a liquid used to dissolve things. Chemists find it helpful to quote the rule "like dissolves like." What this means is that *nonpolar solvents* dissolve non-polar substances while *polar solvents* dissolve ionic and polar compounds.

Water is the commonest solvent. It is a polar solvent and dissolves many ionic compounds. Water molecules also dissolve compounds with which they can form *hydrogen bonds* such as glucose molecules.

Turpentine is a nonpolar solvent consisting of a mixture of *hydrocarbons*. It dissolves oily and greasy materials, including oil paints.

solvent extraction: a technique for separating and purifying substances with a *solvent* which dissolves the product required but leaves all other compounds dissolved in the original solvent. The solvents must not mix so that the two liquids separate after being shaken up together.

The substance being extracted *partitions* itself between the two solvents until the two solutions are in equilibrium. From the *equilibrium law* it is possible to show that it is more efficient to extract with two or three smaller volumes of solvent than to add all the solvent at once in a single extraction.

Solvent extraction is used in the *perfume* industry to extract fragrant oils from chopped up plant blossoms. Solvent extraction is also used to decaffeinate coffee. In both these processes it helps to use a liquefied gas as the solvent. One possibility is carbon dioxide under pressure. After the extraction the solvent is easily removed by lowering the pressure. The solvent turns back to a gas at a low temperature. This allows the use of a nontoxic solvent. It also means that there is no need for heating to distil off the solvent. Heating can easily destroy organic compounds.

Use of a tap funnel for solvent extraction. Caffeine is more soluble in dichloromethane than in water. The other chemicals in tea are much more soluble in water. After solvent extraction the caffeine is recovered by distilling off the dichloromethane.

s-orbitals: *see atomic orbitals.*

space-filling models: atomic models which show the space taken up by atoms in molecules or crystals. They show the sizes of atoms, molecules or ions and how they pack together. They do not show the bond angles or the numbers of bonds between atoms in molecules as clearly as *ball and stick models*.

A space-filling model of benzene, C_6H_6

species: a useful collective noun used by chemists to refer generally to the atoms, molecules or ions taking part in a chemical process. A pure chemical species is a collection of identical chemical *entities*.

specific heat capacity: the energy transferred per unit mass when the temperature of a material changes by one degree kelvin. The SI unit is J /kg·K. When working on a small scale with a *calorimeter* chemists often work with values in J /g·K.

$$\text{energy transfer (J)} = \text{specific heat capacity (J/g·K)} \times \text{mass (g)} \times \text{temperature change (K)}$$

In symbols this becomes:

$$q = mc\Delta T$$

spectator ions: ions in a solution during a reaction which do not take part in the chemical change. For clarity, chemists omit spectator ions from *ionic equations*. Adding silver nitrate to a solution of potassium chloride produces a precipitate of insoluble silver chloride. Both silver nitrate and potassium chloride are ionized in solution.

$$Ag^+(aq) + NO_3^-(aq) + K^+(aq) + Cl^-(aq) \longrightarrow AgCl(s) + NO_3^-(aq) + K^+(aq)$$

The potassium and nitrate ions remain in solution unchanged. They are the spectator ions left out of the ionic equation:

$$Ag^+(aq) + Cl^-(aq) \longrightarrow AgCl(s)$$

spectroscopy: a range of practical techniques for studying the composition, structure and bonding of elements and compounds. These instrumental techniques have been developed in the last 80 years and continue to become more powerful. Spectroscopic techniques are now the essential "eyes" of chemistry.

The instruments used are variously called spectroscopes (emphasizing the uses of the techniques for making observations) or spectrometers (emphasizing the importance of measurements).

Spectroscopy uses the full range of the electromagnetic spectrum to study atoms, molecules, ions and the bonding between them:

- **radio waves** in *nuclear magnetic resonance spectroscopy*
- **microwaves** to study the rotations of polar molecules
- **infrared radiation** in *infrared spectroscopy*
- **visible and UV radiation** in *atomic absorption spectroscopy, atomic emission spectroscopy* and *ultraviolet spectroscopy*
- **X-ray** spectroscopy to study electron jumps between the electron shells in atoms.

spin: the property of electrons which accounts for their behavior in a magnetic field. Electrons behave like tiny magnets. In a magnetic field electrons either line up with the field or against the field.

An *atomic orbital* can hold only two electrons and they must have opposite spins. Arrows pointing up or down represent electrons in *energy level* diagrams.

If all the electrons in molecules or ions are paired with opposite spins the substance is diamagnetic. Elements and compounds with unpaired electrons are *paramagnetic*.

spontaneous reaction: a reaction which tends to happen. In thermochemistry, spontaneity has the same meaning as a *feasibility*. So strictly speaking any reaction which naturally tends to happen is spontaneous even if it is very slow.

The control of spontaneous reactions is of vital importance in *metabolism*. The hydrolysis of *ATP* is spontaneous. However, if all the ATP in cells were to react rapidly with water, the energy from respiration would be wasted and life would cease. Hydrolysis happens only with *enzymes* to speed up the reaction. With enzymes the energy from hydrolysis can be harnessed to growth and movement.

In practice chemists sometimes use the word spontaneous in its everyday sense to describe reactions which not only tend to occur but also go fast when the reactants are mixed at room temperature. Here is a typical example:

The hydrides of silicon catch fire spontaneously in air unlike methane which has an ignition temperature of about 500°C.

The reaction of methane with oxygen is also spontaneous in the thermodynamic sense, even at room temperature. However, the *activation energy* for the reaction is so high that nothing happens until the gas is heated with a flame. A mixture of methane and oxygen at room temperature is kinetically inert (see *inertness of chemicals*).

stability of benzene: the greater stability of *benzene* because of *delocalization* of electrons. Benzene is more stable than expected for a compound which is often shown with three double bonds as in the *Kekulé formula*.

A measure of the greater stability of benzene comes from a comparison of the experimental enthalpy change on adding three moles of hydrogen to benzene (*hydrogenation*) with three times the enthalpy change on adding a mole hydrogen to cyclohexene.

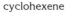

benzene + $3H_2$ $\longrightarrow$ (cyclohexane) $\Delta H° = -208$ kJ/mol

benzene

cyclohexene + H_2 --> (cyclohexane) $\Delta H° = 120$ kJ/mol

cyclohexene

Enthalpy changes for hydrogenating benzene and cyclohexene

Real benzene is more stable than might be expected by about 150 kJ/mol.

$C_6H_6 + 3H_2$

Kekule benzene — $\Delta H° = -360$ kJ/mol
(Estimated enthalpy change for hydrogenating 3 double bonds)

$C_6H_6 + 3H_2$

benzene — $\Delta H° = -208$ kJ/mol
(Measured enthalpy change for benzene)

C_6H_{12}

cyclohexane

Comparing the enthalpy change for hydrogenating real benzene with the calculated enthalpy change for hydrogenating a ring compound with three normal double bonds

stability of compounds: compounds are stable if they do not tend to decompose into their elements or into other compounds. A compound which is stable at room temperature and pressure may become more or less stable as conditions change (see *thermal stability*).

Chemists often use standard *enthalpy changes* as an indicator of stability. Strictly they should use standard *free energy change*, $\Delta G_f°$, values but in many cases $\Delta G_f° \approx \Delta H_f°$.

When discussing the stability of a compound it is important to specify the decomposition reaction. At a high enough temperature, for example, *calcium carbonate* becomes unstable relative to decomposition into calcium oxide and carbon dioxide. The *nitrogen oxides*, however, are unstable relative to decomposition into the elements.

A compound such as the gas N_2O is thermodynamically unstable; the compound tends to decompose into its elements. The expressions $\Delta H_f^\circ (= + 82$ kJ/mol) and $\Delta G_f^\circ (= +104$ kJ/mol) are both positive. The decomposition reaction is exothermic but the rate is very slow under normal conditions. Chemists sometimes say that N_2O is kinetically "stable." It is better to use a different word and to refer to the kinetic inertness of N_2O. (See also *inertness of chemicals*.)

ΔG	Activation energy	Change observed	
positive	high	no reaction	reactants stable relative to products
negative	high	no reaction	reactants unstable relative to products but kinetically inert
positive	low	no reaction	reactants stable relative to products
negative	low	fast reaction	reactants unstable relative to products

standard conditions: the conditions used in *thermochemistry* to define standard enthalpy and free energy changes. These conditions are the temperature 298 K and a *pressure* of 100 000 N/m² = 10^5 Pa (1 bar).

When standard electrode potentials are being measured the standard conditions are the temperature 298 K, solutions at a concentration of 1.0 mol/L and a pressure of 10^5 Pa if gases are involved.

standard electrode potentials: the basis of the *electrochemical series* for predicting the direction of *redox reactions*. They are also used to calculate the emfs of *electrochemical cells*. The standard electrode potential for a *half-cell* is measured relative a standard *hydrogen electrode* under *standard conditions*.

standard enthalpy changes: see *enthalpy change*.

standard form: the form which mathematicians and scientists use to write very large or very small numbers. Standard form is based on powers of 10.

10^{-9} = 0.000 000 001	10^1 = 10
10^{-6} = 0.000 001	10^2 = 100
10^{-3} = 0.001	10^3 = 1 000
10^{-2} = 0.01	10^4 = 10 000
10^{-1} = 0.1	10^6 = 1 000 000
10^0 = 1	10^9 = 1 000 000 000

Worked example

The dissociation constant for acetic acid, K_a
$$= 0.000\ 017 \text{ mol/L}$$
$$= 1.7 \times 0.000\ 01 \text{ mol/L}$$
$$= 1.7 \times 10^{-5} \text{ mol/L},$$

which is standard form

Worked example

The *Faraday constant* = 9 6480 C/mol = 9.648 × 10 000 C/mol

= 9.648 × 10⁴ C/mol which is standard form.

standard free energy changes: see *free energy change.*

standard hydrogen electrode: see *hydrogen electrode.*

standard molar entropy, $S°$: the *entropy* per mole for a substance under *standard conditions*. Chemists use values for standard molar entropies to calculate entropy changes and so predict the direction and extent of chemical change.

The units for standard molar entropy are joules per kelvin per mole ($J·K^{-1}·mol^{-1}$).

solids	$S°/J·K^{-1}·mol^{-1}$	liquids	$S°/J·K^{-1}·mol^{-1}$	gases	$S°/J·K^{-1}·mol^{-1}$
carbon (diamond)	2.4	mercury	76.0	hydrogen	130.6
magnesium oxide	26.9	water	69.9	carbon dioxide	213.6
copper	33.2	ethanol	160.7	propane	269.9

The table shows standard molar entropies at 298 K. Gases have higher standard molar entropies than solids. The number of ways of distributing particles and energy in a system of gas molecules is much higher than in an ordered crystalline solid. The more rigid and regular a crystal, the lower its entropy.

Liquids have intermediate values of standard molar entropies. The more atoms in the molecules the greater the opportunities for vibration and rotation so the higher the entropy values.

standard solution: a solution with an accurately known concentration. The direct method for preparing a standard solution is to dissolve a weighed sample of a *primary standard* in water and to make the solution up to a definite volume in a graduated flask.

Standard solutions are used in *titrations* to determine the concentrations of other solutions. They are also needed for the *calibration* of instruments such as colorimeters.

standard state: the stable state of an element or compound under the *standard conditions* which apply in thermochemistry. *Standard enthalpy changes* and *free energy changes* are defined for substances in their standard states.

When an element such as carbon has two *allotropes*, diamond and graphite, the standard state of carbon is graphite because it is the more *stable* form.

Standard states normally refer to substances in their stable physical states under standard conditions. Carbon dioxide and methane are gases but water is a liquid.

starch: a *carbohydrate* which consists of long chains of glucose units. It is a polysaccharide. Starch is insoluble in cold water. In hot water, starch gelatinizes forming a colloidal dispersion. Starch solution gives an intense blue–black color with iodine. The solution is used as an indicator to detect the end point in *iodine–thiosulfate titrations.*

state functions: measurable properties which depend only on the state of a system, not on how the system got to that state. Examples are *pressure*, *volume* and temperature, as are thermochemical quantities such as enthalpy, *entropy* and *free energy.*

Changes in state functions depend only on the initial and final states of the system and not on the pathway from one state to the other. This is the basis of *Hess's law.*

Equations of state show the mathematical connections between state functions. An important example is the *ideal gas* equation which relates the pressure, volume and temperature of an amount of a gas.

states of matter: the solid, liquid and gaseous states (see *changes of state*) as well as intermediate states such as *liquid crystals* and mixtures, including aqueous solutions and the colloidal state (see *colloids*).

State symbols in equations indicate the states of the chemicals: (s) solid, (l) liquid, (g) gas, (aq) aqueous (dissolved in water).

stationary phase: see *chromatography*.

steady state systems: systems in which the concentrations of chemicals stay constant because they are being supplied or formed as fast as they are removed or destroyed. The temperature is constant too because energy is transferred to the system as fast as it is lost to the surroundings.

The blue flame of the laboratory burner is a steady state system. There is a constant supply of gas and oxygen which burn to release energy. The products of burning and energy are constantly lost to the surroundings.

Steady state systems are important in *environmental chemistry*. The *atmosphere* approximates to a steady state, with gases such as oxygen being added to the air by photosynthesis but removed by respiration and burning.

Problems arise when human activity on a large scale upsets one or more of the processes in a steady state system. *Ozone*, for example is naturally formed and destroyed in the stratosphere:

rate of formation of ozone = rate of destruction of ozone

The release of chlorine compounds such as *CFCs* has increased the rate of destruction of ozone, upsetting the steady state and lowering the ozone concentration, especially over the poles, hence the "hole" in the ozone layer.

steam cracking: see *thermal cracking*.

steam distillation: a useful technique for separating oils from plant materials such as rose petals, cloves, lavender, thyme or fennel. It is an important technique in the *perfume* industry. Steam distillation makes it possible to separate compounds which decompose if heated at their boiling points. The technique works only for compounds which do not mix with water.

Steam distillation is also sometimes used to separate products of organic preparations, leaving behind reagents and products which are soluble in water.

Steam distillation works because the *vapor pressure* of a mixture of *immiscible liquids* is the sum of the separate vapor pressures when they are shaken up together. The mixture boils when the total vapor pressure equals atmospheric pressure. So the mixture of steam and oily product distils at a little below the boiling point of water and well below the boiling point of the oil.

steam reforming: the reaction of steam with methane (or other hydrocarbon) in the presence of a nickel oxide *catalyst* at 800°C under pressure.

$$CH_4(g) + H_2O(g) \rightleftharpoons 3H_2(g) + CO(g)$$

It is followed by further reaction with excess steam. This "shift reaction" converts the CO into CO_2. The reaction happens in the presence of an iron(III) oxide catalyst at 400°C.

$$H_2O(g) + CO(g) \rightleftharpoons H_2(g) + CO_2(g)$$

This mixture of hydrogen and carbon dioxide can be converted directly to *methanol*. Alternatively, absorbing the carbon dioxide in alkali provides hydrogen for *ammonia manufacture*.

steels: *alloys* of iron with carbon and often other metals.

Carbon steel contains about 0.2% carbon as iron carbide. Crystals of iron carbide in the metal structure make the steel strong and yet it is still *malleable*. Mild steel is used for car bodies. As the carbon content increases the steel becomes stronger and harder so that it is suitable for rail and streetcar lines.

Steel is made by the basic oxygen steelmaking (BOS) process which removes impurities from *iron extraction* in a blast furnace. During the process a blast of oxygen converts impurities in the liquid metal, such as carbon, silicon and phosphorus, into their oxides. Carbon dioxide escapes as a gas. The oxides of the nonmetals silicon and phosphorus are *acidic oxides*. They are converted to a molten *slag* by adding the *basic oxides* of calcium and magnesium. The slag floats on the surface of the liquid steel and can be poured off separately

$$SiO_2 + CaO \longrightarrow CaSiO_3(l)$$

Alloy steels consist of iron with small amounts of carbon together with up to 50% of one or more of these metals: aluminum, chromium, cobalt, manganese, molybdenum, nickel, titanium, tungsten, and vanadium. The presence of other metals distinguishes alloy steels from carbon steels. Examples of alloys steels are:

- stainless steels, which include chromium and nickel

- tool steels with tungsten or manganese which make the alloy harder, tougher and keep their properties at higher temperatures so that they are suitable for drill bits and cutting tools.

stereochemistry: the study of molecular shapes and the effect of shape on chemical properties. Stereochemistry is especially concerned with the study of the contrasting properties of *stereoisomers*.

Smell and taste seem to depend on molecular shape. One mirror-image form of limonene smells of oranges while the other form smells of lemons.

Orange   Lemon

The enantiomers of limonene and their smells. Asterisks mark the chiral centers.

Molecular shape can also subtly alter the physiological effects of drugs as tragically illustrated by the two enantiomers of thalidomide (see *chiral compounds*).

The *antibiotic* penicillin destroys bacteria by disrupting their cell walls. Penicillin molecules resemble the shapes of molecules used to make the membrane around bacterial cells. The resemblance is enough for enzymes to build penicillin molecules into the structure. But penicillin lacks the part of the natural molecules which *cross-links* the structure. The walls are fatally weakened so that the bacterial cells break open and die.

Molecular shape affects the *physical properties* of materials such as polymers. This is illustrated by the differences between poly(propene) as an *isotactic polymer* and as an *atactic polymer*.

Steric factors can also affect the mechanisms of reactions and the rates of chemical change.

stereoisomers: distinct compounds with the same molecular formulas and structural formulas but with different three-dimensional shapes. There are two kinds of stereoisomerism: *geometrical isomerism* and *optical isomerism*.

stereoregular polymer: a polymer with a regular three-dimensional shape. Polypropylene can be an *isotactic polymer*. This is the useful stereoregular form. Or it can be an *atactic polymer* which is irregular and does not have useful properties.

stereospecific reactions: reactions which involve one of the *optical isomers* of a *chiral* compound but not the other. All the biochemical processes involving *amino acids* are stereospecific. The enzymes involved can act on the L-amino acids but not on the mirror-image D-forms.

Enzymes act only on molecules which fit their *active sites*. They are so selective that they can pick out from a *racemic mixture* those molecules that are either left-handed or right-handed.

Since the thalidomide tragedy, the pharmaceutical industry has become much more aware of the importance of chirality and the need to test mirror image forms of chiral compounds separately. This has encouraged chemists to develop stereospecific reactions to produce particular optical isomers.

steroids are *lipids*. An important steroid is *cholesterol*. Cholesterol can be converted to biologically active steroids such as the sex hormones estrogen and progesterone.

stick structure: outline formulas for carbon compounds which are a useful shorthand for complex molecules such as many natural products. The formulas need careful study because they represent only the hydrocarbon part of the molecule with lines for the bonds between carbon atoms, leaving out the symbols for the carbon and hydrogen atoms. *Functional groups* are included. (See diagram on page 331.)

stiffness: the opposite of flexibility. It is a very important property of materials. Materials scientists seek to develop materials which are both stiff and strong while not being too dense.

These examples help to explain the difference between strength and stiffness:

> A cookie is stiff but weak, steel is stiff and strong, nylon is flexible and strong, raspberry jelly is flexible and weak.

Stiffness is measured by the Young modulus for the material.

Stick structure *for vitamin A compared to its full structural formula*

stoichiometry: a stoichiometric equation is the *balanced equation* for a reaction. It shows the amounts, in moles, of the substances involved in a reaction.

A stoichiometric compound has a composition which corresponds exactly to its formula. A stoichiometric reaction is one which uses up reactants and produces products in amounts exactly as predicted by the balanced equation.

The word stoichiometry sounds mysterious but is simply based on Greek words meaning "element-measure." Stoichiometry is the basis of *quantitative analysis* where amounts are measured in moles.

storage cells: see *lead–acid cell, lithium cells* and *nickel–cadmium cells*.

STP (standard temperature and pressure): the standard conditions for describing the properties of gases. The standard temperature is 273 K (0°C) and the standard *pressure* is 101.3 kPa (1 atmosphere).

The molar volume of an ideal gas at STP is 22 400 cm³ (22.4 L).

Note the distinction between these values and the values for defining *standard states* in thermochemistry.

strength is measured by the stress needed to deform and break a material in tension or compression. Stress is the force per unit area of the cross section and is measured in N/m². Defining stress in this way allows for the fact that it is easier to stretch a thin sample than a thick one.

Metals such as steel have high tensile strength. Ceramics are weak in tension but have high strength in compression.

For many applications engineers look for materials which combine high strength with high *stiffness*.

strong acids: *acids* which are fully ionized when they dissolve in water. Examples are *hydrochloric acid, nitric acid* and *sulfuric acid*.

strong bases: *bases* which are fully ionized when they dissolve in water. Examples are the hydroxides of *sodium* and *potassium*.

structural formulas show the arrangements of atoms and *functional groups* in molecules.

Sometimes it is enough to show structure in a condensed form, such as $CH_3CH_2CH_2COOH$ for butanoic acid. Often it is clearer to write the full structural formula, showing all the atoms and bonds. (See examples on page 332.)

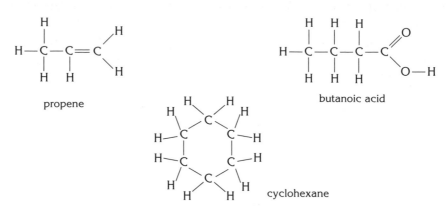

propene

butanoic acid

cyclohexane

Examples of structural formulas. Drawn like this the formulas do not show the true shape in three dimensions.

structural isomers: see *isomers.*

sublimation: the change of a solid directly to a gas on heating. Heating iodine crystals makes them sublime to a purple vapor which condenses to shiny crystals on a cold surface. This process is used to purify iodine.

Another substance which sublimes is solid carbon dioxide ("dry ice") because it turns to gas at minus 78°C without melting.

substitution reactions: reactions which replace an atom or group of atoms by another atom or group of atoms. An example is the reaction of 1-butanol with hydrogen bromide (from sodium bromide and concentrated sulfuric acid).

$$CH_3CH_2CH_2CH_2OH + HBr \longrightarrow CH_3CH_2CH_2CH_2Br + H_2O$$

In organic chemistry, substitution reactions are characteristic of *haloalkanes* (see *nucleophilic substitution*) and *arenes* (see *electrophilic substitution*).

The *ligand substitution* reactions of complex ions are examples of inorganic substitution reactions (see *ligands*).

substrates, in biochemistry, are the molecules on which *enzymes* act as they catalyze change. An enzyme is specific because it acts only on the substrate which fits into its *active site.*

Literally the word "substrate" means a lower layer on which something else can form or grow. One way of creating a large surface area for an expensive *catalyst* is to spread it over the surface of an inert substrate. This helps to keep down the cost of *catalytic converters* in car exhausts.

sugars: water soluble *carbohydrates.* Sugars vary in their *sweetness.* Sugar molecules have many —OH groups and can *hydrogen bond* with each other and with water. This means that they are solid at room temperature and very soluble in water. (See also *reducing sugars.*)

Ribose and deoxyribose sugars are important in the formation of nucleic acids such as *DNA* and *RNA.*

sulfates: salts of *sulfuric acid*. Soluble sulfates such as blue copper(ii) sulfate and green iron(ii) sulfate are familiar laboratory reagents.

Some insoluble sulfates are natural minerals, including calcium sulfate which exists in a hydrated form as gypsum and in an anhydrous form called anhydrite. Heating gypsum produces *plaster of Paris*. Barytes is barium sulfate, called "heavy spar" because of its density. (See also *solubilities of salts* and *anion tests*.)

sulfides of metals: salts of hydrogen sulfide, H_2S. Some valuable metals occur as sulfide ores, including copper, silver, mercury and iron (see also *roasting*). Sodium sulfide is soluble in water but most metal sulfides are insoluble.

sulfites: salts of sulfurous acid which form when sulfur dioxide dissolves in water. Sulfurous acid is unstable and exists only in solution.

$$SO_2(g) + H_2O(l) \longrightarrow H_2SO_3(aq)$$

The most important sulfite is sodium sulfite (Na_2SO_3), made by dissolving sulfur dioxide in sodium hydroxide solution.

Sulfur dioxide and the sulfite ion are *reducing agents*. They reduce chlorine, iron(iii) ions, dichromate ions and permanganate ions.

Sulfur dioxide and sulfites are used as preservatives They are antioxidants used in foods such as lemon juice. They inhibit the growth of bacteria. (See also *anion tests*.)

sulfonamide drugs: see *chemotherapy*.

sulfonation of benzene: an *electrophilic substitution* reaction which takes place in the presence of fuming sulfuric acid – a solution of sulfur trioxide in concentrated sulfuric acid. The product is benzene sulfonic acid

The *electrophile* is sulfur trioxide. A sulfur atom attached to three more *electronegative* oxygen atoms is electrophilic.

Sulfonation of benzene

benzene sulfonic acid

Sulfonation of arenes is important because it helps to produce a range of useful products including *surfactants*, *ion-exchange resins*, *dyes* and sulfonamide drugs.

sulfur (S): a yellow crystalline solid which normally consists of S_8 molecules. Sulfur is a *nonmetal*, coming below oxygen in group VI of the periodic table. Its electron configuration is $[Ne]3s^23p^4$.

Sulfur has two solid *allotropes*. Under atmospheric pressure, rhombic sulfur is the stable form of the element at room temperature and up to 95.5°C. Above this temperature the stable form is monoclinic sulfur. At the transition temperature the two allotropes are in equilibrium. Both forms consist of S_8 molecules. Heating sulfur crystals produces a runny, pale-yellow liquid which darkens as the temperature rises, producing a highly *viscous liquid*. The S_8 rings break and form long, tangled chains of sulfur atoms. Further heating below the boiling point (445°C) makes the liquid more fluid because the chains start to break into shorter lengths. Pouring this dark red liquid into cold water produces an elastic, noncrystalline mass of plastic sulfur. In time plastic sulfur hardens as the sulfur chains gradually reform S_8 rings and produce rhombic sulfur again.

Sulfur is a reactive element but it is a less powerful *oxidizing agent* than oxygen. It combines with most other elements with the exception of nitrogen, iodine, the noble gases and some of the less reactive metals such as gold.

Hydrogen sulfide (H_2S) is a toxic, foul-smelling compound. Despite having a higher relative molecular mass than water, hydrogen sulfide is a gas, because sulfur is not sufficiently electronegative for *hydrogen bonding* between H_2S molecules.

Sulfur forms two *acidic oxides* SO_2 and SO_3, which are both important in the process of *sulfuric acid manufacture*. Sulfur dioxide is a *reducing agent*, as are sulfurous acid and its salts, the *sulfites*.

Bonding in the oxides of sulfur.
Note that the rules for predicting the shapes of molecules apply.

The reactions of *halide ion*s with concentrated sulfuric acid illustrate the range of oxidation states of sulfur.

sulfur extraction: processes which obtain sulfur from underground deposits of the element or from crude oil and *natural gas*.

The Frasch process is an ingenious process for extracting sulfur from underground deposits of the element. Sulfur is a nonmetal with a molecular structure. It melts at 113°C. Superheated steam at 165°C passes down a pipe into the sulfur deposits, where the steam melts the sulfur. Compressed air pumped down a second pipe forces the liquid up a third pipe to the surface where the sulfur cools and solidifies.

Natural gas and crude oil often contain unwanted sulfur compounds which must be removed before the gas and oil are used as fuels or in the chemical industry. The hydrocarbons are mixed with hydrogen under pressure and passed over a *catalyst*. This converts the sulfur to hydrogen sulfide. The hydrogen sulfide is separated from the hydrocarbons and converted to a mixture of sulfur and sulfur dioxide.

Ninety percent of sulfur is used to make sulfuric acid.

sulfuric acid (H_2SO_4) is a highly corrosive but important chemical reagent because it can act as a *strong acid*, a *dehydrating agent*, an *oxidizing agent* and as a sulfonating agent. Pure sulfuric acid is a colorless *viscous liquid*. The molecules of this *oxoacid* are attracted to each other by *hydrogen bonding*.

Structure of sulfuric acid molecules showing hydrogen bonding

In its different roles sulfuric acid acts as:

- **a strong acid** – sulfuric acid ionizes fully as it dissolves in water

$$H_2SO_4(l) + H_2O(l) \longrightarrow H_3O^+(aq) + HSO_4^-(aq).$$

  Dilute sulfuric acid neutralizes bases. It forms two series of soluble salts called *sulfates* and hydrogensulfates. Hydrogensulfates are *acid salts*.

- **a powerful dehydrating agent** – concentrated sulfuric acid can:

  - remove the elements of water from *sugars* and other *carbohydrates* so that they char and turn to black charcoal

  - dehydrate *alcohols* turning them into *alkenes*

  - act as a drying agent to remove water from moist gases

- **an oxidizing agent** – it reacts with *halide ions*, especially bromide and iodide ions

- **a sulfonating agent** – it is used for the *sulfonation of benzene* and other *arenes*.

sulfuric acid manufacture: the contact process for making sulfuric acid from sulfur which produces over 150 million metric tons of the acid in the world each year.

The process operates in three main stages:

Stage 1: Burning *sulfur* to make sulfur dioxide

$$S(s) + O_2(g) \longrightarrow SO_2(g)$$

This is a highly exothermic reaction. The hot gas is cooled in *heat exchangers* which produce steam used to generate electricity. Sulfuric acid plants do not have electricity bills. Electricity and steam exported from the process help to make a sulfuric acid economic.

Outline flow diagram for the manufacture of sulfuric acid

Stage 2: Conversion of sulfur dioxide to sulfur trioxide

$$2SO_2(g) + O_2(g) \rightleftharpoons 2SO_3(g) \quad \Delta H = -297 \text{ kJ/mol}$$

This is an exothermic, reversible reaction which takes place on the surface of a vanadium(v) oxide *catalyst*. The *temperature effect on equilibrium* means that raising the temperature lowers the percentage conversion to SO_3 but the temperature must be high enough to make the reaction go fast enough. The catalyst is not active below 380°C and works best at a higher temperature.

Increasing the pressure would increase the conversion to sulfur dioxide but the extra cost is not usually justified.

Typically the gas mixture passes through four catalyst beds. Between each bed the gas mixture is cooled in a heat exchanger and cold air with more oxygen is added to the mixture. After the third bed the level of conversion is 98%. To ensure conversion of the remaining SO_2, sulfur trioxide is absorbed from the gas stream and more air is added before the gases flow through the fourth bed of catalyst. So three factors contribute to converting as much of sulfur dioxide as possible to sulfur trioxide: cooling, adding more of one of the reactants and removing the product.

Stage 3: Absorption

$$H_2O(l) + SO_3(g) \longrightarrow H_2SO_4(l)$$

Sulfur trioxide cannot be dissolved in water directly because the violent reaction produces a hazardous mist of acid. Instead the gas is absorbed in 98% sulfuric acid. Sulfur trioxide passes up a tower packed with pieces of ceramic, down which the concentrated acid trickles. The circulating acid is kept at the same strength by drawing off product while adding water.

Environmental legislation requires very low emissions of acid gases into the air. A modern plant converts 99.7% of the sulfur dioxide into sulfuric acid in stages 2 and 3.

Sulfuric acid is needed on a large scale to make many other chemicals, including phosphate *fertilizers*, *paints* and *pigments*, *detergents*, *plastics*, *fibers* and *dyes*.

sunscreens protect the skin from harmful ultraviolet radiation from the sun. Suntan lotions include chemicals such as 4-aminobenzoic acid which absorbs UV radiation.

Sunblocks are heavily pigmented so that they form a barrier to sunlight by reflecting and scattering light. The white pigments are zinc oxide and *titanium(IV) oxide*.

superoxides: compounds with oxygen containing the O_2^- ion (see *potassium*).

supersaturated solution: a solution which contains more dissolved solute than a saturated solution. A supersaturated solution is unstable. Crystals form rapidly when the edge of the container is scratched or a tiny seed crystal is added.

The way to make a supersaturated solution is to cool a hot, saturated solution of a salt such as sodium thiosulfate or sodium acetate. Sometimes an unwanted supersaturated solution forms when a solid product is purified by *recrystallization*.

Some "instant hand warmers" consist of a supersaturated solution of sodium acetate in a plastic bag. Flexing a metal disc in the side of the bag starts crystallization which is exothermic and heats the bag to about 60°C. The hand warmer can be reused after heating in boiling water to redissolve the salt and then allowing it to cool to room temperature so that the solution is once more supersaturated.

surface tension: the tendency of surfaces to contract to the minimum surface area. Surface tension accounts for the spherical shape of soap bubbles.

Surface tension arises because of *intermolecular forces*. Molecules inside the liquid are pulled in all directions by surrounding molecules. Molecules near the surface experience a net pull inwards. This means that as many molecules as possible leave the surface which tends to shrink.

The surface tension of many organic liquids is in the range 15 to 30 mN/m. The surface tension of water is higher because of strong *hydrogen bonding*. For water the value is about 73 mN/m. *Surfactants* lower the surface tension of water so that it can wet greasy surfaces (see *wetting*).

The surface tension of the soap film makes the slider move to the left unless balanced by a pull on the thread

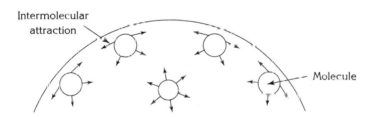

Intermolecular forces acting on molecules inside the liquid and at the surface

surfactants: surface-active agents. They are chemicals with molecules which seek out the boundary surface between two liquids or between a liquid and a gas. One of the important effects of surfactants is that they change the *surface tension* of liquids.

A surfactant molecule has an ionic or polar group attached to a hydrocarbon chain. The polar group is water-loving (hydrophilic). The hydrocarbon chain is water-hating (hydrophobic).

The hydrophobic chains of surfactant molecules tend to escape from water by moving to the surface or by forming *micelles*. Surfactant molecules at the surface lower the surface tension of the water so that it wets surfaces more effectively. Surfactant molecules adsorbed at the boundary between air and water stabilize bubbles of foam by stopping all the water draining away.

Nonpolar;
hydrophobic "tail"

Ionic;
hydrophilic "head"

Structure of a surfactant molecule

Surfactants help to separate greasy dirt from surfaces. They keep dirt dispersed in water so that it rinses away. They also help to prevent dirt from reattaching itself to the surface of fabrics.

Surfactants come in three kinds based on the type of hydrophilic group:

- anionic surfactants such as *soap*, $CH_3(CH_2)_{16}COO^-Na^+$
- nonionic surfactants made from *epoxyethane*, $CH_3(CH_2)_{11}(OCH_2CH_2)_6OH$
- cationic surfactants such as *quaternary ammonium salts*, $CH_3(CH_2)_{11}N^+(CH_3)_3Br^-$

Synthetic anionic surfactants are widely used. They are included in bubble bath, shampoos and dishwashing liquids. They make stable foams with water.

Nonionic surfactants are used in many household cleaners because they allow smooth drainage and leave no deposit even when they are not fully rinsed away. Nonionic surfactants also make less stable foams so they are included in detergents for dishwashers and washing machines.

Cationic surfactants are used in fabric softeners and hair conditioners.

surroundings: see *system*.

suspension: particles of a solid suspended by shaking or stirring in a liquid or gas. In time the particles of a suspension settle out, unlike the smaller particles in *colloids*.

sweetness: a taste sensation on the tip of the tongue produced by *sugars* and a number of synthetic chemicals such as saccharin, cyclamate and aspartame. Sugars vary in their sweetness. Fructose is sweeter than sucrose, which is sweeter than glucose.

Most synthetic sweeteners have been discovered when chemists have accidentally tasted chemicals made for other purposes. Synthetic sweeteners are taken in small amounts especially by people trying to cut the quantity of energy foods in their diet.

Chemical formulas of two sweeteners. Saccharin is 300 times sweeter than sucrose when equal quantities are compared.

saccharin

cyclamate

Cyclamates are banned in the US and UK because some people metabolize the sweetener to another chemical which causes bladder cancer if fed in large doses to rats.

Aspartame is an artificial sweetener which is about 200 times sweeter than *sucrose*. It consists of the methyl *ester* of a dipeptide consisting of two *amino acids* linked by a *peptide* bond.

aspartame (Nutrasweet®)

Structure of aspartame

Aspartame has no aftertaste but cannot be used in cooking because *hydrolysis* on heating destroys its sweetness. There is a warning on the labels of soft drinks and other foods sweetened with aspartame that they contain a source of phenylalanine. This can harm some people with a genetic disorder. Hydrolysis of aspartame produces phenylalanine.

synthesis: a process of making compounds from simpler starting materials. Synthesis puts things together. It is the opposite of analysis which takes things apart to see what they are made of.

The Haber process is an example of synthesis: two elements (nitrogen and hydrogen) combine to make a compound (ammonia). This is the method of *ammonia manufacture* on a large scale in industry.

Synthesis of more complex molecules often takes several reaction steps.

synthetic routes: see *organic routes*.

system: a term used in *thermochemistry* to describe the material or mixture of chemicals being studied. Everything around "the system" is "the surroundings" which includes the apparatus with maybe a waterbath, the air in the laboratory – in theory everything else in the universe.

An open system can exchange energy and matter with its surroundings. Most chemical reactions in laboratories take place in open systems.

A closed system cannot exchange matter with its surroundings. Energy can transfer in or out of a closed system but not the reactants or products.

An isolated system cannot exchange energy or matter with its surroundings. A mixture of chemicals in a vacuum flask with a stopper comes close to being an isolated system.

temperature effect on equilibria: the shift in the position of equilibrium which happens when the temperature changes. The effect depends on the enthalpy change for the reaction. *Le Chatelier's principle* predicts that raising the temperature makes the equilibrium shift in the direction which is endothermic. In *sulfuric acid manufacture*, for example, raising the temperature lowers the percentage of sulfur dioxide at equilibrium.

$$2SO_2(g) + O_2(g) \rightleftharpoons 2SO_3(g) \qquad \Delta H = -98 \text{ kJ/mol}$$

The equilibrium shifts to the left as the temperature rises because this is the direction in which the reaction is endothermic.

These shifts happen because equilibrium constants vary with temperature.

$$K_p = \frac{P_{SO_3}^{\,2}}{P_{SO_2}^{\,2} P_{O_2}}$$

At 500 K, $K_p = 2.5 \times 10^{10}$ atm^{-1}, but at 700 K, $K_p = 3.0 \times 10^4$ atm^{-1}. The value of the equilibrium constant falls as the temperature rises. With a smaller value of K_p the proportion of $SO_3(g)$ falls while the proportions of $SO_2(g)$ and $O_2(g)$ rise at equilibrium.

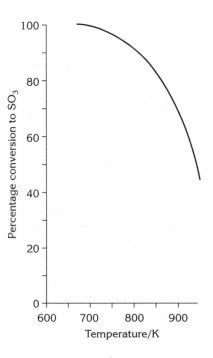

Effect of rising temperature on the equilibrium between SO_2, O_2 and SO_3

temperature effects on reaction rates: see *activation energy* and the *Arrhenius equation*.

teratogens: substances which may cause harm to an unborn child if inhaled, swallowed or absorbed through the skin of the mother.

termination: a step in a *radical chain reaction* which tends to stop the process by removing radicals. In a termination step two radicals combine to form a molecule.

tertiary organic compounds: see *primary, secondary and tertiary organic compounds*.

theoretical yield: see *yield calculations*.

thermal cracking breaks up bigger *hydrocarbon* molecules into smaller molecules by heating them with steam at a high temperature. The process is especially useful for converting the *naphtha* fraction from crude oil distillation into starting materials for synthesis in the petrochemical industry.

Thermal cracking with steam converts ethane to *ethene* which is a very important starting compound for chemical synthesis in industry.

A mixture of the hydrocarbon vapor and steam passes through tubes in a furnace where they are heated to about 1000°C. Thermal cracking involves *radical chain reactions*, unlike *catalytic cracking* which has an ionic mechanism. The first step in the thermal cracking of ethane is for the molecules to split into methyl radicals.

thermal decomposition: a reaction in which a compound decomposes on heating. An important example for industry and agriculture is the thermal decomposition of *calcium carbonate* (limestone):

$$CaCO_3(s) \longrightarrow CaO(s) + CO_2(g)$$

Sometimes heating causes decomposition because a compound is stable at room temperature but becomes unstable at a higher temperature. This is the case with the calcium carbonate and the other carbonates of *group II* metals.

Sometimes heating causes decomposition of a compound which is unstable at room temperature but does not decompose because the rate of reaction is so slow. This is true of the *nitrogen oxides* which, at room temperature, are examples of unstable but *inert chemicals*. They all tend to decompose into nitrogen and oxygen but they do so only on heating.

thermal dissociation: a reversible process which splits a compound into fragments as the temperature rises, but re-forms the starting material on cooling.

The brown gas which forms on heating some metal *nitrates* contains an equilibrium mixture of N_2O_4 and NO_2.

$$N_2O_4 \rightleftharpoons 2NO_2$$
colorless   brown

The mixture darkens on heating as more colorless N_2O_4 dissociates into NO_2 which is brown. The color fades on cooling as N_2O_4 reforms.

thermal stability of carbonates: the ease with which the carbonates of metals decompose on heating.

Chemists relate differences in the properties of the compounds to two factors:

- the charges on the metal ions
- the sizes of the metal ions

A B C D E F G H I J K L M N O P Q R S T U V W X Y Z

Group II carbonates and nitrates are generally less stable than the corresponding group I compounds. This suggests that the larger the charge on the metal ion the less stable the compounds. Down either group I or group II the carbonates become more stable. This suggests that the larger the metal ion the more stable the compounds.

Chemists explain the trend in thermal stability by analyzing the energy changes. Two of the energy quantities they take into account are:

- the energy needed to break up the carbonate ion into an oxide ion and carbon dioxide
- the energy given out (–lattice energy) as the positive and negative charges get closer together when the larger carbonate ions break up into smaller oxide ions and carbon dioxide.

It turns out that all the carbonates are thermally stable at room temperature but become less stable as the temperature rises. The key factor is the energy released as the ions get closer together. This is greater when the metal ion is small than when the metal ion is large.

Detailed analysis of the energy changes is complex, so chemists find it convenient to correlate the stability of compounds such as carbonates and nitrates with the *polarizing power* of the metal ions. Generally, the greater the polarizing power of the metal ion the less stable the carbonates and the more easily they decompose to the oxide.

thermite reaction: a highly *exothermic reaction* used for welding rails and in some incendiary devices. The reaction is similar to the process for chromium extraction but operates on a smaller scale. A magnesium fuse heats and sets off the reaction in a mixture of iron(III) oxide and aluminum. The aluminum reduces the oxide to a glowing mass of molten iron.

$$Fe_2O_3(s) + 2Al(s) \longrightarrow 2Fe(s) + Al_2O_3(s)$$

thermochemistry: the study of energy changes during chemical reactions, including *enthalpy changes*, *free energy changes* and *entropy* changes. It is a major part of chemical thermodynamics. Thermochemistry is important for theory because it helps chemists to explain the *stability of compounds* and to predict the likely direction of chemical change. With the help of thermochemistry, chemists can decide on the *feasibility* of reactions.

Practically, thermochemistry is important because of the significance of:

- calculating the energy from burning fuels
- keeping large scale reactions under control in the *chemical industry*
- estimating the impact of energy changes in the environment.

Thermochemistry was developed mainly in the nineteenth and early twentieth centuries.

thermodynamic control operates where there is a choice of possible products and the main product is the one that is most *stable* (according to thermodynamics). This contrasts with *kinetic control* which operates when the main product is the one that forms fastest.

thermodynamic stability: see *stability of compounds*.

thermodynamics: see *thermodynamics (laws of)* and *thermochemistry*.

thermodynamics (laws of): the laws which govern energy transfers and the direction of change. The laws originally grew out of the study of heating, mechanical working and steam engines. They have now been given a molecular interpretation which makes them applicable to chemical systems.

The first law of thermodynamics says that the total quantity of energy for a system and its surroundings is the same before and after any change. In other words, energy cannot be created or destroyed. This is the law of conservation of energy. Chemists are interested in energy transfers:

- by heating, as energy flows from a higher to a lower temperature
- by working as a system expands or contracts against atmospheric pressure
- by the flow of an electric current
- by the absorption or emission of radiation.

The value of the first law of thermodynamics is that it provides a check that all energy transfers have been accounted for.

The second law of thermodynamics shows which changes can happen (are *feasible*). The law states that a change tends to happen if the total entropy of the system plus the surroundings increases. The *entropy* change, ΔS_{total}, is positive. Chemists often find it more convenient to work with *free energy changes*. The rule is that a process tends to happen if the free energy change, ΔG is negative.

Thermodynamics indicates the possible direction of change but says nothing about the rate of change. A reaction which tends to happen may in practice be so slow that it is *kinetically inert*.

thermoplastic polymers are solid when cold but they soften and become *plastic* on warming so that they can be molded. Most *addition polymers* are thermoplastics. They consist of long chain molecules with no cross-linking.

Thermoplastics such as polythene, polypropene and PVC are supplied by the chemical industry as small chips or as powders which can be melted and forced under pressure into closed molds. Plastic buckets, crates, combs and many other articles are made by this process of injection molding.

Thermoplastics can also be extruded by squeezing the heat-softened material through a die to create a continuous strip with a constant cross section. This is the technique for making car trim, draft seal, curtain track and pipes.

Extrusion through the fine holes of a spinneret converts a thermoplastic such as nylon into *fibers*. Cold drawing the fibers by stretching them when they are cold lines up the molecules and increases the tensile *strength*.

thermosetting polymers: *resins* which set to strong, stiff material once they have been heated in a mold. Heating *cross-links* the polymer chains with strong covalent bonds. Once formed, a thermosetting *plastic* cannot be melted again.

Bakelite is a cross-linked thermoset made from phenol and methanal (formaldehyde)

Saucepan handles and electrical fittings are made from thermosets because these plastics do not melt on heating and are electrical insulators.

thin-layer chromatography (TLC) is a type of *chromatography* in which the stationary phase is a thin layer of a solid supported on a glass or plastic plate while the mobile phase is a liquid. The rate at which a sample moves up a TLC plate depends on the *equilibrium* between *adsorption* on the solid and solution in the solvent. The position of equilibrium varies from one compound to another so the components of a mixture separate.

Thin-layer chromatography is quick and cheap. Only a very small sample is needed. The technique is widely used both in research laboratories and in industry. It can be used quickly to check that a chemical reaction is going as expected and making the required product. After an attempt has been made to purify a chemical, TLC can show whether or not all the impurities have been removed.

Colored compounds are easy to see on a TLC plate. However, the technique is often used with colorless compounds as well. A quick way of finding the position of colorless organic spots is to stand the plate in a covered beaker with iodine crystals. The iodine vapor stains the spots.

Solvent tank – atmosphere saturated with solvent vapor by lining it with paper soaked in the solvent

Solvent rises up plate by capillary action

Thin layer of powdered solid on plastic or glass plate

Spots of solutions applied here, including one or more reference compounds

Solvent

Apparatus for thin-layer chromatography

An alternative is to use a TLC plate impregnated with a fluorescent chemical. Under a UV lamp the whole plate glows, except in the areas where organic compounds absorb radiation, so that they show up as dark spots.

R_f *values* can be used to record the distances moved by chemicals in a mixture relative to the distance moved by the solvent.

thio compounds: sulfur compounds with similar formulas and structures to equivalent oxygen compounds. The thiosulfate ion, $S_2O_3^{2-}$, for example, can be thought of as a sulfate ion, SO_4^{2-}, with one oxygen atom replaced by a sulfur atom.

$$S_2O_3^{2-}$$

thiosulfate

$$\begin{array}{c} H_2N \\ \diagdown \\ C=S \\ \diagup \\ H_2N \end{array}$$

thiourea

$$SO_4^{2-}$$

sulfate

$$\begin{array}{c} H_2N \\ \diagdown \\ C=O \\ \diagup \\ H_2N \end{array}$$

urea

Examples of thio compounds with their oxygen equivalents

Oxygen and sulfur are in the same group of the periodic table with the same number of electrons in the outer shells of their atoms. They therefore form compounds with similar formulas and structures.

tin (Sn): a shiny metal with a long history. Tin was valued as an ingredient of a range of *alloys* including pewter (with antimony), solder (with lead) and *bronze* (with copper). The main use of tin today is for coating the steel for tin cans. The layer of tin stops the iron from corroding. Tin's Sn^{2+} ions are not toxic. Small traces of dissolved tin contribute to the characteristic taste of some tinned foods such as canned fruit and tomatoes.

At room temperature tin has a metallic structure but below the *transition temperature* 13.2°C the stable form is grey tin which has the diamond structure.

above 13.2°C

grey tin $\rightleftharpoons$ metallic tin

below 13.2°C

It takes a long time for atoms in a solid to rearrange themselves. So metallic tin has to stay cold for a long time before it shows the symptoms of "tin plague" as it gradually alters to the crumbly, brittle, grey form.

Tin, with the *electron configuration* $[Kr]4d^{10}5s^25p^2$, comes below germanium but above lead in *group IV* of the periodic table.

titanium (Ti): a very strong metal which is much less dense than steel. It melts at the very high temperature of 1675°C. It does not corrode because, like *aluminum*, it is protected by a thin layer of oxide on the surface of the metal. The difficulties of titanium extraction mean that the metal is not as widely used as might be expected considering its properties.

Titanium is a *d-block* element with the electron configuration $[Ar]3d^24s^2$. Titanium forms compounds in the +2, +3 and +4 states but only the +4 state is common. Titanium(IV) chloride is a colorless liquid formed as an intermediate in *titanium extraction* and in the manufacture of titanium(IV) oxide.

Smoke grenades produce dense clouds of titanium(IV) oxide by the rapid *hydrolysis* of titanium(IV) chloride.

titanium extraction: a process for extracting the metal from its ores which are rutile (TiO_2) and ilmenite ($FeTiO_3$). Titanium is the fourth most abundant metal in the Earth's crust and it would be more widely used if the methods of extraction were less difficult and expensive. In theory it should be possible to extract titanium from its oxide with carbon but in practice some of the titanium reacts with carbon forming carbides which make the metal brittle.

After the ore has been purified it is heated with carbon in a stream of chlorine gas at about 1275 K.

$$TiO_2 + 2Cl_2 + C \longrightarrow TiCl_4 + CO_2$$

The titanium(IV) chloride condenses as a liquid which can be purified by *fractional distillation*.

The *reducing agent* for producing titanium metal from its chloride is magnesium. This is a *batch process*.

In the process hot titanium(IV) chloride passes into a steel reactor inside a furnace containing molten magnesium in an atmosphere of argon at 800 K. A series of exothermic reactions heats the reactor to 1100 K producing titanium(III) and titanium(II) chlorides. Magnesium slowly reduces these chlorides to the metal as the furnace is raised to a temperature of 1300 K. Overall:

$$TiCl_4 + 2Mg \longrightarrow Ti + 2MgCl_2$$

After 36–50 hours the reactor is allowed to cool for at least four days. The solid product is crushed and then *leached* with dilute hydrochloric acid to remove magnesium chloride. The titanium metal is purified by vacuum distillation at a high temperature.

Titanium is used mainly to make aircraft engines and airframes. Other major uses are the production components of chemical plants such as *heat exchangers*.

345

titanium(iv) oxide: a brilliant white *pigment* which is usually known as titanium dioxide. In has two crystalline forms: anatase and rutile. Millions of metric tons of titanium(iv) oxide are made each year. The main use of the pigment is in *paint* but it is also a surface coating for paper, a filler in plastics and an ingredient of cosmetics and toothpaste. The oxide makes a good white pigment because it has a very high refractive index but absorbs almost no light in the visible part of the spectrum. When ground to a fine powder it is both intensely white and very opaque so as a pigment it has excellent covering power. The oxide is also inert and nontoxic, which is why it has replaced lead oxide in many applications.

There are two processes for producing titanium(iv) oxide from ilmenite: the long established sulfate process and the newer chloride process.

titration: a *volumetric analysis* technique for finding the concentrations of solutions and to investigate the amounts of chemicals involved in reactions. Titrations are widely used because they are quick, convenient, accurate and easy to automate.

The procedure only gives accurate results if the reaction is rapid and is exactly as described by the chemical equation. So long as these conditions apply, titrations can be used to study *acid–base*, *redox*, *precipitation* and *complex*-forming reactions.

The measured volume of the unknown is transferred to a flask with a *pipet*; then the standard solution is added carefully from a *buret*. The *end point* is determined by adding a few drops of an indicator or by using an instrument such as a pH meter, a *colorimeter* or a conductivity meter.

Sometimes no indicator is needed because the excess reagent itself produces a permanent color change at the end point. This happens in *potassium permanganate* titrations.

The diagram shows the apparatus for a titration involving a solution A which reacts with solution B. Suppose the equation for the reaction takes the form:

$$n_A A + n_B B \longrightarrow \text{products}$$

which means that n_A moles of A reacts with n_B moles of B.

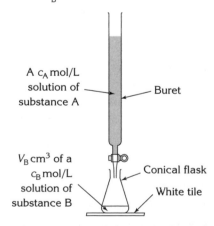

A c_A mol/L solution of substance A — Buret

V_B cm^3 of a c_B mol/L solution of substance B — Conical flask / White tile

Apparatus for a titration

In the laboratory volumes of solutions are normally measured in cm^3 but they should be converted to L in calculations so that they are consistent with the units used to measure concentrations. (1L = 1000 cm^3)

The concentration of B in the flask is c_B mol/L and its volume is V_B L.

The concentration of A in the buret is c_A. A volume V_A of the solution A is added until the indicator shows that the end point has been reached.

The amount of B in the flask at the start $= V_B \times c_B$ mol

The amount of A added from the buret $= V_A \times c_A$ mol

The ratio of these amounts must be the same as the ratio of the amounts shown in the equation:

$$\frac{V_A \times c_A}{V_B \times c_B} = \frac{n_A}{n_B}$$

In any titration, all but one of the values in this formula are known. The one unknown is determined from the results.

(See *acid–base titrations* and *iodine–thiosulfate titrations* and *complex-forming titrations* for worked examples.)

Tollen's reagent: a test reagent used to distinguish *aldehydes* from *ketones*. Warming Tollen's reagent with an aldehyde produces a precipitate of silver which coats clean glass with a shiny layer of silver so that it acts like a mirror.

The reagent consist of an alkaline solution of diamminesilver(I) ions, $[Ag(NH_3)_2]^+$. Aldehydes reduce the silver ions to metallic silver. Ketones do not react.

toughness: a property of materials which measures how much energy is needed to break them. Tough materials are hard to break. Metals and polymers are tough. Ceramics and glass are not tough – they are brittle.

toxic substances are labelled with the skull and cross-bones *hazard warning sign* because they can lead to serious, acute or chronic health risks or even death. Toxic substances can cause harm if inhaled, swallowed or absorbed through skin.

Toxic substances include poisons, *carcinogens*, *mutagens* and *teratogens*.

trace elements: the micronutrients which plants need in small amounts from the soil. Examples are cobalt (Co^{2+}), copper (Cu^{2+}), iron (Fe^{2+} or Fe^{3+}), manganese (Mn^{2+}) and zinc (Zn^{2+}). Plants need these nutrients in much smaller amounts than the nitrogen, phosphorus and potassium supplied by NPK *fertilizers*.

tracers: chemicals used to keep track of chemicals in the environment, in living organisms, in a chemical process or in many other circumstances where chemicals are on the move. The tracer is chosen so that it does not interfere with the changes to be studied. For this reason radioactive isotopes are often chosen as tracers because they are chemically identical with the substances being tracked but easily detected from the alpha, beta or gamma particles emitted during radioactive decay. Examples of radioactive tracers include the use of:

- iodine-131 to study the behavior of the thyroid gland

- hydrogen-3 (*tritium*) to investigate the metabolism of drugs in the human body

- phosphorus-32 to explore the uptake of phosphate fertilizers from the soil by plants.

transition metals: The *d-block elements* which represent a transition from the highly reactive metals of the s-block to the much less reactive metals in group III. In the d-block series from scandium to zinc the only element that does not use its d-electrons in compound formation is *zinc*. For this reason zinc is not classified as a transition metal.

Zinc, $[Ar]3d^{10}4s^2$, only forms compounds in the +2 state. Losing two electrons gives an ion Zn^{2+} with the electron configuration $[Ar]3d^{10}$ in which all the *d*-energy levels are full.

Transition metals share a number of common features. They:

- are *metals* with useful mechanical properties and high melting points
- form compounds in more than one *oxidation state*
- form *colored compounds*
- form a variety of *complex ions*
- act as *catalysts* either as metals or as compounds.

Many transition metal salts are *paramagnetic* because of the presence of unpaired electrons in the partially filled inner *d* energy levels.

					+7					
			+6							
		+5								
	+4	+4		+4						
+3	+3	+3	+3		+3	+3				
				+2	+2	+2	+2	+2	+2	
								+1		
Sc	Ti	V	Cr	Mn	Fe	Co	Ni	Cu	Zn	

Main oxidation states of the elements Sc to Zn. Note that scandium and zinc form ions only in one state.

The chemistries of *chromium, cobalt, copper, iron, manganese, nickel, titanium* and *vanadium* provide many examples of these features.

transition state: the state of the reacting atoms, molecules or ions when they are at the top of the activation energy barrier for a reaction step. Transition states exist for such a brief moment that they cannot be detected or isolated, unlike the *intermediates in reactions* formed between two steps.

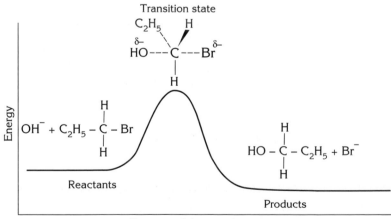

Energy changes during a reaction with a one-step mechanism showing the transition state

The combination of reacting atoms, molecules or ions in a transition state is sometimes called an *activated complex*.

transition temperature: the temperature at which two *allotropes* of an element are in equilibrium. One form is stable below this temperature, the other is stable above the transition temperature. (See *sulfur* and *tin*.)

transmittance measures the extent to which a sample in a spectrometer absorbs radiation at a particular wavelength. Printouts from some spectrometers show transmittance on one axis.

Transmittance, T, is defined as the ratio of the intensity of the radiation leaving the sample, I, to the intensity of the radiation entering the sample, I_0.

$$T = \frac{I}{I_0}$$

For a particular wavelength of radiation, transmittance is low when a sample absorbs strongly. It is 100% if the sample does not absorb at all.

trend: a term often used by chemists to describe the way in which a property increases or decreases along a series of elements or compounds. In the periodic table the term can describe the variations of a property down a *group* or across a *period*. In organic chemistry the term may describe the changes in a property from one member of an *homologous series* to the next.

triglycerides: *esters* of *fatty acids* with *glycerol* (1,2,3-propanetriol) which is an alcohol with three —OH groups. Animal *fats* and *vegetable oils* are examples of triglycerides. Triglycerides, like other *lipids*, are soluble in organic solvents.

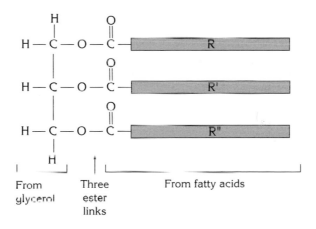

General structure of a triglyceride. In natural fats and vegetable oils, R, R' and R'' may all be the same or they may be different.

triiodide ion: the ion that forms when *iodine* dissolves in aqueous potassium iodide.

$$I_2(s) + I^-(aq) \rightleftharpoons I_3^-(aq)$$

Iodine is only very slightly soluble in water. It dissolves in potassium iodide solution because it forms $I_3^-(aq)$. A reagent labelled "iodine solution" is normally $I_2(s)$ in $KI(aq)$.

The $I_3^-(aq)$ ion is a yellow-brown color which explains why aqueous iodine looks quite different from a solution of iodine in a *nonpolar solvent* such as hexane.

trioxygen: the systematic name for *ozone*.

triple bond: three covalent bonds between two atoms as in *nitrogen*, *alkynes* and the cyanide ion. With three electron pairs involved in the bonding there is a region of high electron density between two atoms joined by a triple bond.

$$\overset{\times}{\underset{\circ}{{}^{\circ}_{\circ}N{}^{\times}_{\circ}N}}{}^{\times}\qquad\qquad H{}^{\bullet}_{\bullet}\overset{\times}{\underset{\circ}{C{}^{\times}_{\circ}C}}{}^{\bullet}_{\bullet}H$$

Examples of molecules with triple bonds $\qquad N{\equiv}N \qquad\qquad H-C{\equiv}C-H$

The molecular orbital model for a triple bond shows that one of the bonds is a normal σ-*bond* while the other two bonds are π-*bonds*.

triple point: the unique combination of temperature and pressure at which the solid, liquid and gaseous forms of a substance are at equilibrium. The triple point of water is at 611 N/m² (0.006 atmospheres) and 273.16 K (0.01°C).

Three lines meet at the triple point on a *phase* diagram for a pure substance.

tritium: the *isotope* of hydrogen, $^{3}_{1}H$. It is radioactive and emits beta particles with a *half-life* of 12.3 years.

ultraviolet radiation, UV, is *electromagnetic radiation* with shorter *wavelengths* (and so higher frequencies) than the violet end of the visible spectrum. The energy of UV photons is high enough to excite the electrons in covalent bonds to higher energy levels. Excited molecules may then split into atoms (dissociate). Thus UV light can start chemical reactions by forming *radicals*.

Ozone forms in the upper *atmosphere* (the stratosphere) where UV light from the Sun shines on oxygen molecules, splitting them into oxygen atoms which then combine with other oxygen molecules to make ozone. In the laboratory, UV light can start the reaction of chlorine with alkanes by splitting chlorine molecules into chlorine atoms. This is an example of a *free radical chain reaction*.

ultraviolet (UV) spectroscopy is particularly useful for studying colorless organic molecules with unsaturated *functional groups* such as $C = C$ and $C = O$. The molecules absorb UV radiation at frequencies which excite shared electrons in double bonds. A UV spectrometer records the extent to which samples absorb UV radiation across a range of *wavelengths*.

In organic molecules with *conjugated systems* the UV absorption peak moves to longer wavelengths as the number of alternating double and single bonds increases. The maximum absorption by ethene is at 185 nm; this shifts to 220 nm for buta-1,3-diene. These are both colorless compounds.

The longer the conjugated chain the stronger the absorption. Beta-carotene with eleven conjugated carbon–carbon double bonds absorbs strongly with a peak shifted so far that it is not in the UV region but lies in the blue region of the visible spectrum (450 nm) so carotene is bright orange (see *colored compounds*).

UV spectrometers make it possible to extend the techniques of *colorimetry* to colorless compounds. In the *pharmaceutical industry*, for example, scientists use UV spectroscopy to check that medicines contain the correct amounts of drugs and that the products do not deteriorate in storage.

uncertainty of measurement refers to variations in analytical results due to factors which the analyst cannot control. Uncertainty in *titrations* can arise from:

- features of the measuring devices – such as the difficulty of reading a buret if the meniscus lies between two of the graduations

- aspects of the procedure – such as changes in the laboratory temperature which affect the volume of volumetric glassware which is calibrated at a particular temperature

- the behavior of the analyst – such as the detection of the end point by observing the color change of an indicator which is a matter of judgement.

Examples of measurement uncertainty for laboratory glassware:

- 50 cm³ class A buret; ±0.03 cm³ when used to deliver 25 cm³
- 25 cm³ class A pipet; ±0.03 cm³
- 50 cm³ measuring cylinder; ±0.6 cm³.

Measuring the percentage uncertainty makes it possible to compare the uncertainties due to each measuring device used in an analytical procedure. This makes it possible to identify the measurement that contributes most to the overall uncertainty in the calculated value.

$$\text{percentage uncertainty} = \frac{\text{uncertainty in the result}}{\text{result}} \times 100\%$$

unit cell: the smallest unit of a crystal structure which, when piled up in three dimensions, gives a whole crystal.

The unit cell of a cubic crystal is a minute cube. An atom or ion inside the unit cell entirely belongs to that cell. An atom or ion at a corner is shared between the eight unit cells which meet in three dimensions at a corner.

Unit cell for the cesium chloride structure

The unit cell for cesium chloride consists of one cesium ion and $8 \times \frac{1}{8}$ of a chloride ion. So the composition of the unit cell still corresponds to the *empirical formula* CsCl. (See *cesium chloride structure*.)

universal indicator: a mixture of several *acid–base indicators* which changes color through the colors of the spectrum from red to indigo as the *pH* rises from 1 to 11. Narrower range indicators are also available which cover a smaller range of pH.

unsaturated compounds: compounds containing one or more double or triple bond between atoms in their molecules. The term is often applied to the *hydrocarbons alkenes* and *alkynes* which typically undergo addition reactions. The term is also commonly used to describe unsaturated *fats* and *fatty acids* which have double $C = C$ bonds in their hydrocarbon side chains.

urea: a white crystalline solid. It is an end product of the metabolism of proteins, and excreted as a waste product in urine.

Structure of urea.

Urea slowly hydrolyzes, releasing ammonia, making it a useful nitrogen *fertilizer*. It is manufactured by the reverse process of heating ammonia and carbon dioxide under pressure.

Urea and formaldehyde are used to manufacture a range of *thermosetting polymers* which are used as adhesives in chip board, and for producing electric sockets and plugs.

vacuum distillation: distillation under reduced pressure which makes it possible to distil liquids with very high boiling points below the temperatures at which they start to decompose. Lowering the pressure lowers the temperature at which liquids boil.

The process is used in oil refineries to separate the *hydrocarbons* in the residue from fractional distillation at atmospheric pressure. Vacuum distillation separates lubricating oils and waxes from other hydrocarbons fed to the plant for *catalytic cracking*.

Vacuum distillation is also used on a small scale to separate products of synthesis which would decompose if heated to their boiling points at atmospheric pressure.

vacuum filtration: a procedure for filtering solutions especially when isolating and purifying solid products of reactions by *recrystallization*. A pump (usually a water pump) lowers the pressure inside the flask so that a pressure difference across the filter paper speeds up filtration.

Apparatus for vacuum filtration

valency theory covers all the theories of chemical bonding. The outer electrons of an atom are its valency electrons which take part in chemical bonding. Electron transfer leads to *ionic bonding* or electrovalency. Electron sharing leads to *covalent bonding*.

Chemists use a variety of ways of describing or predicting the number of bonds formed by the atoms of an element. These include:

- the pattern of charges on the ions in a group of element – *group II* metals, for example, form 2+ ions

- the number of covalent bonds which atoms normally form in molecules – carbon, 4; hydrogen, 1; oxygen, 2; nitrogen, 3; and the halogens, 1

- the *coordination numbers* in crystals and *complex ions* – giving the numbers of nearest neighbors bonded to an atom

- *oxidation numbers* – which provide a formal code for keeping track of the numbers of electrons taking part in bonding.

None of these are rigid guidelines and there are exceptions to all the rules.

van der Waals forces: weak *intermolecular forces*, including all the weak attractions between molecules other than *hydrogen bonding*.

Johannes van der Waals (1837–1923) was a Dutch scientist who studied *real gases* and their deviations from *ideal gas* behavior. He devised a modified gas equation for real gases. His equation includes two correction factors based on *kinetic theory* – one to allow for the existence of intermolecular forces and the other to allow for the fact that gas molecules have a definite volume.

van der Waals radius: the effective radius of an atom when held in contact with another atom by weak *intermolecular forces*.

Atoms do not have a definite size. The apparent size of an atom depends on the way it is bonded to a neighboring atom, so the *ionic radius* and the *covalent radius* of an atom are not the same. Generally, the stronger the bonding the smaller the effective radius. Intermolecular forces are much weaker than covalent bonds, so van der Waals radii are relatively large.

van der Waals radii determine the effective size of a molecule as it bumps into other molecules in a liquid or gas, and when it packs together with other molecules in a solid.

Covalent and van der Waals radii for bromine

vanadium (V): a *d-block* metal with the electron configuration $[Ar]3d^34s^1$. The metal is used to make alloy *steels* which are strong and *tough* making them suitable for machine tools and parts of engines. Vanadium is a typical *transition metal*: it forms colored compounds in several *oxidation states*, it forms *complex ions* and has compounds which can be used as *catalysts* such as vanadium(V) oxide in *sulfuric acid manufacture*.

In solution, vanadium forms ions in the +2, +3, +4 and +5 oxidation states. The +5 state is available as the yellow solid ammonium vanadate(V).

$$
\begin{array}{ll}
+5 & VO_2^+(aq) \ \text{yellow} \\
+4 & VO^{2+}(aq) \ \text{blue} \\
+3 & V^{3+}(aq) \ \text{green} \\
+2 & V^{2+}(aq) \ \text{violet} \\
+1 & \\
0 & V
\end{array}
$$

Oxidation states of vanadium showing the colors of the ions

Standard *electrode potentials* help to identify reducing agents to reduce vanadium(v) to the succession of lower states. (See *electrochemical series.*) Iodide ions reduce vanadium(v) to vanadium(vi).

$E°/V$

$$VO_2^+(aq) \ + \ 2H^+(aq) \ + \ e^- \ \rightleftharpoons \ VO^{2+}(aq) \ + \ H_2O(l) \qquad +1.00$$

$$I_2(aq) \ + \ 2e^- \ \rightleftharpoons \ 2I^- \qquad\qquad\qquad +0.54$$

Half-equations and electrode potentials for the reduction of vanadium(v) to vanadium(iv)

Copper metal reduces vanadium(v) to vanadium(iii).

Zinc in acid will reduce vanadium(v) all the way to vanadium(ii). The $V^{2+}(aq)$ ion is a strong *reducing agent*.

$$V^{3+}(aq) \ + \ e^- \ \rightleftharpoons \ V^{2+}(aq) \qquad E° = -0.26 \ V$$

$$Zn^{2+}(aq) \ + \ 2e^- \rightleftharpoons Zn(s) \qquad E° = -0.76 \ V$$

Half-equations and electrode potentials for the reduction of vanadium(iii) to vanadium(ii)

vapor pressure: the pressure of a vapor in a closed container with its own liquid. The full term is either:

- **equilibrium vapor pressure** – a reminder that in a closed container a liquid and vapor reach a state of dynamic equilibrium

- **saturated vapor pressure** – a reminder that the atmosphere above the liquid holds as much vapor as it can at equilibrium at a particular temperature.

The vapor pressure of a liquid rises as the temperature rises. A liquid starts *boiling* when its vapor pressure equals the external pressure.

A solution of an involatile substance in a liquid has a lower vapor pressure than the pure solvent. This follows from *Raoult's law*. Because of this effect the solution has to be hotter before it starts to boil at a given pressure. Dissolving an involatile salt in water raises its boiling point slightly.

The lowering of vapor pressure depends on the mole fraction of dissolved particles. It does not depend on the chemical nature of the particles. (So the lowering of vapor pressure is an example of a *colligative property*.)

vapors: gases formed by evaporation of substances which are usually liquids or solids at room temperature. So chemists talk about oxygen gas but water vapor.

Vapors are easily condensed by cooling or increasing the pressure because of relatively strong *intermolecular forces*. Vapors therefore tend to deviate markedly from *ideal gas* behavior.

Physicists sometimes broaden the definition of a vapor to includes gases such as butane, ammonia and carbon dioxide which can be liquefied at room temperature by increasing the pressure. (These are gases below their *critical temperature*.)

vat dyes: dyes such as indigo which are insoluble in water but can be converted to a soluble form to dye cloth. Chemical reduction converts indigo to an almost colorless, water-soluble dye which soaks into cloth such as denim for jeans. Oxidation converts indigo back to the insoluble blue form which precipitates in the fibers.

vegetable oils: liquids extracted from plants which are esters of fatty acids with glycerol. They are *triglycerides* which belong to the broad class of *lipids*. The *fatty acids* in vegetable oils contain a higher proportion of *unsaturated* fatty acids than animal *fats*. Examples of vegetable oils are olive oil, sunflower oil and palm oil.

Triglycerides with unsaturated fatty acids have a less regular structure than saturated fats. The molecules do not pack together so easily to make solids, so they have lower melting points and have to be cooler before they solidify.

vibration of bonds: the spring-like vibration of covalent bonds as they bend and stretch. A vibrating *polar covalent bond* is an oscillating *dipole* which can interact with infrared radiation. Like other energy changes involving atoms and electrons, energy is gained or lost in fixed amounts (quanta). According to *quantum theory*, $\Delta E = h\nu$. It turns out that the sizes of the energy jumps for vibrating bonds correspond to frequencies in the infrared region of the spectrum. Each type of bond has a particular value for the energy jumps so it absorbs at a characteristic frequency. This is the basis of *infrared spectroscopy*.

viscous liquids: thick and sticky liquids such as syrups and molasses. Some liquids are viscous because they consist of a tangled mass of long chain molecules. Lubricants from crude oil consist of *hydrocarbons* with chains of over 25 carbon atoms in the molecules.

Other liquids are viscous because of extensive *hydrogen bonding*. Glycerol is a sticky liquid for this reason. It flows much more slowly than propan-1-ol. Hydrogen bonding also contributes to the high viscosity of concentrated solutions of sugars in water.

On an atomic scale toffee and *glass* have disordered structures like liquids. They are, however, so viscous and flow so slowly that they are effectively solids.

visible radiation: *electromagnetic radiation* with *wavelengths* between 400 nm and 700 nm. This band of radiation is visible because it can bring about reversible chemical changes in cells of the retina. These chemical changes lead to electrical impulses in the nerve cells of the eye interpreted in the brain as colors.

Colored compounds absorb radiation in the visible region of the spectrum. Chemists study these compounds using visible spectroscopy or *colorimetry*.

vitamins: a group of chemically unrelated organic compounds which are needed in very small amounts in the diet for healthy growth and body functions. The B vitamins and vitamin C are soluble in water. The other vitamins are fat soluble.

The B vitamins are essential for the activity of some *enzymes*. Lack of vitamin B_{12} leads to a form of anemia. One of the triumphs of twentieth-century organic chemistry was the analysis and total *synthesis* of vitamin B_{12}. Vitamin C is *ascorbic acid*.

volatile organic compounds (VOCs) include a wide variety of carbon compounds which pollute the atmosphere. They are gases or liquids which evaporate easily. Motor vehicles are one source of VOCs, including *hydrocarbons* from the evaporation of fuels and from unburnt *fuel* released from exhausts. Another source is leakage from pipelines and storage tanks. The organic solvents used in the chemical industry and in chemical products are yet another source of this type of air pollution. Many *paints*, varnishes, and wax contain organic solvents, as do many cleaning, disinfecting, cosmetic and degreasing products. One of the aims of *green chemistry* is to find new formulations of these products based on water instead of non-aqueous solvents.

volatile substances: solids or liquids which evaporate easily. A volatile substance easily turns into a *vapor*. Iodine is an example of a volatile solid. Most familiar volatile compounds are molecular liquids at room temperature. Examples are *water, hexane, ethanol* and *acetone*.

The equilibrium *vapor pressure* of a substance is a measure of its volatility at a particular temperature.

Makers of *perfumes* take advantage of differences in volatility.

voltmeters measure the potential difference between two points in a circuit in volts. Chemists use voltmeters to measure the *EMFS* of electrochemical cells.

The value needed when determining *standard electrode potentials* is the EMF when no current is flowing between the electrodes. These values are measured with an accurate voltmeter with a very high internal resistance.

volume: the amount of space taken up by a sample. The *SI unit* of volume is the cubic meter (m^3). Gas volumes are converted to m^3 before substituting values into the *ideal gas* equation.

Chemists generally measure volumes in liters (L) or cubic centimeters (cm^3). There are 1000 cm^3 in a liter, so 1 mL (milliliter) = 1 cm^3.

Measuring volumes of gases and solutions is an essential part of quantitative chemistry. (See *gas volume calculations, molar volume, reacting volumes of gases* and *volumetric analysis*.)

volumetric analysis: a very important aspect of quantitative analysis because volumetric methods can be very *accurate*.

VSEPR theory: the letters VSEPR stand for "valence shell electron pair repulsion theory." This is the theory which helps to predict the *shapes of molecules* from the number of bonding pairs of electrons and lone pairs of electrons in the outer (valence) shell of the central atoms in a molecule.

wastes: unwanted solids, liquids and gases discarded from homes, commerce, industry, agriculture and the public services. They can help to conserve resources and limit the amount of waste dumped in *landfill* by:

- **reduction** – the chemical industry has an extensive program of research and development to introduce new processes which give higher yields and create less waste so that the disposal problem is reduced

- **reuse** – this depends on solvents and detergents which help to clean products or containers so that they can be used again

- **recycling** – a variety of physical and chemical processes recover metals, polymers and other materials so that they can be added to the raw materials used to make new products; in some industries, such as the steel industry, recycling is well established

- **recovery** – this includes the recovery of the energy resources tied up in materials; *incineration* not only disposes of the waste but provides energy to generate electricity. However, incineration can be hazardous if not carried out at a high enough temperature to ensure that harmful chemicals do not escape into the air.

water (H_2O): a familiar liquid with remarkable properties. Despite the small size of its molecules, water is a liquid at room temperature because of *hydrogen bonding*. The $O — H$ bonds in water molecules are highly polar because oxygen is so *electronegative*. Water molecules too are polar because they are not linear thanks to the two lone pairs of electrons which help to determine the *shape of the molecule*.

As a *polar solvent*, water can hydrate ions and dissolve salts. Water can also dissolve some organic compounds because it can form hydrogen bonds with hydroxyl groups in alcohols, sugars or carboxylic acids and with $— NH_2$ groups in amines.

Water plays an important part in many reactions:

- **hydration reactions** – water forms aquo *complex ions* with metal ions

- **acid–base reactions** – water acts as both an acid and as a base because it is an *amphoteric compound*

- **redox reactions** – reactive metals such as *group I* elements reduce the hydrogen in water to hydrogen gas

- **hydrolysis reactions** – this includes the *hydrolysis of nonmetal halides* and of *organic compounds* such as esters (*acid catalysis* or *base catalysis*).

water cycle: the cycling of water in the environment between the oceans, the atmosphere and the crust of the Earth.

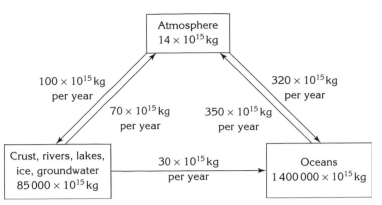

The water cycle, showing the amounts of water in each of the three main reservoirs and the flows between them

The specific latent heat of vaporization of water is relatively large so huge amounts of energy are transferred by the water cycle. Energy from the Sun evaporates water in the tropics. Winds carry the warm moist air to higher latitudes where the energy is released as the water condenses and falls as rain.

water ionization constant: see *autoionization constant of water*.

water of crystallization: water molecules which make up part of the crystal structure of a compound. There are five molecules of water of crystallization for each $CuSO_4$ unit in blue copper(II) sulfate crystals, $CuSO_4.5H_2O$.

Heating hydrated crystals drives off the water as steam, leaving the anhydrous salt. Stronger heating may then lead to other changes (see *nitrates* for example). Other salts with water of crystallization include:

- hydrated sodium carbonate (washing soda crystals) – $Na_2CO_3.10H_2O$
- hydrated magnesium sulfate (Epsom salts) – $MgSO_4.7H_2O$

water treatment provides water for drinking, water for industry and water for medical uses.

Some of the chemical processes used to provide water for homes include:

- **coagulation** – the addition of aluminum(III) or iron(III) salts to add positive ions which help to coagulate negatively charged *colloid* particles so that finely divided solids clump together and settle out
- **adsorption** – removing organic chemicals on the surface of activated *charcoal*
- **disinfection** – use of chemicals to kill microorganisms which might otherwise cause disease (see *chlorine water treatment*). Increasingly *ozone* or *ultraviolet radiation* are being used in place of chlorine because they are less hazardous and do not react with organic impurities to produce chlorinated hydrocarbons which may be harmful
- **water softening** – removing calcium and magnesium ions from *hard water* by precipitating them as insoluble carbonates.

Ion exchange can remove all the ions from water to give very pure water needed industrially and commercially.

wavelength: the distance between the peaks (or troughs) of a wave. Wavelenghs of *electromagnetic radiation* vary from about 1000 m for radiowaves down to about 10^{-3} nm for gamma rays.

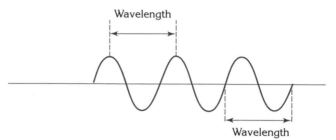

A waveform showing the wavelength

Light in the *visible region* of the spectrum has wavelengths from about 400 nm at the blue end to 700 nm at the red end.

All electromagnetic radiation travels at the same speed, c, in a vacuum. The *frequency*, v, wavelength, λ, and speed are simply related by $c = v\lambda$.

waxes: materials which can be molded when warm but are hard and brittle when cold. They are insoluble in water and water repellent. Waxes are ingredients of polishes and are used to waterproof cloth and leather. Natural waxes such as beeswax from honeycombs and lanolin from wool are esters of *fatty acids* with alcohols with only one hydroxyl group. This distinguishes them from *fats* and *vegetable oils*. The formula of beeswax is $C_{15}H_{31}CO_2C_{30}H_{61}$. Carnauba wax, an ingredient of polishes, varnishes and lipstick, comes from the leaves of a palm tree which grows in Brazil.

Hydrocarbon wax, such as the paraffin wax used for candles, is one of the products from the *vacuum distillation* of the residue from the *fractional distillation* of crude oil.

The chemical industry makes synthetic waxes by polymerizing *epoxyethane*.

weak acids: only slightly ionized when they dissolve in water. In a 0.1 mol/L solution of acetic acid, for example, only about one in a hundred molecules reacts with water to form *hydronium ions*. The more dilute the solution the greater the degree of ionization. The *acid dissociation constant*, K_a measures the strength of an *acid*.

Note the important distinction between strength and concentration. Strength is the extent of ionization. *Concentration* is the amount of acid present in mol/L. Note that it takes as much sodium hydroxide to neutralize 25 cm³ of 0.1 mol/L of a weak acid (such as acetic acid) as it does to neutralize 25 cm³ of 0.1 mol/L of a strong acid such a hydrochloric acid.

Measuring the *pH* of a solution of acid is not enough to show whether or not the acid is strong or weak. A solution of an acid with pH 4 might be a very dilute solution of a strong acid or a concentrated solution of a weak acid.

The salts made from weak acids and strong bases are alkaline in solution (see *hydrolysis of salts*).

weak bases are only slightly ionized when they dissolve in water. In a 0.1 mol/L solution of *ammonia*, for example, only about one in a hundred molecules reacts with water to form ammonium ions. The *base dissociation constant*, K_b measures the strength of a *base*.

As with *weak acids*, it is important to distinguish between strength and concentration.

The salts made from weak bases and strong acids are acidic in solution (see *hydrolysis of salts*).

weathering: a process which breaks down rocks physically and chemically. Physical weathering cracks rocks into smaller fragments increasing the surface area exposed to chemical attack. During chemical weathering rocks react with water, oxygen and carbon dioxide. The types of reaction involved are *hydrolysis* and *oxidation*. Weathering releases soluble ions which plants need for growth. It also forms insoluble *clay minerals*.

weight, measured in newtons (N), is the pull of the Earth's gravity on an object. A laboratory balance responds to weight but is calibrated to give readings in grams or kilograms which measure the mass of the sample. On the surface of the Earth the pull of the Earth is proportional to mass.

$$\text{weight} = \text{mass} \times g \qquad \text{where the constant } g = 9.8 \text{ N/kg}$$

So chemists use "weighing" to determine the mass of a sample. This accounts for the continuing use of concentration units such as weight/volume percent (w/v) which should strictly be the mass/volume per cent.

wetting happens when water spreads out and covers the surface of a solid. Water wets clean glassware but breaks up into separate droplets on greasy glassware. This makes it easy to see if pipets and burets are clean.

Wetting is an important step in getting materials clean but water does not wet fabrics which are contaminated with greasy dirt. One of the functions of *surfactants* in *detergents* is to lower the *surface tension* of water so that it will wet dirty surfaces.

Surfactants are added to pesticides so that when sprayed onto plants the solution spreads over the leaves and is absorbed. With no surfactant the solution would break up into tiny droplets on the surface of the leaves.

A B C D E F G H I J K L M N O P Q R S T U V W X Y Z

X-ray crystallography: a technique used to determine *crystal structures* by interpreting the diffraction patterns formed when *X-rays* are scattered by the electrons of atoms in crystalline solids. Interference between the scattered X-rays produces patterns recorded on photographic film. Crystallographers have worked out techniques for interpreting the patterns so that they can determine:

• the arrangements of atoms, molecules or ions in crystals

• the sizes of atoms and ions

• the *shapes of molecules.*

Since X-rays are scattered by electrons it is possible to draw up *electron density maps* from the diffraction patterns showing the positions of atoms. Small atoms such as hydrogen with few electrons do not show up well, so their position can be hard to determine.

The use of X-ray crystallography to determine the structures of crystals was pioneered by Lawrence Bragg (1890–1971) who developed the theory, and his father William Bragg (1862–1942) who designed the instruments.

Among the early successes of X-ray crystallography were the determination of the structures of:

• the protein α-helix by Linus Pauling in 1951

• myoglobin and hemoglobin and John Kendrew and Max Perutz in the 1950s

• DNA by Rosalind Franklin, Maurice Wilkins, Francis Crick and James Watson in 1953

• vitamin B$_{12}$ by Dorothy Hodgkin in 1956.

X-rays: electromagnetic radiation with very short wavelengths in the region 1×10^{-9} m (1 nm). X-rays have wavelengths of the same order of magnitude as the lengths of atomic bonds. As a result it is possible to use X-ray diffraction to investigate *crystal structures* and the *shapes of molecules* including long chain molecules such as *proteins* and *nucleic acids.*

xenon (Xe): a noble gas which makes up less than 0.1% of the atmosphere. Xenon is separated from liquid air by *fractional distillation*. For a long time chemists thought that the noble gas elements of *group VIII* were completely unreactive so they called them the "inert gases." This changed in 1962 when the British-born Canadian chemist Neil Bartlett was working on the properties of platinum fluorides. Based on the accidental discovery that platinum hexafluoride could combine with oxygen molecules, he used values for *ionization energies* to predict the formation of a compound between platinum hexafluoride and xenon and produced the yellow solid $Xe^+PtF_6^-$. Since then chemists have produced a range of xenon compounds but only with the most reactive elements oxygen and fluorine. Examples are XeF_2, XeF_4 and XeF_6.

yield calculations are used to assess the efficiency of a chemical synthesis. A perfectly efficient reaction would convert all of the starting material to the desired product. This would give a 100% yield.

Few reactions are completely efficient and most reactions, especially organic reactions, give lower yields. There are several reasons why the overall yield may be low:

- the reaction may be incomplete (perhaps because it is slow or because it reaches an equilibrium state) so that a proportion of the starting chemicals fails to react

- there may be side-reactions producing by-products instead of the required chemical

- recovery of all the product from the reaction mixture is usually impossible

- some of the product is usually lost during transfer of the chemicals from one container to another when the product is separated and purified.

The "theoretical yield" is the mass of product assuming that the reaction goes according to the chemical equation and the synthesis is 100% efficient.

The "actual yield" is the mass of product obtained.

The "percent yield" is given by this relationship:

$$\text{percent yield} = \frac{\text{actual yield}}{\text{theoretical yield}} \times 100\%$$

Worked example

What is the theoretical yield of aspirin when 15.5 g salicylic acid reacts with excess acetic anhydride? What is the percent yield if the actual yield of aspirin is 7.25 g?

Notes on the method
Start by writing the equation for the reaction. This need not be the full balanced equation so long as the equation includes the *limiting reactant* and the product.

Since the acetic anhydride is in excess the limiting reactant is the salicylic acid. This means that the acetic anhydride can be ignored during the calculation.

Answer
The equation:

The molar mass of salicylic acid, $C_7H_6O_3$ = 138 g/mol

The amount of salicylic acid at the start of the synthesis

$$= \frac{15.5 \text{ g}}{138 \text{ g/mol}} = 0.112 \text{ mol}$$

1 mol of the acid produces 1 mol of aspirin.

The molar mass of aspirin, $C_9H_8O_4$ = 180 g/mol

The theoretical yield of aspirin = 0.112 mol × 180 g/mol = 20.2 g

Percent yield = $\frac{7.25 \text{ g}}{20.2 \text{ g}}$ × 100% = 35.9%

zeolites: sodium aluminum *silicates* in which the three-dimensional structures of the silicon and oxygen atoms form tunnels and cavities into which ions and small molecules and ions can fit.

Model of a zeolite crystal structure

Synthetic zeolites can make excellent *catalysts*. They can be developed with active sites to favor the required reactions by acting on molecules with particular shapes and sizes. They are used for *catalytic cracking*.

Natural and synthetic zeolites are *ion exchangers*. Synthetic zeolites are used to soften water by swapping the calcium ions in *hard water* with sodium ions. For this reason many laundry detergents contain up to 25% of zeolites.

Zeolites can be very good drying agents; they can absorb water molecules selectively because the water molecules fit into the holes in their crystal structure.

Similarly, zeolites can acts as "molecular sieves," sorting out molecules by size. Some zeolites will absorb nitrogen from the air while letting oxygen pass through. This is one of the methods used to separate *oxygen* from the *air* on a large scale. Once the zeolite is saturated with nitrogen it can easily be regenerated by lowering the pressure and releasing the waste nitrogen back into the air.

zero order reaction: a reaction is zero order with respect to a reactant if the rate of reaction is unaffected by changes in the concentration of that reactant. The concentration term for this reactant is raised to the power zero in the *rate equation*. (See diagrams on page 366.)

$$\text{Rate} = k[X]^0 = k \text{ (a constant)} \qquad \text{since } [X]^0 = 1$$

(Any term raised to the power zero equals 1.)

The rate of reaction of iodine with acetone in the presence of acid is zero order with respect to iodine.

$$\text{Rate} = k[\text{acetone}]^1[\text{hydrogen ion}]^1[\text{iodine}]^0$$

Zero order reactions *Variation of concentration of a reactant plotted against time for a zero order reaction. The gradient of this graph measures the rate of reaction. The gradient is a constant so the rate stays the same even though the concentration of the reactant is falling. (See page 365.)*

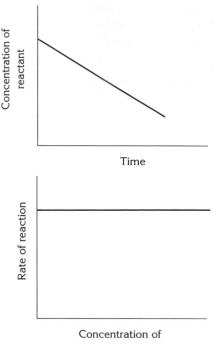

Zero order reactions *Variation of reaction rate with concentration for a zero order reaction (See page 365.)*

Ziegler–Natta catalysts speed up the polymerization of *alkenes* such as ethene and propene making it possible to produce *addition polymers* at relatively low temperatures and pressures. Ziegler–Natta catalysts are made from titanium(IV) chloride and aluminum alkyls (such as triethyl aluminum) in a hydrocarbon solvent.

There is little chain branching in the polymers, so the poly(ethene) chains formed by this method can pack closely together, forming the high density form of poly(ethene). Poly(propene) made with a Ziegler–Natta catalyst has a regular arrangement of the methyl side chains. The molecular structure is *isotactic*. The long chains can pack together more closely because of this regularity, allowing stronger *intermolecular forces* between the chains and a strong polymer with the ability to form useful fibers.

zinc (Zn): a metallic element; the last in the *d-block* series of the fourth row of the periodic table. Zinc is sometimes not regarded as a *transition metal* because both its atoms, $[Ar]3d^{10}4s^2$ and its Zn^{2+} ions, $[Ar]3d^{10}$, have full *d*-subshells.

Zinc and its compounds show few of the typical properties of transition metals. The metal has lower melting and boiling points than transition elements. It forms only one series of compounds in the +2 oxidation state and these compounds are colorless. Examples are zinc oxide (ZnO), zinc chloride ($ZnCl_2$) and zinc sulfate ($ZnSO_4$).

Zinc oxide and hydroxide are *amphoteric* which means that they will dissolve in both acids and alkalis. Adding sodium hydroxide to a solution of zinc ions produces a white precipitate of the hydroxide which will redissolve either on adding acid or on adding excess alkali.

$$Zn(OH)_2(s) + 2H^+(aq) \longrightarrow Zn^{2+}(aq) + 2H_2O(l)$$

$$Zn(OH)_2(s) + 2OH^-(aq) \longrightarrow [Zn(OH)_4]^{2-}(aq)$$

Zinc resembles transition elements by forming *complex ions* such as $[Zn(NH_3)_4]^{2+}(aq)$. Adding ammonia solution to a solution of zinc ions produces a white precipitate of the hydroxide which redissolves in excess ammonia solution as the ammine complex forms.

$$Zn(OH)_2(s) + 4NH_3(aq) \longrightarrow [Zn(NH_3)_4]^{2+}(aq) + 2OH^-(aq)$$

zinc extraction produces the metal from one of its sulfide ores. The most abundant ore is sphalerite, consisting of zinc blende (ZnS). *Froth flotation* is commonly used to concentrate the ore which is then roasted by heating in air to convert the sulfide to zinc oxide.

$$2ZnS(s) + 3O_2(g) \longrightarrow 2ZnO(s) + 2SO_2(g)$$

A neighboring plant converts the sulfur dioxide into sulfuric acid for making fertilizers.

Most zinc (around 80%) is obtained from the oxide by *electrolysis* of zinc sulfate solution made by dissolving the zinc oxide in sulfuric acid. The electrolysis cells have lead *anodes* and aluminum *cathodes*. Zinc forms at the cathodes.

$$Zn^{2+}(aq) + 2e^- \longrightarrow Zn(s)$$

At regular intervals the zinc is stripped from the anodes, melted and cast into ingots with metal at least 99.96% pure.

Oxygen bubbles off from the anodes where sulfuric acid reforms. Recycled acid is reused to dissolve more zinc oxide.

The main use for zinc is for protective coatings such as the layer on *galvanized* steel. It is also used to make *alloys*.

zwitterions: ions with both a positive and a negative charge. *Amino acids* form zwitterions when the amino groups accept protons and the acid groups gives away protons.

An amino acid forming a zwitterion

In crystals and in aqueous solution, amino acids exist largely as zwitterions. Adding acid to a solution of zwitterions puts a proton on the —COO^- group, turning it into —$COOH$, so that the only charge on the ion is the positive charge on the nitrogen atom. Adding alkali to a solution of zwitterions removes the proton from the —NH_3^+ group, turning it into —NH_2, so the ion has only a negative charge only on the —COO^- group (see *isoelectric point*).